KV-030-988

ISBN 1 85310 179 6

First English edition published 1990
by Swan Hill Press, an Imprint of Airlife Publishing Ltd.

**SWAN HILL PRESS**

An Imprint of Airlife Publishing

101 Longden Road, Shrewsbury SY3 9EB, England.

# IN THE WILD

## Wildlife in Great Britain and Europe

SVEN MATHIASSON AND GORAN DALHOV

ENGLAND

# Contents

# Author's Preface

Animals may captivate us for many reasons. The fisherman has a purpose in seeking his elusive prey, while an insect collector is often delighted by colourful, shapely butterflies and beetles, rather than by flies or ants. But birds and mammals enjoy such a general preference because countless nature enthusiasts throughout history have been fascinated by them. Both these kinds of animals, however, differ greatly as to how they reveal themselves. Most birds are active by day, whereas most mammals are just the opposite. Birds tend to be clad in beautiful colours, while mammals are more sober in appearance.

Observing and studying, collecting and portraying, hunting and fishing — all these pursuits involve making use of what nature has to offer. It is a form of exploitation which, in most parts of the world today, must be regulated in relation to resources. Long ago, and even now among non-industrialised peoples, man hunted for a livelihood. In many places, hunting and fishing remain an integral aspect of land usage. Once there were hunting parks, amounting to protected natural environments, for hunting by royalty and nobility; these have saved numerous species for posterity. Today, the environment is protected mainly in order to ensure the survival of species. We also pay attention to the ecological context. Previously, an individual species was put under protection, while its essential habitat was ignored. Certain species were protected, particularly during the breeding season, while others were lawful game. Today we do the reverse: fauna as such is protected, and hunting seasons are established only for those species which can safely be hunted. The protective regulations also cover nature studies, photography, and ring-marking.

All this will hopefully result in a richer natural world and more animal life. Animals ought to be not only an economic and exploitable resource, but a normal feature of obvious value in a shared living environment. We should never begin to question any species' right to exist. What threatens animals, in the form of pollution and environmental destruction by thoughtless and selfish use of land and resources, has painfully proved to have the same consequences for us human beings.

Experience of animals in their natural surroundings is a source of refreshment and recreation. It may be an encounter with beauty, a pleasure to the eye and ear. It is also a challenge to discover their secrets, whether through research or mere curiosity. In this challenge, too, lie the excitement and charm of hunting — another aspect of the same interest. Personally, I have had the good luck to encounter most of the animals which are presented in this book, and under different circumstances. Polar bears, through binoculars and camera from a sled on the sea-ice north of Baffin Land; elk, with my gun in hand on the hunting grounds of western and central Sweden; captive swan and geese, with my scales and ruler and ring-marking gear in various places around the Arctic; and so on.

The book's contents, though, chiefly reflect the general curiosity about nature which msot of us feel. There is, of course, much more to be said, and perhaps important details have been overlooked. But the aim of this book is to give a measure of insight into the lives and behaviour of the species that play a leading role in our contacts with the animal world.

Comparatively great space is devoted to certain species, because they are regarded as representatives of a way of life which is common to other, more briefly discussed, relatives of theirs. By contrast, phenomena which are peculiar to certain species have been emphasized for these animals. Likewise, species of unusual interest to mankind are given more attention. Biological and technical aspects, as well as general matters of folklore, history, nomenclature and so on, have been treated along with the species which best illustrate them.

The species are considered in terms of the environments where they are primarily encountered as regards their occurrence, seasonal importance, largest populations, and so on. A separate section deals with threatened and introduced species.

For my own opportunities of experiencing these wonders of nature, I thank those who have helped to make it possible — first and foremost, my parents and my family.

**Sven Mathiasson**

# Artist's preface

One of the first days in March finds me by the water down at the north cove. It ripples with long-missed sounds as fragrances are released by the melting spring ice. The lake in this neighbourhood, a favourite spot of mine, is part of everyday nature, and it becomes a measure of what goes on. Squadrons of plummeting ducks are caught in the memory and on the sketch pad. Often only a turn round the shores, where mere footsteps locate the path, is enough to loosen a mental knot in my studio.

The pictures in this book are actually a record of such strolls and moments from my favourite wildlife habitats as well as other expeditions. I remember especially the Great Hungarian Plain's expanses and the bountiful mountains of Portugal, but every place stimulates the pen with its wildlife and its unique identity. Meeting a bear in the Ström Water Valley, or standing eye to eye with the master stag in a park district outside Vienna, rank high among my memories — yet seeing friends again, like the thrushes of Tjörn Island in early April, seems even more enjoyable.

An experience of wildlife in the wilderness is frequently a fleeting impression, complete in an instant but retained on the retina by memory. Whoever takes the time to learn about an environment for some years may, of course, learn what to expect. But the unexpected sights are what excite and ensure. Light and movement against backgrounds in a landscape can, at best, be put in order on the page: figures and surroundings combine to make a complete picture. The artist's aim is to select what the paper should portray; sifting among the signals from eye to hand is the artist's difficulty, and — if you will — his art.

The impressions assembled to illustrate this book communicate my experiences of various species in the wild. I hope that they will not only be useful but, even better, that they may inspire people to take walks and encounter wildlife directly. Pop out to the park now and then — make time for repeated weekend visits to pleasant places that you can get to know through the years. Discover the museums and collections. Awareness of nature around us is undoubtedly vital for our continued coexistence with wildlife on the planet.

The Natural History Museum of Gothenburg has always been a place of pilgrimage for me. Here the details of animals can be studied and, apart from my sheer frustration with museums, it is doubtful whether the book's anatomical drawings could have been done without those collections. Deep gratitude goes to my friends there.

And lastly, thanks to Kerstin.

**Göran Dalhov**

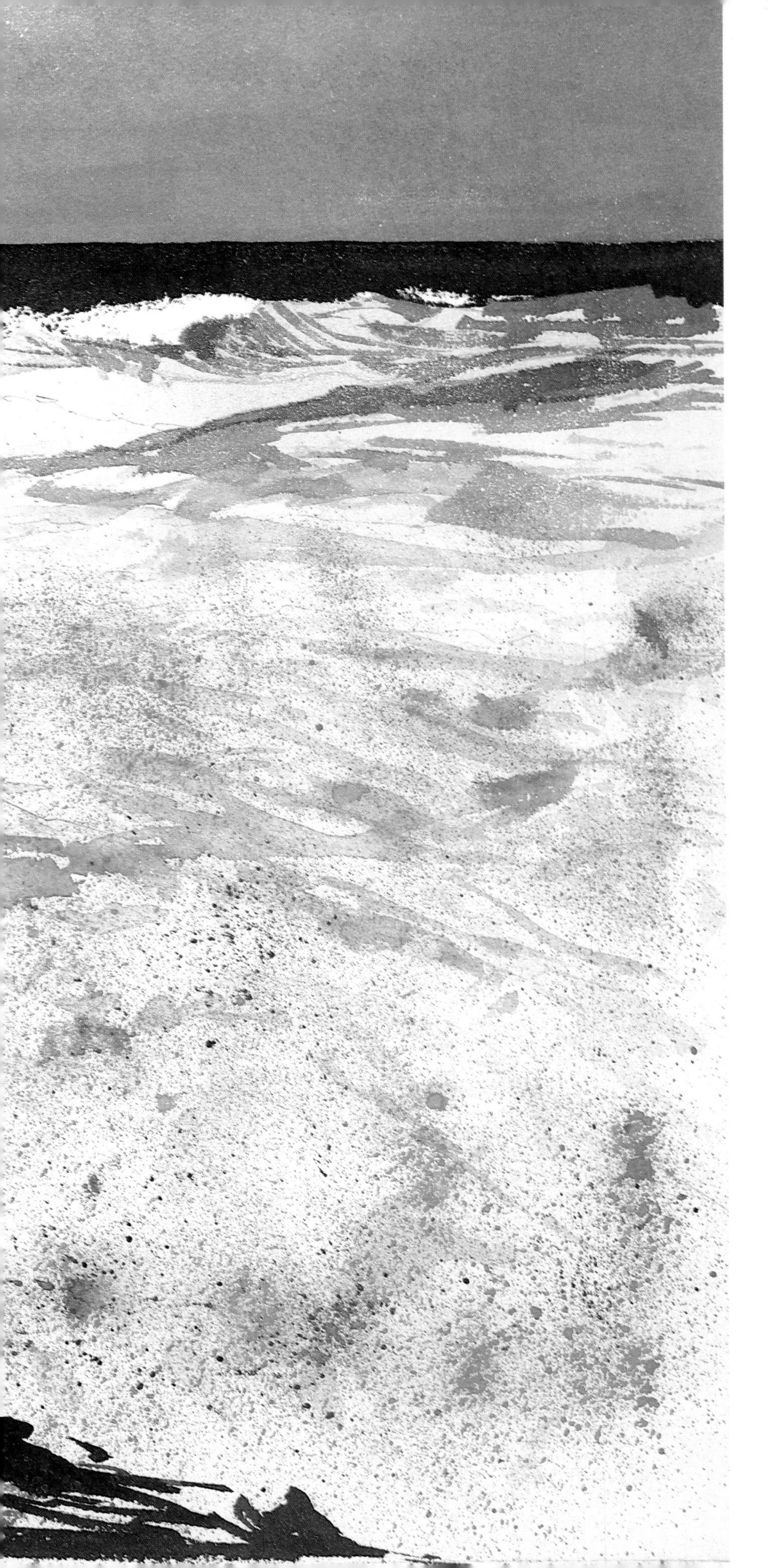

# Coasts and the sea

The sea is a rich natural environment. It is largely free of ice in winter, which gives it obvious attractions for those creatures which are forced to leave lakes and other wetlands which become ice-bound in winter. Sea-dwelling species such as seals, whales and seabirds benefit from the ice-free conditions, although in some northern areas even they are forced to move away owing to sea ice in winter. Some species of seals have adapted to use the sea ice to their advantage, for protection from predators and as breeding grounds. Historically, man has used the presence of the sea ice to exploit some of the larger species, and the age-old practice of seal hunting is an example of this. Elsewhere, subsistence hunters have often taken advantage of dense concentrations of sea birds.

Sadly and alarmingly, many of mankind's waste products threaten the ecological balance of the shallow and very organically rich coastal regions. It should be remembered that although the seas occupy two-thirds of the earth's surface, only 7% of this area consists of shallow water less than 200 metres deep. In particular, oil discharges have caused the deaths of millions of seabirds and other creatures. Individual incidents such as the wrecks of the Torrey Canyon and the Amoco Cadiz have accounted for hundreds of thousands of seabirds and done major damage to marine ecosystems. In addition to such major disasters, there is also a steady leakage of oil and oil products into many rivers and other watercourses which lead to the sea, and it has been estimated that the total amount of oil leakage is in the region of six or seven million tons.

The dumping of commercial waste material at sea is still a normal and widespread practice. Heavy metals such as mercury, and other toxic materials, have a highly poisonous effect when large-scale dumping takes place, polluting the marine environment and passing on into the environment in general. The consequences can be seen in the incidence of disease in marine species, especially among seals and many types of fish. Reduced breeding success and even complete sterility is common among seals. Seabirds have also been affected, with reduced breeding success and poor rates of survival among young birds.

It is therefore rather surprising to find that the sea coast is relatively rich in higher forms of life, and there are rich rewards for the naturalist (and for the hunter) in the coastal environment. In some instances this may be because man's activities and even his waste products have created local conditions which are especially favourable to one or two particular species. For instance, swans have undoubtedly benefited from the organic enrichment of some areas. Some species of sea ducks have also benefited because the populations of bottom-living species have become simplified, allowing the tougher and more resilient species to flourish at the expense of other, more vulnerable creatures. A habitat which enjoys a rich diversity of species is usually found to be thriving and dynamic, with certain species increasing and decreasing in abundance from time to time in relation to natural environmental forces. Pollution and human impact on the habitat can adversely affect this natural dynamism and richness.

# Seals

Since their physical characteristics are clear and distinct, seals are nowadays usually regarded as belonging to a unique order of mammals, the *Pinnipedia.* With few exceptions seals lack external ears, and when these are present they are simple and rudimentary. Seals can nevertheless hear and see well, even under water, although their nose and ear openings are closed when they dive. For the most part seals are active by day. Their fingers are webbed and they can often be seen to lack front teeth. Their molars are all similar in appearance, and all seals have a thick layer of blubber under their skins.

Seals can be divided into three groups: two of these are the *Otariidae*, which have ears and are found in the Pacific Ocean, and the earless seals or *Phocidae*, which are found in both the Atlantic and Pacific oceans. The latter can also use their hind limbs to assist their movements on land, something which the eared seals cannot do. The eared seals also swim largely with the help of their sizeable front paws. From the evolutionary point of view, these two groups are considered to have had different origins and do not stem from any common ancestral species. The third group in the seal family is the *Odobenidae* or walruses.

Thirty-three species of seal are known to exist world-wide. For the most part they occur in areas of cold fresh water and in the sea. However, two species of seal — the Baikal seal and the Black Sea seal — are only found in lakes, and the Ringed seal occurs principally in Arctic waters or in cold lakes such as Ladoga in Russia and Saimen in Sweden. The Common seal is a sea-dwelling species, but it also occurs in certain freshwater lakes in Canada.

In European waters Common seals, Ringed seals and Grey seals occur in large numbers in the north Atlantic, the North Sea and the Baltic. In the same waters the Bearded seal (*Erignathus barbatus*), the Harp seal (*Pagophilus groenlandicus*) and the Hooded seal (*Cystophora cristata*) all occur as rare visitors, but they are more commonly encountered in the European sector of the Arctic Ocean. In the Mediterranean there are monk seals (*Monachus monachus*) and in northern latitudes the walrus (*Odobenus rosmarus*).

## Common seal — *Phoca vitulina*

This species is also sometimes called the harbour seal or the spotted seal. The male Common seal is always some 10%–15% larger than the female, and a large male may attain an overall length of up to two metres, but most adults are considerably smaller — 150–185cm. Individual weights vary with the seal's age and the time of year, as well as with its sex, but they normally weigh within the range 50–120 kg. However, individual specimens have been found to weigh as much as 150 kg., and one Pacific race achieves weights close to 200 kg. Newly born seal pups have an overall length of 70–97 cm. and an average weight of 9–11 kg., but by the time they are three or four months old they will weigh two or three times as much. Seal pups studied on the west coast of Sweden have been found to have lengths in the range 90–100 cm. and weights between 12.5 and 18 kg. in July–August.

The fur colour varies from a light grey to a darkish grey-brown with numerous prominent dark spots, chiefly over the animal's back. Before birth the pups have a pale whitish fur, but at or immediately after birth a darker fur emerges, not dissimilar to that of the adults. Young seals remain paler on their undersides than the adults, however. The Common seal's nose is short, rounded and blunt, with distinct white whiskers. The nostrils form a definite V-shape and the muzzle area between the eyes and the nose is prominently curved and concave. Females are thought to live longer than males, and a female aged 32 years has been recorded, compared to the oldest known wild male at 26 years old. In captivity, however, Common seals have achieved proven lifespans of 40–45 years.

The Common seal feeds on various species of fish, and its diet is dictated largely by what is available in the locality. Along the western coast of Sweden, for example, cod, herring and haddock predominate. The quantities of fish eaten by seals are slight compared with mankind's harvest, and it has been estimated that the fish requirements of 1,000 Common seals corresponds to only 0.7% of the fishermen's catch of cod and 0.5% of the herring catch. Along the North Sea coasts of Germany and the Netherlands young seals have been seen to feed on sand shrimps, and one-third of the diet of adult seals comprises flat-fish. Around British and Irish coasts seals are often accused of high levels of predation on salmon.

The Common seal normally fishes for his food by daylight, and it is believed that they can dive as deep as 100 metres. A seal's normal fishing dive is estimated to last between five and seven minutes.

The gestation period of the Common seal lasts for eleven months, beginning with up to ten weeks delayed development of the foetus, owing to delayed implantation. Although twins occur occasionally, seals normally have only one pup, which is born in early summer and about one month before the adults gather in herds for mating. The pups are born on land but the act of mating takes place in the water. Some seals are monogamous, but promiscuous mating behaviour can also occur, and seals have been observed to mate both before and after the annual moulting period in July–August. The young pups become weaned and independent of their mothers as early as six weeks after birth. The female seal's milk is known to be especially rich in protein and fat, promoting rapid growth of the young, which suckle for a period of 4–6 weeks, both on land and in the water. The pups enter water as early as a few hours after birth, but if they are left undisturbed they prefer to remain on land. While the pups are still suckling they begin to learn to fish. The females become sexually mature at two years old, but do not normally reproduce for a further 2–3 years. The males usually mature a year later than the females, and the adult females produce one pup each year.

The Common seal occurs all around the Arctic Circle, and five distinct local races are known. The nominate form, *Phoca v. vitulina*, is found in northern Europe, while the others occur in North America and Greenland, and in the northern Pacific. *Phoca v. concolor* frequents coastal waters and *Phoca v. mellone* is found in some Canadian freshwater lakes. *Phoca v. richardi* occurs along the American coast, and the large, dark-coloured *Phoca v. stejnegeri* is found in the Kurilerna area of Russia. Around the Bering Strait and southwards to Japan and Korea there are large numbers of Alaskan fur seals, which were at one time thought to be a race of the Common seal.

Of all the various seal species worldwide, the Common seal is the most widespread, and the total world population has been estimated at about 500,000 individuals. In Europe the Common seal has declined considerably in certain areas, and in some places such as the Baltic Sea

its survival is a matter of concern. The chief threat to seal populations along the coasts of industrialised European countries and in inland seas is the effects of environmental poisons. Among other toxins it is suspected that the influence of PCBs — polychlorinated biphenyls — is related to the sterility of about one half of the female seal population in the Baltic. This total population consists nowadays of only about 100 individuals (200 in 1981) and their rate of reproduction is minimal. To the west in the Skaggerack–Kattegat area the Common seal has fared much better, and there have even been local increases following recent seal protection measures. In 1986 this population was estimated at approximately 4,000 animals.

## The Grey Seal — *Halichoerus grypus*

The Grey seal is the largest of all seals which breed in Europe. The head is large and the muzzle is long and concave, and this is the best identifying characteristic of the species, together with the prominent and parallel nostrils. Mature males are often 15%–20% larger than the females, and an adult can measure more than 3 metres and weigh over 300 kg. A British study of the Grey seal has shown that adult males average 207 cm in length and females 180 cm, with average weights of 233 kg and 155 kg respectively. The seals' colour can vary, but the basic colour is grey to yellowish-brown and the body is dotted with dark grey and light grey spots. The males are darker than the females and often have smaller spots. Both are much paler on the belly than on the upper parts.

At birth the pups' fur is yellowish-white, turning to a spotted blueish-grey after three weeks. Grey seal pups remain on land longer than those of the Common seal, usually until they have moulted their juvenile fur, but they can nevertheless swim well from birth, like other seal pups. At birth the pup is about 1 metre long and weighs 10–15 kg. Within 3–4 weeks of birth their weight will have trebled. By the time they are one year old the largest will measure 1.35 metres.

Grey seals live chiefly on fish, especially salmon and cod. Along the British coast their diet also comprises large quantities of octopus. Grey seals often catch more food than they require, and prey may therefore only be partly eaten. Their daily food requirements vary with age and size, but have been estimated at between 7.5 and 12.5 kg. The Grey seals of the Baltic subsist chiefly on Baltic herring and cod. But Swedish studies have revealed that potential prey is abundant, and more than 20 different species of fish occur in the diets of Grey seals in the Baltic. Occasionally they will also seize a seabird resting on the water. They can dive to depths of up to 100 metres and remain submerged for up to 20 minutes.

Grey seals form herds and the males and females form separate groups throughout the year, except at mating time. One female Grey seal is recorded as having lived to the age of 46, while the oldest recorded male is 31 years.

The Grey seals have their pups in small family groups or in larger herds. At mating time the largest males take possession of a territory and attract sexually receptive females, and a female may choose one male in one year and another the next. Female Grey seals are believed to breed at the age of five, but males do not begin to mate until they are several years older. Mature males create harems around their territories, and where the seals are numerous these may consist of up to 20 females in one harem.

Mating takes place when the female comes into season, which is usually about the time when the young pup has become independent and left her, having shed his pup fur. The implantation and development of the foetus is delayed for about three months, and the total period of gestation takes about 11½ months. The single pups (twins are very rare) are born in February–March in the Baltic, while in the North Sea and the Atlantic they are born in the period September–October. Grey seals give birth either on ice or on islands with gently sloping beaches. The pups are suckled for 3–4 weeks before their mothers leave them.

The Grey seal is found in two areas of the north Atlantic, around the coasts of Europe and America, and also in the Baltic. The different geographical groupings are regarded as separate races, and the nominate *Halichoerus g. grypus* is found in the St. Lawrence Bay, around Newfoundland and Labrador, and in southern Greenland. *Halichoerus g. atlanticus* occurs in the White Sea, Spitzbergen and Norway, Iceland and from the Faeroe Islands south to the British Isles, while *Halichoerus g. balticus* is found in the Baltic. The world population of Grey seals has been estimated at around 120,000 animals,
of which an estimated 80,000 are found around the British Isles, with some 10,000 in the remainder of the eastern parts of the north Atlantic. The Baltic population has been estimated at 1,500 animals, and therefore approximately half of the world population is to be found off the coasts of North America.

In the Baltic the Grey seal was very numerous well into the early part of this century. In the period 1886–1900 no fewer than 84,226 Grey seals were recorded as killed by hunters in Finland alone, an average of 5–6000 each year. The total Baltic population at that time has been estimated at about 50,000 seals. In Swedish waters the Grey seal was hunted eagerly in the Ostergotlands archipelago, where the majority of Baltic seals lived. Since hunting has been abandoned, however, environmental pollution has devastated the population. Today the Swedish population is estimated at 750 animals, and the total Baltic population at barely twice that number. Since 1974 the Grey seal has been protected in Sweden and throughout the Baltic states.

## Ringed seal — *Phoca hispida*

The Ringed seal is the smallest of the three species of seal which breed around European coasts. As with other seal species, the male is larger, attaining lengths of up to 1.80 metres and weights of up to 130 kg. The female is about 20% smaller, and when she is suckling a pup her body weight may fall by up to 40%. In other respects, however, the sexes are similar in appearance. The basic colour is greyish-brown, sometimes almost silver-grey, with a paler belly. These seals are dotted with dark spots which are ringed with light greyish-white. Ringed seal pups weigh approximately 5–6 kg at birth. The head is small, rounded and short-nosed, and the rest of the body is plump and rotund. The young pups keep their white juvenile fur for 3–5 weeks, after which the mature coat emerges. The whiskers of the Ringed seal are brown, unlike the white whiskers of the common seal. The teeth are smaller and weaker than those of Common and Grey seals, and they grow at an oblique angle.

Ringed seals live chiefly on fish, principally Baltic herring and to some extent on crustaceans such as crawfish and lobsters. In summer the Ringed seal often lives in bays and inshore coastal waters where it catches whitefish and cottidaue on the sea bottom. The Ringed seal is believed to dive for up to 20 minutes at a time, often to a depth of 100 metres, and there is one recorded instance of a Ringed seal diving to 200 metres. However, it usually seeks its food in much shallower waters. Individual life-spans of up to 46 years have been recorded, and the males tend to live rather longer than the females.

Male Ringed seals become sexually mature earlier than females, which are unlikely to breed before they are at least three years old. The male and females form a pair bond and remain together for one season. New pairings are formed prior to the next mating season. The male defends a territory around the ice cave (there are often several alternative caves or dens) in which the female gives birth to her single pup, and the male also has a cave or den of his own. The pups are born in March and are suckled by their mothers for 5–6 weeks. By the time they are weaned the females are ready for another mating. In common with Grey and Common seals the implantation of the foetus is delayed and the total gestation period lasts from 8–11 months.

At birth the pups can see well, and if danger threatens they can enter the water. But their juvenile fur is not very water-resistant. The importance of the ice cave is not clear, but it apparently aids the survival of the pup, and may also protect the suckling female. However, polar bears and arctic foxes are skilled at locating the caves and sometimes kill a great many pups. Otherwise, the cave acts as important shelter from the wind on the exposed and windswept ice, and they can also prevent the freezing-up of the vital diving holes, of which there are one or two in each cave. The young pups suckle for up to seven weeks, and sometimes for 2–3 weeks longer, after which they leave the caves to assume an independent existence. By this time they have the same fur as the mature seals and have reached an overall length of 60-70 cm.

The Ringed seal is found in both the Old and New worlds, and usually in Arctic waters. In Europe there are isolated populations in the Gulf of Bothnia, and in lakes such as Saema in Finland and Ladoga in Russia. Ringed seals flourish in the area of the pack ice belt, and in summer they tend to be found in shallow bays. The Ringed seals of the Finnish and Russian lakes have been considered as distinct races — *Phoca hispida saimensis* and *Phoca h. ladogensis* — and a separate racial designation, *Phoca h. botnica*, has also been given to the Ringed seal of the Baltic. The populations in the northernmost Atlantic are regarded as belonging to the same race as those of northern Canada and Greenland, the nominate *Phoca h. hispida*. Isolated occurrences of this race have been made as far south as England and the western coast of Sweden, and two further races have been identified in the northern Pacific.

The Ringed seal is regarded as the world's most numerous arctic species of seal, with a total population of at least three million and perhaps many more. In 1981 the Baltic Sea population was estimated at about 10,000 animals, while the Lake Saimen population was reckoned at 100 individuals. As with other species of seals in the Baltic, the Ringed seal is under threat from environmental pollution and toxins, which have already rendered three-quarters of the Baltic seals sterile.

Traditionally, the Ringed seal has been hunted by coastal peoples on both the Swedish and Finnish sides of the Gulf of Bothnia. Hunting expeditions used to go out regularly in spring onto the ice, with specially equipped boats pulled on sleds. The hunting was principally a matter of shooting, but they were also captured by netting, harpooning and clubbing, as well as by hunting with dogs. The annual harvest was sometimes as many as several thousand seals. The Ringed seal is now only hunted to a very limited degree in Finland, and many are accidentally caught and drowned in fishing nets, which is a particular threat to the lake populations.

# Common porpoise

## *Phocaena phocaena*

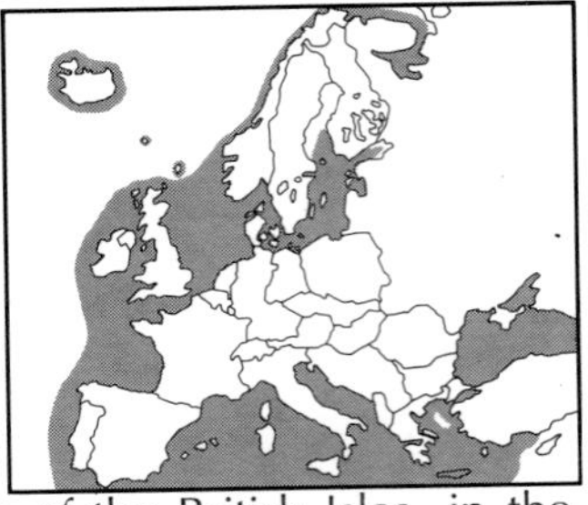

Sixty-six species of Toothed whales and ten species of Baleen whales have been identified world-wide. Seventy of them exist along the coast of Europe, but many are only occasional visitors, and most of them are bound for waters to the south of the British Isles, in the Mediterranean and the Atlantic.

Historically, many species of whale have been of great financial importance to coastal peoples in Europe. The Norwegians, the Faeroese and the Basques have all traditionally hunted whales in their home waters as well as on a large commercial scale in more distant seas. Off the Spanish coast the Biscayan Right whale *Eubalena glacialis* was hunted, while in the Faeroe Islands the Pilot whale *Globicephala melaena* was pursued. Off northern Norway the Lesser Rorgual or Minke whale *Balaenoptera acutorostrata* was the hunters' quarry. The Pilot whale and the Common porpoise, together with the dolphin, have been a more general quarry for many coastal peoples, who caught them in nearby waters and shared out the meat and other by-products among the local villagers.

The hunting of Pilot whales still continues as a tradition in the Faeroe Islands. The proceedings begin with the raising of the alarm by the first person to see the school of whales. The school is then driven inshore into suitable bays, which is where they are killed, and the whole business is carried out in a highly organised and systematic way, ending with the distribution of the meat according to time-honoured rules.

The most widely distributed and best known whale around European coasts is the Common porpoise. In the past it has been the object of large-scale hunting, some of which is still carried on today. Porpoise hunts were carried out in the Black Sea, around Iceland and in the Oresund Strait between Denmark and Sweden. An eighteenth-century Swedish account stated that porpoises were to be hunted because of the damage they did to nets and fishing equipment. Old documents reveal that the right to hunt porpoises was highly regarded and carefully maintained, often causing serious disputes and resulting in legal battles. Possibly the largest European catches were made in the straits between Denmark and Sweden through which tens of thousands of porpoises moved westwards to avoid the winter ice in the Baltic.

The Common porpoise belongs to a group of six closely related species, all of which are small by comparison with other whales, and these occur principally in the north and south Atlantic and in the Pacific and Indian oceans. However, they chiefly occur in the northern hemisphere.

### Characteristics

The Common porpoise is a small-toothed whale. Typically, a fully grown adult will weigh between 50 and 70 kg. The porpoise's total

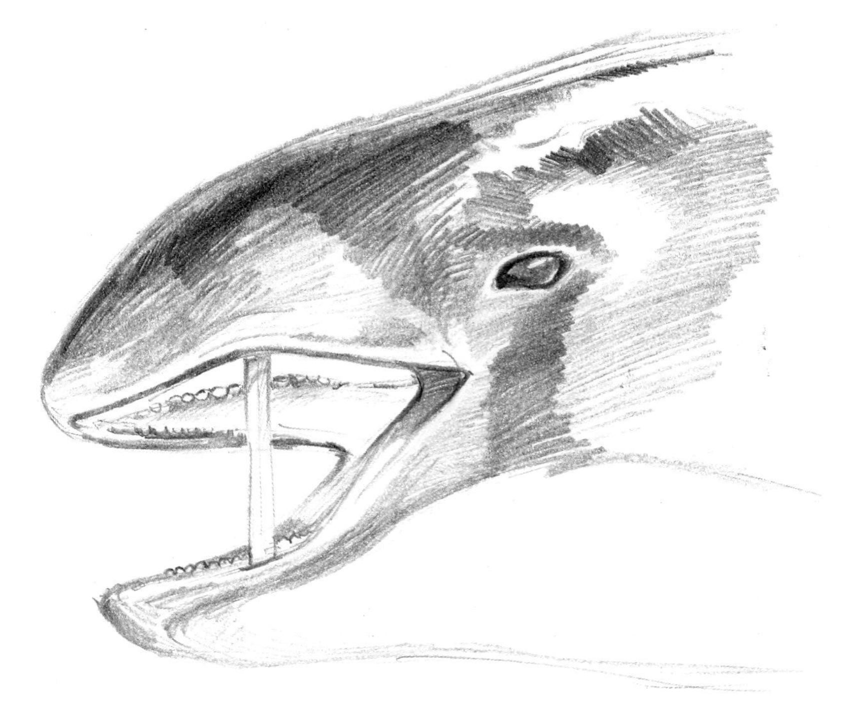

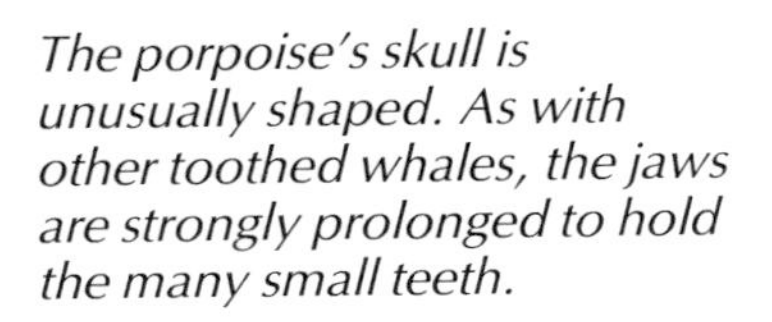

*The porpoise's skull is unusually shaped. As with other toothed whales, the jaws are strongly prolonged to hold the many small teeth.*

body length is in the range 1.5–2 metres. In common with all whales the tail fin is horizontal, and the porpoise's triangular dorsal fin is situated about midway between its nose and its tail. The colouration on the back parts is black, while the under-belly is greyish-white. Although the nose is not very prominent the jaws are long and equipped with a row of small, spatulate teeth, usually between 22 and 27 on both the upper and lower jaws. The sexes are quite similar.

The Common porpoise mostly occurs in small schools, usually close to the coast and sometimes within harbours and river estuaries. There are also some recorded instances where it has ventured upstream into rivers. The Common porpoise feeds on fish, especially herring and cod, but it will also feed on flat-fish and to a lesser extent on crustaceans and molluscs.

Under water porpoises sometimes use echo-location to orientate themselves, but they also enjoy very acute eyesight. They do not leap out of the water, but arch their backs over the water as they dive. The greatest age a porpoise is known to have attained is about fifteen years.

## Reproduction

The Common porpoise is usually sexually mature at about 3–4 years of age, and females normally give birth to one pup each year. The gestation period lasts eleven months and the pup is born in the period May–July. At birth the pup weighs 5–6 kg and has a body length of about 75 cm.

## Distribution and numbers

The European occurrence of the Common porpoise is limited to the northern Atlantic where it exists northwards towards the Baltic and southwards towards tropical waters. Elsewhere it occurs in the Mediterranean and Black seas and in the Baltic Sea. In many places it is the commonest member of the whale family, but its numbers have fallen sharply in recent times. This is due partly to excessively intensive commercial fishing, and local populations are seriously threatened by environmental pollution and by the incidence of drowning in fishing nets.

The Baltic population has been protected under Danish law since 1967, and it is also protected under Swedish law. Nevertheless, 3–4000 Common porpoises are reckoned to die each year in Danish waters as a result of becoming entangled in fishing nets. A further 100 are estimated to die annually in this way in Swedish waters. However, some increase in local numbers has been observed in recent years, including along Sweden's western coast. Yet this increase need not necessarily reflect actual changes in the population: it may only signify that these little whales follow different migration routes than formerly. The Common porpoise regularly migrates to escape ice coverings; unlike the Common seal, the porpoise lacks the ability to form air holes in the ice for breathing and can therefore be drowned when extensive and rapid ice cover develops.

Sharks and large-toothed whales are natural predators of the Common porpoise. In the five-metre-long stomach of one killer whale there were found 13 Common porpoises, in addition to 13 seals!

# Small Sea Geese

The family *Branta* to which the Canada goose and its relatives belong consists of a total of five species. Three of these — the Canada goose, the Barnacle goose and the Brent or Brant goose — are regularly found in large numbers in parts of the British Isles and northern Europe. The red-breasted goose is a rare visitor from its eastern haunts, and its breeding grounds are situated in northern Siberia; and the truly wild Hawaiian goose exists only in the islands from which it takes its name.

The Barnacle goose *Branta leucopis* is a bird of the Arctic, which breeds on the tundra within the Arctic Circle from east Greenland eastwards through Spitzbergen to the archipelago of Novaya Zemlya. The western or Greenland population winters in western Ireland and Scotland, where the largest numbers occur in the Inner Hebrides and especially on the island of Islay. The smaller population of Barnacle geese which breeds in Spitzbergen winters in one location only, on the inner parts of the Solway Firth in south-west Scotland, and migrates there via the Baltic Sea. The eastern populations which breed in Novaya Zemlya winter in the Netherlands and Germany, with occasional numbers to be found in south-east England.

The Barnacle goose got its scientific name fairly late, and even its breeding grounds have only been determined relatively recently. The breeding population in Spitzbergen was only identified as recently as 1907, and prior to this the east Greenland population had been identified in 1891–92, in both cases by German scientific expeditions. These geese are more easily counted than most other species of waterfowl, and the world population of Barnacle geese was established at about 93,000 in the early 1980s, and of these 35% are known to winter in the British Isles, which therefore has an international responsibility for this species.

It is interesting to note that the Barnacle goose has established a few small breeding haunts well to the south of its Arctic range, especially in Sweden. It is considered that these breeding populations have stemmed from birds which have escaped from zoos or wildfowl collections.

The Barnacle goose got its common name in early times owing to a very curious misunderstanding of its method of reproduction. The traditional story, stemming from early medieval times, was that these small black and white geese were engendered from shellfish or barnacles which encrusted fallen trees by the sea shore and large pieces of wooden flotsam. This belief led some medieval churchmen to regard these birds as the offspring of shellfish and therefore permissible to be eaten in Lent and on other days of fasting when meat and poultry were forbidden.

The Brent goose *Branta bernicla* is another Arctic species which is found to breed on the Arctic tundra. Brent of the subspecies *Branta b. bernicla* belong to the dark-bellied variety, which breed on the Russian tundra and winter in Europe. The light-bellied subspecies, *Branta bernicla hrota*, breeds in Greenland and in northern Canada. Some remain to winter along the Atlantic coast of America, but the main wintering grounds are in Ireland. There is also a north-eastern Asiatic subspecies, *Branta bernicla nigricans*, which is darker than the previous two subspecies, and its wintering quarters are along the Pacific coasts of Asia and North America. It is only a rare visitor to Europe.

While the Barnacle goose feeds principally by grazing on salt marshes, the Brent goose searches for its food along the shoreline and in shallow water. The Barnacle goose eats land grasses, while the Brent selects sea grasses, *Ulva lactula* and *Enteromorpha intestinalis*. The European wintering population of Brent geese declined sharply in the 1930s, owing to a fungal disease of the eel-grass which was their staple diet, but numbers have recovered steadily since then, although only the dark-bellied *B. b. bernicla* subspecies has fully recovered to its pre-1930s level. Of the world total of approximately 200,000 Brent geese, some 45% winter in the British Isles.

The third member of the *Branta* family to be found in Europe is the Red-breasted goose, *Branta ruficollis*, although it is a very rare visitor to western Europe. Its principal wintering areas are around the Black Sea, in Bulgaria and Romania.

All these small, northern-breeding geese are faced with annual hazards which can jeopardise successful breeding. The percentage of young birds in the winter flocks varies from 5%–31% for the barnacle goose, and from 0%–50% in the case of the Brent goose. Both species require between 70 and 80 days for the laying of their eggs, for incubation and for the rearing and fledging of the goslings. This is almost exactly the same length of time during which their nesting sites are free from ice and snow. In years when the arrival of the geese coincides fairly exactly with the unfreezing of their breeding grounds breeding success can be very high, but a small variation in time can result in a very much reduced level of breeding and in bad years the birds may not succeed in breeding at all.

There are other hazards in the form of predators like the Arctic fox, the Raven and various large gull species, all of which will take eggs and goslings. Both species of geese get some measure of collective protection by nesting in fairly large colonies. The Red-breasted goose breeds in smaller groups which are often close to the nests of falcons, which help to drive away their mutual enemies.

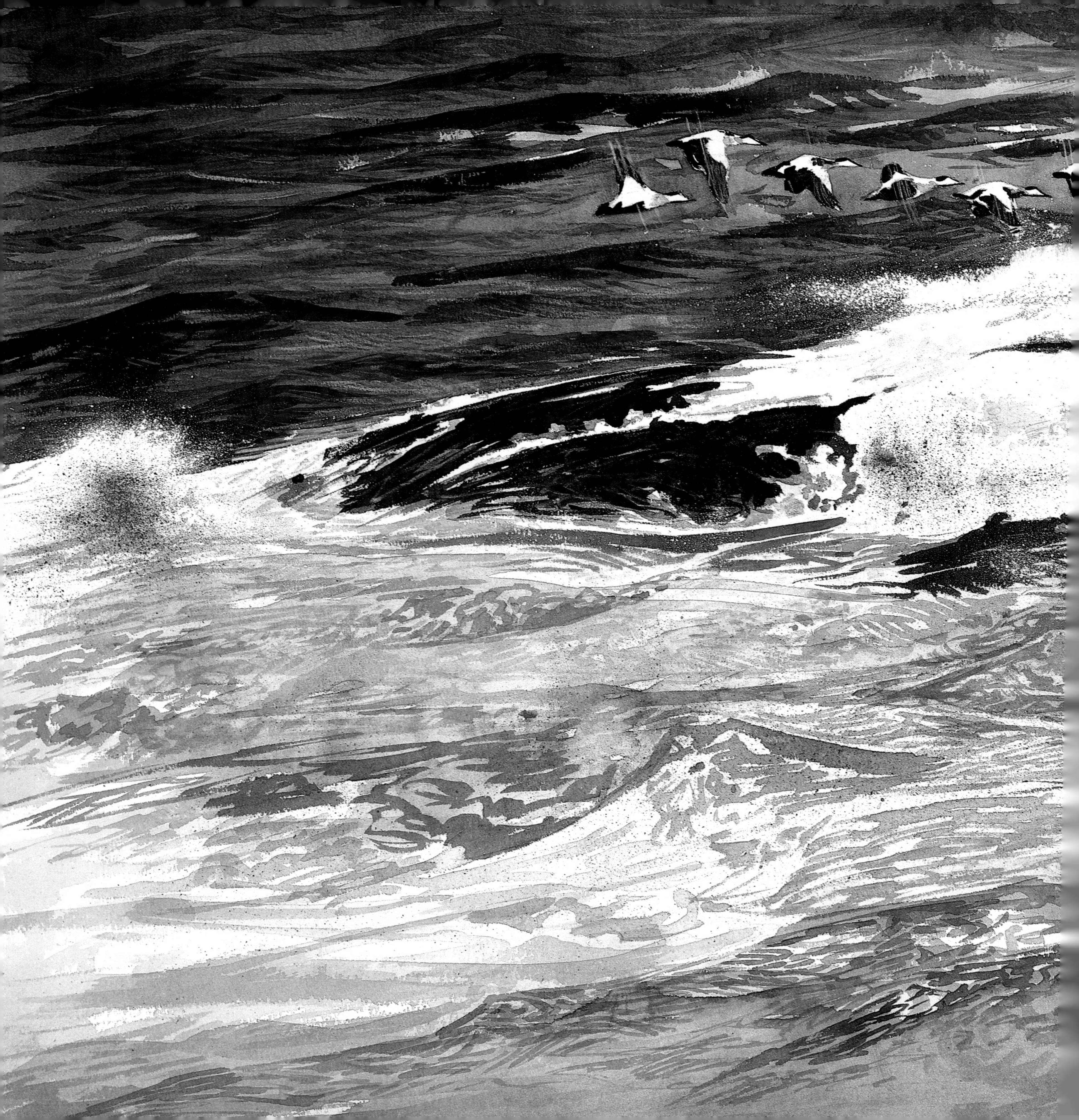

# Common eider

## *Somateria mollissima*

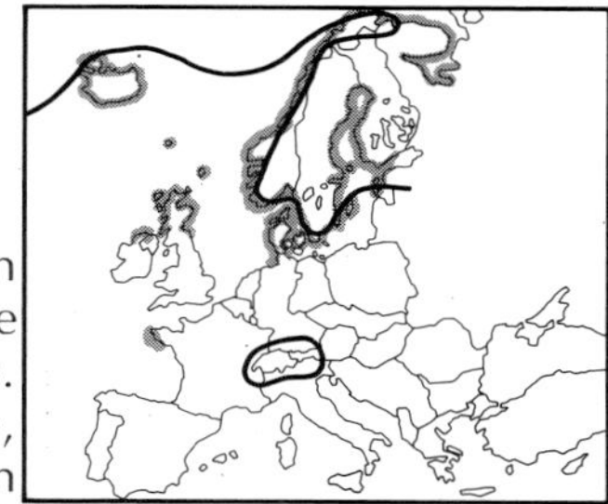

Archaeological excavations have shown that the Eider was a source of food for the people of Stone Age coastal settlements. The oldest known non-fossil bird bone, which was found in Denmark, is from an Eider, and this bird is believed to have lived during the period before the last Ice Age, long before the appearance of man in western Europe.

At present the Eider breeds in northern Scotland and in Iceland. Elsewhere it is a bird of the Nordic countries, and especially Denmark, Sweden, Finland and Norway. In all four countries there is a long historical tradition of hunting the Eider duck and coastal peoples have also exploited the Eider population by gathering the down from its nests and also collecting its eggs.

From the earliest times, writers on natural history praised the Eider's down for its exceptional softness and warmth, and for its high commercial value. This was especially recognised in the Nordic countries such as Sweden, where the birds were given an important measure of special protection by royal decrees from the eighteenth century onwards, which were publicly proclaimed from the pulpits of every country church, with substantial fines for those who transgressed. In 1951 a controversial ban was placed on the traditional spring hunting of Eider in Sweden, and observers noticed a significant rise in the Swedish Eider population after this date.

### Characteristics

The Eider is a large and heavy duck with a short neck and a large, elongated head with a wedge-shaped, sloping bill. The newly hatched young are covered in greyish-black down, after which they gradually assume a darker plumage not dissimilar to that of the adult female. The plumage of the mature male is very different from that of the female, and the young males do not attain their full adult plumage until they are four years old. The adult male in mature plumage has a black top to his head and a dark bill, black flanks and tail, with a breast colour of subtle pink and whitish patches on either side of its rump. On the sides of the head and on the back of the neck there are flashes of greenish-hued feathering. Immature males may be distinguished from mature males and from females by their dark, blackish-brown plumage on their backs.

The iris of the eye is brown, and the male's bill is a greyish-green or olive colour shading to a yellowish-green; the bill also has a pale tip to the upper mandible. The feet are yellowish-green.

The Eider rides quite deep in the water, its tail pressed downwards onto the water's surface and its short neck retracted so that the head appears almost to rest on the upper part of the bird's back. The angle of the bill is slightly downward. Eider are usually found in flocks, and the birds' flight is low across the water with heavy wingbeats.

## Reproduction

Eider females do not lay their first clutch of eggs until they are between two and four years old, and the males are not sexually mature until they are three years old. The females are observed to return year after year to the same breeding islands and to lay their eggs in the same places; they remain very faithful to the breeding locality and to their nesting places. eiders are monogamous, but the pair bond only lasts for one season at a time.

Eiders nest on small islands within larger archipelagos, and the birds tend to nest widely throughout a scattering of such islands. The Eider also nests on sandy islets and occasionally on salt marshes. Nests are placed close to one another, thus forming nesting colonies, and this occasionally leads to more than one female laying eggs in one nest. The nest site consists of nothing more than a simple depression or scrape in the ground or in a bank of seaweed. The nest tends to be lined with whatever materials are close to hand, but all nests are warmly and softly lined with the birds' rich down. Nests are usually quite exposed, but they can occasionally be found under bushes or among their exposed roots.

A typical clutch comprises 4–6 eggs, which are large and oval in shape and greyish-green in colour. The egg size in within the range 67–88 mm x 47–55 mm, and weigh 102–110 grams. Incubation begins when the last egg has been laid and lasts for 25–26 days. It is believed that average clutch sizes tend to be larger among the more northerly populations. The clutch size may also be influenced by the birds' physical condition as dictated by the weather conditions of the preceding winter, and mature birds lay larger clutches than young ones.

The chicks are not fully fledged for 9–11 weeks after hatching, but often become independent some weeks earlier. The breeding cycle from the start of nest building to eventual fledging takes 3–3½ months.

By the time the females hatch their eggs, the males are usually far away and undergoing their annual moult, when all the body and flight feathers are renewed. The females have sole responsibility for building the nests, brooding the eggs and young, and taking care of the growing ducklings. Females with young often band together in groups which typically comprise 20–25 young and 3–4 females, and it is believed this strategy aids chick survival.

Although almost all the eggs are fertile and will hatch, many clutches of eggs are lost and mortality among the young chicks is high. It is estimated that between 5% and 35% of clutches are lost to gulls and crows, and also because of disturbance and inundation of the nests by rising water levels. This varies from place to place and from season to season. Further studies have shown that from 30%–90% of recently hatched young are lost each year. In the Baltic Sea area the survival rate of the chicks is sufficiently high to provide a surplus of young birds, and this fully compensates for the mortality among unfledged youngsters and in the adult population. In the North Sea, however, it is estimated that only 10% of the chicks survive to fledging.

In certain years there can be localised and also widespread mass mortality among Eider chicks, and this is partly attributed to the effects of parasites. Mortality in Baltic and North Sea breeding colonies can be as high as 90%. Should such mortality occur for a number of seasons in succession it can decimate Eider populations.

## Adaptations and behaviour

The nesting eider sits very tightly on her nest. Her camouflaged plumage makes her hard to spot, and the nesting birds normally emit little or no scent which could reveal their whereabout to predators. When chased off her nest, she squirts a malodorous excrement over the eggs, which makes them less appetising to predators. This excrement may also have a certain camouflage value, but excrement can have the effect of chilling the eggs by evaporation if they are left unattended for a long time, and it may also block up the pores of the eggs. When an undisturbed female leaves her nest, she does not normally excrete, but covers the eggs carefully with down, which conceals them and keeps them warm.

The newly hatched downy chicks remain in the nest until all the eggs have hatched, and on the water the chicks stay close together. Usually female Eider with chicks gather together into groups, and females without chicks may attach themselves to such groupings. When the chicks are threatened, perhaps by gulls, the females respond by swimming towards the attackers and counter-attacking with outstretched necks. However, broods which are attacked often split up, and it is easy for gulls to catch those chicks which have become widely separated from the group, even though the chicks often attempt to save themselves by diving. One means of reducing predation by gulls is for several female Eider to band together and jointly fend off an attack by a single gull.

Female Eider may travel long distances with their chicks, and this has the effect of reducing local competition for limited supplies of food. The dispersal of the colonies also reduces the levels of predation on the young chicks, and helps minimise the spread of infectious diseases.

Like most male ducks the Eider drake has a bright and striking plumage. This can be seen when the annual moult is completed, between September and November, and the full plumage is retained until the following June. Like most other species of duck, eider form their breeding pairs during the winter months. However, if the ducks belong to a migratory population, pair formation may not take place until the birds have returned to their spring nesting grounds.

The Eider duck is a salt water bird, which eats salty foods and drinks salt water. As an adaptation to this, it has well developed salt-filtering glands located in front of its eyes.

During the nuptial displays in winter and early spring eider are noisy. The displaying drakes utter a "a-oooh, a-oooh" call which can be heard a long way off. Often an entire group of up to ten males will display around a solitary female, and smaller groups of drakes can also be seen to display within large mixed flocks. As with most ducks, the display motions are stereotyped and mechanical, and repeated unchanged in several sequences. The neck is thrown back over over the bird's back and then stretched upwards and forwards, like many other duck species. Other movements include head-turnings, preening, wing-flapping with the body held upright, etc.

## Population structure and distribution

Adult female Eider often breed for the first time in either their first or second year, and the males sometimes take a little longer. Recent studies indicate that 25% of females breed in their first year, while about 50% wait until their third year. The remainder may not mate

until even later in life. Since there is an estimated 20%–40% rate of mortality among juvenile and old birds, and since the returning frequency among surviving females is high, one can make a fairly accurate estimate of the general age structure among the breeding population.

A survival rate of 1.5 chicks per adult female (i.e. 0.75 young females) will result in an increase in the population, and it indicates that one-third of the breeding population of females consists of birds aged 3 or more years old, with one-year-old females making up about 15% of the breeding population. The most productive age groups are those between one and three years old.

In the early 1970s the total population of Eider in the Baltic area was estimated at 300,000 breeding pairs. Finnish and Swedish Eider lay clutches averaging 4–5 eggs, which means the total annual egg production of the Baltic population is approximately 1.5 million. Finnish studies show that 75%–95% of the eggs hatch, while 25%–30% of the chicks survive to fledge. Thus 19%–31% of eggs laid result in fledgeling Eiders. With an estimated annual mortality of 20%–40% this rate of reproduction provides a large surplus to augment the population, or to allow increased levels of predation or shooting. Shooting accounts for much of the mortality among adult birds. About a quarter of a million birds are shot annually around the Baltic, especially in Denmark. In certain areas both young and adult Eiders have suffered from parasitic disease and also from pollution. The primary cause of mass deaths has been oil discharges. But the worst breeding results allied to the highest mortality rate would still maintain the total numbers of birds. The Eider is therefore a relatively successful breeding bird, which is holding its own and even increasing in numbers in the Baltic Sea area.

This has been reflected in the expansion of breeding areas. Finnish data indicates that the Baltic population doubled in size during the 1970s, and by 1980 the population was estimated at 600,000 breeding pairs. North Sea populations are also reckoned to have increased, although to a lesser extent. The total northern European breeding population of Eider ducks is probably of the order of 1-1½ million pairs.

The Eider is part of the Arctic avifauna, and it is probably one of the species which first inhabited European coasts after the Ice Age. They have also recently expanded southwards and at the same time they have become more numerous, as have some other seabird species including the Fulmar, the Gannet, Kittiwakes, Razorbills and the Herring and Greater Black-Backed gulls.

## Environment and food

Although the Eider is a sea duck, in certain places it has colonised lakes. Normally Eider will spend the winter along the sea coast, but some birds have also been found inland. During the 1980s numbers of Eider were found to over-winter on Swiss lakes, where they feed on certain species of shellfish (*Dreissena polymorpha*) which have become very common. One Eider, ringed as a young bird at Oland, was found among those wintering in Switzerland.

During the breeding season Eider live mainly among scatterings of small islands, and small to medium-sized islands are especially attractive to them. But they also breed along the beaches of larger islands with low sand dunes, and they can also be found inland on some of the higher grassy islands of the Atlantic, although in smaller numbers. Eider are also occasionally found along the coastal salt marshes of the mainland. The birds' principal sources of food are mussel beds, and along the coast they will also eat crabs and other small crustaceans found in shallow water. Eider are active by day and night, and can dive to depths of 10–15 metres.

Females with young prefer shallow sea bays where there is an ample supply of small crustaceans and mussels. Even when the chicks are very small the females and their young will gather together along the tideline where there is plenty of food. When they are fully grown, Eider often feed on mussel beds which are far from land.

The Eider's bill is strong and sharp, and it uses this to tear mussels off the rocks. Mussel beds which contain plenty of young, small mussels are especially attractive to Eider, which cannot swallow the very largest mussels. On the little islets where Eiders rest their droppings can be seen to contain purplish-blue and white fragments of shell, which reveals what they have been eating. Mussels and sand crabs are swallowed whole, and the powerful muscular stomach grinds their hard shells. When feeding on crabs, Eiders can often be seen to shake and hit the crabs against the water's surface to break off the straggly legs. In addition to mussels, Eiders also eat other small shellfish and small shoreline crustaceans.

## Relations and races

The Eider and its closest relatives have been difficult to place as a family. At present they are often assigned, along with the mergansers, to the *Mergini* group of merganser-like birds. Equally often, however, they are placed in their own distinct group, the *Somateriini* or Eider birds. Besides the common Eider this group also includes the King Eider and Steller's Eider. The birds' breathing system and the structure of the bronchia distinguishes the Eider from other species of diving ducks and brings them closer to dabbling ducks. The downy chicks are also different from the young of diving ducks. The three species of the *Somateria* family all occur in north Eurasia and North America. The Common Eider is therefore part of the holarctic avifauna, occurring all around the pole. However, there is virtually no contact between the Eider populations of Europe and Asia, and those of North America. The Pacific Eiders are distinctly isolated, for the species is absent over a large part of the mid-Asiatic area of the Arctic Sea. Nor do the Eider of Greenland mix with those of the eastern Atlantic, and in these circumstances one might expect to find separate races, which is the case. In North America and Greenland,

and in the north Pacific area, four races occur. In Europe and the eastern Atlantic, however, there are only two races. The larger of these is the nominate form with the scientific name *Somateria m. mollissima*, and the smaller race is the Faeroe Island Eider, with its small and dark females, and is known as *Somateria m. faeroeensis*.

## Migration

Many of the North Sea Eiders are sedentary, as also are the Eiders of Iceland, Norway and the Faeroe Islands. However, the Eiders of the Baltic and of the west coast of Sweden are mostly migratory.

Baltic Sea Eiders leave their breeding grounds to find winter quarters in the south-western Baltic. Some migrate via Schleswig-Holstein in Germany and Denmark to winter in the North Sea. Most stay around the Danish islands.

Eiders leave their wintering grounds at the beginning of March, and by the beginning of April most have returned to their breeding areas. By early June the drakes' moult begins, and hundreds of thousands of Eider drakes fly to their favourite moulting grounds, which are principally in the western Baltic. Individual flocks can comprise as many as 100,000 birds.

*The Eider has relatives which resemble each other in female and juvenile plumage, and in male eclipse plumage. But the King Eider* (Somateria spectabilis) *is smaller, with wing length 249–295 mm for males and 244–291 mm for females. Its featherless wedge in front of the eye is shorter, and the feathered area on the bill side is rounded (pointed on the Eider). Steller's Eider* (Polysticta stelleri) *is even smaller, with wing length 209–217 mm for males and 210–215 mm for females. It has no featherless patch, a rather short grey to blue-grey bill, and a dark blue wing speculum. Below are shown, besides the Eider (top left, with female below and subspecies* borealis *to her right), some northern diving ducks related to the Eider. At bottom left, the Surf Scoter* (Melanitta perspicillata)*; to its right, Harlequin Duck* (Histrionicus histrionicus)*; above this, Steller's Eider and King Eider. In next row, Velvet Scoter, King Eider and Long-tailed Duck; top right, Common Scoter. All are males.*

# Goldeneye

## *Bucephala clangula*

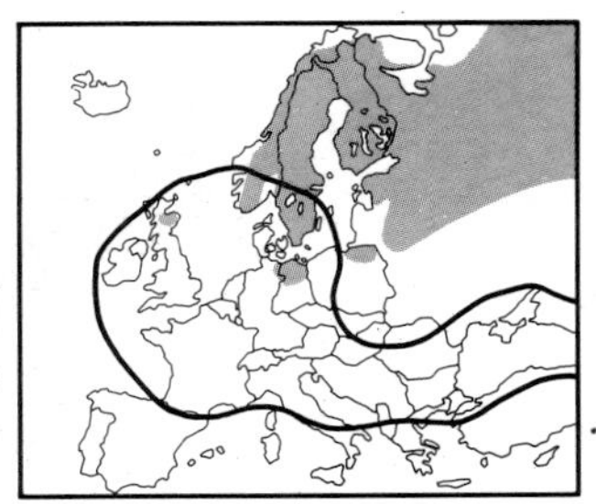

The first word of a Latin name for a species, written with a capital initial letter, states the family of which the species is a member. *Bucephala*, which means "ox-head" and refers to the round, high and powerful head of the Goldeneye, is the family name of this species. *Clangula*, which means sonorous or loud, alludes to the whining wing sound of the flying drake. In the family *Bucephala* are two more species, Barrow's Goldeneye (*B. islandica*) and the American Bufflehead (*B. albeola*).

Because the second name, *clangula*, tells us it is the Goldeneye and not any other member of the family, the Latin name therefore confers on this species a clearly defined and internationally recognised identity. It can otherwise be doubtful quite which species is meant. Compare the different national and vernacular names for the Goldeneye: English — Goldeneye; French — Canard garrot or Garrot commun; German — Schellente; Danish — Hvinand etc.

The Latin names also show family relationships. Species within a family are naturally more closely related than those from other families. Latin names have been developed one stage further, for in addition to binary nomenclature — i.e. one family name and one species name — there is also trinary nomenclature, which means that the species has been divided into two or more subspecies or races.

The concept of a species depends on the fact that the individual members of one species do not normally breed with individuals belonging to any other species. Isolation and separate development have led to such complete differences in behaviour and genes that these different species can live side by side without interbreeding. However, this is not the case with members of different races within the same species. If the individuals belonging to a certain population of one species can be distinguished from the individuals of another population of the same species, then it can be worthwhile and meaningful to give them different names to indicate this. But the differentiating features are not so major that individuals from different populations cannot interbreed if their paths should cross. The race is maintained by genetic isolation, which is usually achieved by geographical factors. Separate races often develop on islands, for example. But where the races meet, hybrid zones can occur.

The Goldeneye is considered to have two races: one is Eurasian and the other is American.

### Characteristics

The Goldeneye is perhaps the easiest of the smaller diving ducks to identify with certainty. Both sexes are distinguished by a high forehead, conspicuous white wing patches and a bright yellow eye. Seen in the wild, Goldeneye give the impression of a plump and compact bird: the neck appears short, as does the bill, and the tail is depressed against the surface of the water. These ducks take off with a short dash along the water's surface. The drake is distinguished by his bright plumage with the prominent white cheek-spot and his distinctive, whining wingbeat. The female is also unusually attractively coloured, with a brown head and a white collar. The eclipse

plumage of the males and the juvenile birds is similar in appearance to that of the females.

Goldeneye chicks are born at the end of May and in early June, and they have a richly coloured downy plumage, with white cheeks, white belly and small white spots on the lower back contrasting with the darker upper parts and the dark chest band.

The male is larger than the female, and the drake's wingspan is between 209–235 mm, compared with the female's, which is between 193 mm and 205 mm. A mature bird normally weighs between 0.8 and 1.1 kg.

The Goldeneye occurs in pairs in the summer, but small flocks of non-breeding birds are often to be seen near the nesting areas. These ducks form flocks in winter, usually of 20–30 birds and sometimes as many as 100.

The Goldeneye is vocally silent except during mating time when pairing takes place. The drake throws his head back and utters a creaking *knirr*, while the female gives a rough-sounding *garr-gra*. When the chicks are in down their bills are grey in colour, later becoming olive-brown among the juvenile birds and greyish-black among the mature adults. As they develop the drake acquires his superb plumage, while the female develops a yellowish band inside the point of her upper bill. The eyes of the young chicks are brownish, but gradually change through greyish-white to become a bright yellow in the adult stage. The feet of the young birds are greyish-yellow, but those of the mature adults are orange-yellow.

The Goldeneye is usually active by day, but these birds migrate by night. The oldest recorded Goldeneye lived to be 17 years old.

## Reproduction

The female Goldeneye usually comes into breeding condition in her second year, but many do not breed until they are older. The males are slower to mature sexually than the females.

The female goldeneye returns year after year to the same nesting place, and young females return to breed in the areas where they were born. However, the males tend to emigrate, and pairing only lasts for one season. This tendency naturally contributes to their genes being widely mixed.

By the time the birds have returned to their nesting areas the competition for mates is usually over, and pairing has taken place. However, females often compete for nesting sites. Goldeneye build their nests in holes, sometimes in tree trunks, in nesting boxes and even in the chimneys of cottages. Normally several females will lay their eggs together in one place, and while this may be partly due to a shortage of suitable nesting places there may also be a more serious biologicial reason. There is an obvious advantage for any female which is spared the stress and exertion of incubating, brooding and feeding her chicks, but it is more difficult to see what benefits there are for the female who does all the work.

The female Goldeneye nests in April–June and usually lays a clutch of of between 6 and 11 blueish-green eggs. These measure 52–64 mm x 41–46 mm, and the normal weight is 56–64 grams. If the eggs are destroyed or removed a replacement clutch is normally laid, a fact which egg collectors exploited in the past. The female takes sole responsibility for the incubation and brooding, and the eggs hatch after a period of 27–32 days. The chicks quickly become active and will jump unassisted from the nest, which may be as much as 10 metres off the ground. Within eight or nine weeks of hatching the chicks will be fully fledged, and during that time the drakes play no part in looking after them or feeding them. Sometimes several broods of Goldeneye chicks will join together to form a larger juvenile group.

European studies have shown that the hatching success of the eggs is variable, with anything from 22%–94% of eggs hatching successfully. However, the majority of clutches hatch with 50%–80% success. A high proportion of unhatched eggs were found not to have been fully incubated, and a small proportion were not fertilised. It would appear that those clutches of eggs which are incubated by a number of different females are often abandoned. The rate of chick survival after hatching is variable, and as many as half the chicks may fail to survive to the fledging stage. All in all, the Goldeneye's breeding success seems to vary a great deal from place to place and from year to year. In particular the onset of egg laying and the survival rate of the chicks appear to depend on the weather, in addition to being influenced by the age structure of the breeding population.

## Special adaptations and behaviour

The Goldeneye performs various ritual courtship movements which are similar to those of other species of diving ducks. These include throwing their heads backwards over their backs, repeated diving and landing and taking off repeatedly from the water. However, the Goldeneye differs from the majority of diving ducks, including the Scaup, the Tufted duck and the Pochard, by its distinctively shaped skull and bill, by the structure of the bronchia, the colour of the eggs and by the downy colour of the young chicks. The above diving ducks have yellowish-green chicks, while the Long-tailed duck, the Smew and the Common Scoter have dark coloured chicks, similar to the Goldeneye. It has also been observed that the structure of the chest bone corresponds very closely to that of the mergansers.

Goldeneye show no reluctance to interbreed and produce hybrids with the smallest of the Merganser family, the Smew. The Smew drake undertakes the the same head-tossing displays at mating time as the Goldeneye. When mating, the Goldeneye mates with the female in the water, and he flutters his half-opened wings. This mating behaviour is unique to the Goldeneye and the Smew.

A Goldeneye-Smew hybrid was first fully described from a single specimen shot in Scandinavia in November 1881. Prior to this it had been designated by the name *Mergus anatarius*, from a specimen which had been found in Germany in 1825, and another individual was shot in Scandinavia in 1843 and given the name *Anas clangula mergoides*. Such hybrids appear now to be commoner than hitherto and are called *Bucephala clangula x Mergus albellus*.

Breeding Goldeneye females moult together with their chicks in the vicinity of the nesting area, whereas drakes and non-breeding juveniles, together with non-breeding females, fly off to separate moulting grounds. Some small moulting flocks can be found on inland lakes but larger flocks are to be found in suitable coastal areas.

## Population structure and numbers

Since the Goldeneye is principally a second-year breeder, the number of young birds in each different age-group depends upon the breeding success of the preceding years. The survival rate of the older birds is also an important factor, and annual mortality among breeding females may be as high as 37%. It is presumed that there is a general predominance of younger birds among the breeding females, as juvenile females mature and begin to breed.

Goldeneye appear to be increasing in numbers in most of the species' European haunts, and their range has also been extended. The Swedish and Baltic population in the mid-1970s was put at an estimated 100,000 breeding pairs, and the Goldeneye was estimated to be the commonest duck species breeding in this area. It is thought that the species has benefited from the erection of nesting boxes, and also that some of the effects of acid rain have been to the benefit of the Goldeneye. The international mid-winter census of seabirds has revealed an estimated 200,000 Goldeneye, although the actual population is believed to be very much larger.

## Occurrence and food

The Goldeneye belongs to the holarctic group of birds, which indicates that it occurs in the northern parts of both the Old World and North America. Its chief stronghold is in the lakes amid the northern conifer forests, but the erection of nesting boxes and the planned release of ducks, allied to natural dispersal and expansion of range, has led to the species spread well to the south of the region of coniferous forests. During the twentieth century the species has taken up permanent residence in the British Isles, Poland and Czechoslovakia and has also bred occasionally in Denmark, Switzerland, Rumania and Yugoslavia.

The Goldeneye is chiefly found in lakes and will also breed by flowing water. The actual nesting site may be situated at some distance from the water, and the birds do not require reeds, rushes or marine vegetation in which to nest.

Most feeding is done by diving and feeding along the bottom at a depth of up to 5 metres, but the Goldeneye is quite capable of diving to depths of 10 metres and more. This species is omnivorous, but it is chiefly carnivorous and feeds avidly on molluscs, small mussels and shrimps in salt water. When breeding close to fresh water their diet comprises a large proportion of insects and invertebrate larvae. In winter more vegetation is consumed, and this includes the seeds of pondweed and bulrushes, the leaves of Water Milfoil and Water Lobelia. In certain areas the Goldeneye is believed to have benefited from the effects of acid rain, which has killed off various species of fish which previously competed with it for insect foods.

## Relations and races

Although the Goldeneye is distinct in several respects from other small diving ducks, the various members of the Goldeneye family are very much alike. Three species occur worldwide. The closest relative of the European Goldeneye is the Barrow's Goldeneye (*B. islandica*). As this name implies, this species is found chiefly in Iceland, but it also occurs widely in North America from Greenland and Labrador and westwards to Alaska. The adult male can be distinguished from the Goldeneye by the half-moon shape of the cheek spots and by the more purplish-blue tones of the plumage of the head, which is also more elongated. In addition, the neck feathers also grow more directly outwards. The female Barrow's Goldeneye closely resembles the European Goldeneye, but both sexes of the Barrow's Goldeneye have fewer white wing markings. This species occurs occasionally in western Europe, but only in very small numbers, and the Icelandic-Greenland birds are quite sedentary.

The other related species is the American Bufflehead (*B. albeola*), which is found across North America from Ontario to Alaska. Both the male and female of this species can be quite clearly distinguished, for the males have a large white triangular marking on the back of the

head, while the female has an elongated oval spot. The drake's head is coloured from reddish-purple to green, and the birds' feet are pink. The Bufflehead occurs from time to time in the British Isles, but is otherwise rare outside its North American range.

The familiar European Goldeneye also occurs in North America, where it breeds continuously from Newfoundland westwards to Alaska. The American race (*B. c. americana*) is slightly larger than the Eurasian race, but is otherwise almost identical.

## Migration

The Goldeneye of Russia and Scandinavia leave their northern breeding grounds in October–November, and occasionally in September, to winter in milder parts of western Europe. Birds breeding in north-eastern and eastern Europe overwinter in the vicinity of the Black Sea.

The species' most important winter quarters are in the south-western Baltic, and other important wintering populations are found around the coasts of Scotland and the north of Ireland, in Holland and in Switzerland. Birds which have bred in Fenno-Scandia have been found from ringing returns to winter in all these areas, but their principal winter quarters are along the Danish coast.

As is the case with most ducks, the more northerly wintering populations have been found to be dominated by large numbers of mature drakes, while larger proportions of females and juvenile birds occur in the more southerly wintering areas.

In summer Goldeneye drakes, non-breeding females and young birds undertake a moulting migration. Certain moulting grounds in Denmark attract 10–20,000 birds, and around Scandinavian coasts flocks of up to 1000 birds are to be found. The moulting flocks gather in late July and August. Moulting birds come from east of the Ural mountains to moult on the lakes in the wooded northern areas of Siberia.

The spring or pre-breeding migration takes place in March–April, and the flocks of Goldeneye follow the retreating edge of the ice on their northward journey, and arrive early on ice holes, ponds and other open waters around the breeding grounds.

## Shooting and hunting

Goldeneye are gregarious and form large winter flocks, and this habit has been exploited by commercial hunters and latterly by sportsmen through the use of decoys. At one time these were made of carved wood, but modern plastics have replaced them and the skillfully carved and hand-painted wooden decoys of old are now valuable collectors' items.

Hunting and shooting with the help of decoys has been carried on world-wide, with decoys for wildfowl, waders, pigeons and small passerine birds. Diving ducks used to be the special quarry of those who used decoys for shooting along the coastline, and in certain northern countries spring shooting was an annual tradition. The birds were hunted as they returned northwards from their wintering grounds to their breeding areas. The old style of hand-painted wooden decoys tended always to show the birds in full breeding plumage.

Decoying requires that the decoys be placed close to the birds' accustomed flight lines and in places where the birds normally alight, and it is rare for fewer than ten decoys to be used. At one time sea ducks were hunted by stretching nets across narrow coastal channels and at other critical points on the birds' flight lines. However, the art of decoying was stimulated by the growth of shooting and the development of shotguns. Among sea ducks the Goldeneye and the Long-tailed duck were regarded as the two which could most easily be lured to within range of the waiting guns.

*Barrow's Goldeneye male (left) and Common Goldeneye*

# Long-tailed duck

## *Clangula hyemalis*

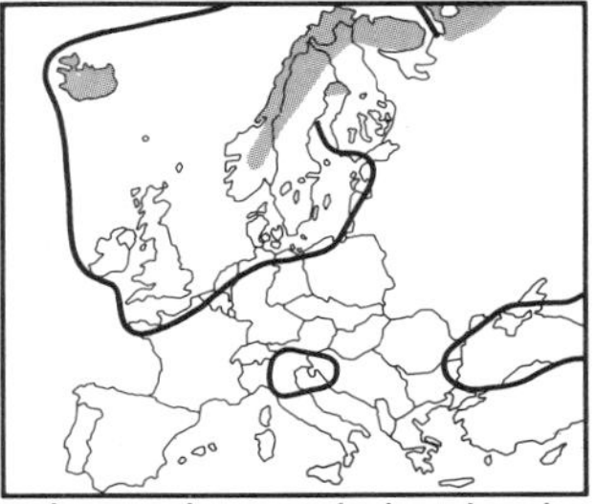

The Long-tailed duck has been given many local and vernacular names. Most of these are onomatopoeic, seeking to imitate the sound of the birds' call, and most end with *awl* or *all*. Sometimes the names include a prefix which refers to the colour of the birds' plumage, for this species varies significantly in appearance depending on the sex and the time of year.

The calls of the Long-tailed duck at dawn — a soft and melodious chorus from hundreds or thousands of voices — imparts an inimitable atmosphere to the scene. But Long-tailed ducks tend to remain far out at sea, usually close to shallows and banks where they can find mussels to eat. Hunting and shooting the Long-tailed duck has usually involved going out to sea in a boat. Among Scandinavian sportsmen and commercial hunters the Long-tailed duck has traditionally been highly prized, together with the Eider and the Goldeneye. The nineteenth-century Swedish naturalist and sportsman Gustav Kolthoff described how in severe winter weather Long-tailed ducks would gather in enormous numbers at holes in the sea ice or where currents kept the water open, and there they were shot in large numbers. The birds were shot either by stalking them in a boat or decoying them within range.

The greatest and most recent destruction of Long-tailed ducks has been caused not by shooting but by oil discharges at sea. From the 1940s onwards it was an annual event to find tens of thousands of Long-tailed ducks floating in to shore, either dead or close to death, with their feathers soaked in oil. This was largely due to small freighters and other vessels flushing out their tanks at sea as they made their way to winter drydocks in the Baltic harbours. In February 1976 some 30,000 oil-soaked Long-tailed ducks were found off Sweden, their plight entirely due to oil discharges from ships cleaning their tanks. During the 1980s this grave situation improved considerably as a result of international conventions forbidding oil discharges.

### Characteristics

The Long-tailed duck is a small, plump diving duck with a rounded head, a high forehead and a short bill, and of all ducks it exhibits the most variation in plumage. In summer the plumage of the drake is dark and mainly brown, but in winter this is much more magnificent looking, consisting chiefly of black and white, with a white crown to the head and a large sooty brown patch to the rear of the cheek. The drake also has a special eclipse plumage, which characterises it during the post-breeding moulting period. At this time its plumage is not dissimilar to the brownish summer plumage, but the bird has a pale greyish-white head, and the elongated central tail feathers which give this bird its name are missing at this time of year. Young birds resemble the appearance of the mature females in winter, with a pale head and neck which turns darker in the summer. No other duck species displays such plumage variations among the females. The young chicks in down are coloured blackish-brown with grey-white cheeks and light spots above, below and to the rear of the eyes, and their bellies are pale coloured.

*Male in winter plumage*

The Long-tailed drake is larger than the female, with a wingspan in the range 209–240 mm, while that of the female is 180–222 mm. The body weight of the adult birds is within the range 580–930 grams, and varies according to the time of year. The Long-tailed duck is very sociable and gregarious, and forms very large flocks in winter. Their flight is usually low and fast over the water, with a rapid wing beat. In posture they are similar to the dabbling ducks, for they ride high in the water and that long tail points obliquely upwards. They can take flight quickly from the water, and dive with great agility. In appearance and behaviour they most closely resemble the Goldeneye.

## Reproduction

Like most duck species the Long-tailed duck is monogamous, with pairs forming in the winter and lasting for one breeding season. The drakes court the females on the water, with groups of birds displaying in a formal and stylised way. Most birds are believed to breed in their second year. More than most other ducks, the male Long-tailed ducks remain in the vicinity of the nesting sites while incubation is in progress, and even while the female is brooding the young chicks. However, the choice of nesting site, the construction of the nest and the task of incubation and brooding are all the sole responsibility of the female. The nest is in a depression in the ground, and is usually sited quite close to the water. It is lined with dark, lightly spotted down and the clutch of eggs, from five to nine in number, is laid in the period June–July, although some early clutches are complete in May. The eggs are greyish-green and an elongated oval in shape. The egg weights are in the range 50–60 grams and the average dimensions are 50–57 mm x 35–41 mm. A number of Long-tailed ducks occasionally lay their eggs together in one nest, as do some other species of diving ducks, and this can lead to clutches of up to 10–17 eggs. Egg-laying usually begins in the period from late May–July, but laying takes place later in the most northerly and easterly parts of the species' breeding range.

Incubation lasts for 24–28 days, and the chicks are fledged and ready to fly five or six weeks after hatching. At hatching the average weight of a chick is 29 grams, but by the time they are fledged and flying their body weights will have increased to almost 500 grams.

## Distribution and food

The Long-tailed duck breeds around the coastal beaches and bays of northern Norway, and also on the inland tundra lakes, and in Scandinavia most breeding grounds are north of latitude 60°. In northern Norway these ducks nest at sea level, but in Sweden they occur mainly in lakes more than 500 metres above sea level, and in Finland it nests close to the upper tree line. In northern Norway the breeding density can reach four pairs per 100 km$^2$ but in southern Norway and Sweden the density is more like one breeding pair per 100 km$^2$.

The Long-tailed duck is claimed to have bred on several occasions in northern Britain, and there are records from Shetland in 1848 and 1887, and from Orkney in 1911. During the period 1969–72 pairs of Long-tailed ducks were recorded in the Outer Hebrides and in Shetland, and breeding was strongly suspected at that time. Since 1954 this species has enjoyed total protection in Britain.

In winter this species is very much a sea duck and tends to remain out on open waters, especially over suitable shallow banks which afford ample food. Long-tailed ducks are believed to be able to dive to depths of 30 metres, and a feeding flock in search of food moves in much the same way as Mergansers. The ducks at the front of the flock dive just as those at the rear of the flock have surfaced, to fly forward and dive again. In smaller flocks Long-tailed ducks will often dive in a more haphazard way, more like Goldeneye, Eider and Tufted ducks.

Their winter diet comprises chiefly mussels, although the limited gape of the birds' bills puts a limit on the size of mussels which can be swallowed. During the breeding season in high mountain lakes their diet comprises insects and invertebrate larvae, freshwater crayfish and molluscs, and also the seeds of some aquatic plants.

## Distribution and numbers

The Long-tailed duck belongs to the holarctic fauna, which means it occurs in all the area around the North Pole. It is just as common in North America as in northern Europe and Asia. This species' habit of remaining well out to sea has made accurate winter censuses difficult to carry out, but an estimated 50,000 individuals winter off the Norwegian coast and a further 350,000 in the Baltic Sea area, and it has been estimated that the total winter population of Long-tailed ducks in northern Europe is approaching 500,000 individuals. Those birds which winter around the coasts of Britain and Ireland are at the south-western extremity of the species' range, and they are chiefly found around the coasts of Shetland and Orkney and along the north-eastern coast of Scotland. The average British and Irish wintering population has been estimated at 20,000 birds.

Several hundred thousand pairs of Long-tailed ducks breed in Iceland, but the species is generally sedentary there, and those birds which do migrate tend to move to southern Greenland. Thus their numbers do not swell the total wintering population in northern Europe.

## Migration

Studies of ringed birds have shown that the birds which winter in the Baltic Sea have mainly come from breeding areas well to the east of the Ural mountains, and other birds from the same areas may also swell the numbers of wintering ducks off the Norwegian coast. It is assumed that those Long-tailed ducks which breed in the upland areas of Norway and Sweden move to the Atlantic coasts in winter, while some of the Icelandic birds move to winter quarters around the coasts of southern Greenland.

Ringing returns also show that there is a good deal of movement of breeding populations around the northern latitudes, which makes it difficult to determine specific local races, and it has never been possible to distinguish physical characteristics to indicate different races.

Autumn migration takes place in the period from mid-September to November. During the early winter local migrations are dictated by the weather, and particularly by the first appearance of sea ice in certain areas. As the winter progresses the Baltic Sea population is pushed further to the south and west.

The spring migration occurs in the period from March to May, and occasionally lasts into June. The last ducks usually leave their wintering areas by early May, but flocks of Long-tailed duck have also been seen off the west coast of Sweden as late as June. The ducks usually arrive at their Russian breeding grounds in late May and June and egg-laying may be delayed as late as June or early July.

## Disease and mortality

The Long-tailed duck has been affected by environmental changes both in its breeding grounds and as itt winter haunts, as has happened with many other duck species. In earlier times shooting accounted for a large annual harvest of this species, which was regarded as easily approached and shot in large numbers owing to the birds' gregarious habits. In certain areas the winter populations were very heavily shot.

Decreasing numbers of breeding Long-tailed ducks in Sweden is largely attributed to changes in the availability of suitable food. Various species of fish stocked artificially in the lakes have provided serious competition for freshwater feeding such as aquatic invertebrates and insects which are essential for the early growth and survival of the ducklings.

The large winter flocks of Long-tailed ducks have also been ravaged by oil spillages, and for many decades the mortality due to oil has far exceeded that which has been caused by shooting.

*Long-tailed ducks in summer plumage*

# Shelduck

## *Tadorna tadorna*

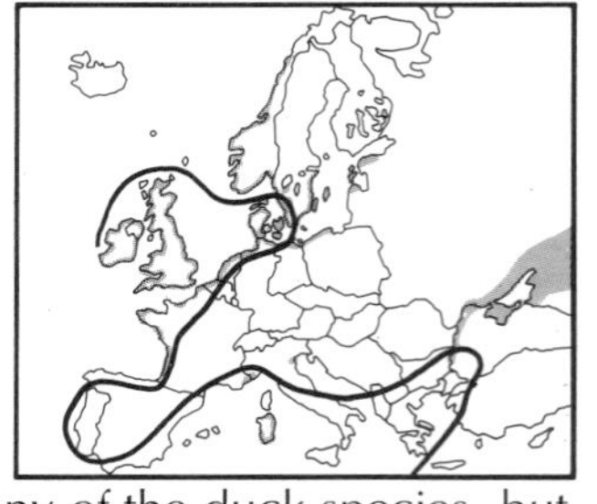

The Shelduck is regarded as occupying a position somewhere between geese and ducks, for the Shelduck and its five close relatives have certain of the characteristics of both ducks and geese. In appearance the Shelduck is colourful, as are many of the duck species, but the male and female are very similar in appearance, which is more characteristic of geese. Their plumage is largely the same in both winter and summer. This duck's legs are comparatively long, like those of geese, and this is especially noticeable in the young ducklings. In terms of behaviour Shelduck are like geese, for the pairs remain together during the breeding season. Although the Shelduck drake does not take any direct part in the incubation of the eggs or the brooding of the young, he remains close by and rejoins the family at a later stage. Most of the Shelduck family like to feed by grazing on grassy fields, like geese, and so the Shelduck species are sometimes referred to as "half-geese", and they are assigned to a special subfamily among the ducks, the *Tadorni*.

### Characteristics

The mature Shelduck has an unmistakable appearance, and the drake can be distinguished from the female by his more distinct colouring and his larger, redder bill with a prominent tubercule. The female tends to have a conspicuous white feathery area around the base of the bill. The juvenile birds are more palely marked in grey, greyish-black and white, and they lack the chestnut-coloured breast marking

of the adults. The duckling in down has a characteristic black and white colour. In their eclipse plumage the adult birds have paler and less distinct colours. The chestnut-brown breast colouring and the blackish-brown belly are heavily marked with white feathering.

The drake is considerably larger than the female; he weighs about 20%–25% more and his wingspan is in the range 317–337 mm, while that of the female is within the range 295–311 mm.

The calls of Shelduck are quite unlike those of geese and ducks. The drake utters a whistling sound, together with his bobbing, rolling head movements which are part of his courtship display, and the females' calls are harsh and somewhat goose-like cacklings which can be heard at a considerable distance, as well as a softer, hoarser *korrr*. However, at most times the Shelduck is a silent bird.

Shelduck are most active by day, and unlike many species of dabbling and diving ducks it does not make regular morning and evening flights to and from its feeding grounds. It seeks its food by searching in shallow water and by sifting along the water surface and probing along the bottom, filtering the water through the tooth-like serrations or *lamella* of the edges of its bill. The Shelduck's food comprises small invertebrates, molluscs, mussels, marine worms and various marine insects and their larvae. Although the Shelduck appears to feed on whatever is available, it has been noticed that one specific type of food may predominate where it is especially abundant, and these birds are therefore capable of adapting to a specialised and preferred diet when conditions are suitable.

It is not known how long Shelduck can live, but birds of at least 16 years of age have been recorded. It is surprising that although a large proportion of Shelduck will not breed in any given season, the annual mortality among the adults is high, estimated at 30%–38%.

Shelduck have been known to form pairs as early as the second season of life, but the female will not normally lay any eggs for a further season, and the drakes do not normally begin to breed until their second or third year. This results in the formation of groups of non-breeding birds, and there may be a ratio of several hundreds of non-breeders to comparatively few breeding pairs.

When one considers the generally high mortality rate, the population ought to be falling in many areas, but this does not appear to be the case, and a steady increase in Shelduck numbers have been noted in many parts of Europe. However, British studies have indicated that the breeding success of some populations cannot be sufficient to maintain existing population levels.

The Shelduck is monogamous, but the pair formation is generally considered to last for one season only. When the ducklings hatch both parents take care of them, and in areas where there are several breeding pairs broods of the same age will often join together into larger units. At this time some of the adults leave their young, but the females are more inclined to remain than the drakes. Occasionally it is possible to find an extended brood of up to 20 ducklings led by only one adult bird. Otherwise, however, both parents share in the responsibility of caring for their brood until the young have fledged.

The female Shelduck often returns year after year to the same nesting place. The nests are made in holes — in old fox earths and rabbit warrens, in mounds of stones, beneath boat houses and in haystacks. The average size of the clutch will be 8–12 eggs, which are creamy-white, oval and rather large. The dimensions of the eggs are within the ranges 60–70 mm x 45–50 mm, and the newly hatched chick weighs an average of 85 grams. Some very much larger clutches of up to 20 eggs have been reported, but these are almost certainly made by two or more females laying their eggs together.

Egg-laying takes place in May–June, and in the North Sea area it can take place as early as March. The female has sole responsibility for incubation, and this lasts for 28–31 days. The young ducklings usually fledge after 42–50 days, but may sometimes take up to 60 days before they are fully fledged. The whole breeding cycle from egg-laying to the fledging of the young therefore takes about 3½ months, and there is therefore no time for a second brood to be reared.

## Migration and distribution

Shelduck in northern Europe are usually migratory, especially around the Baltic Sea. By contrast, most of the North Sea populations are sedentary. However, the migration movements of the Baltic birds are not long, and small flocks winter in Denmark and in south-west Sweden while the majority move to the North Sea. The autumn migration takes place in the period from August to October and the return passage in the period February to April.

The Shelduck makes a special migration to its moulting grounds, of which the biggest and best known is on the German coast between the estuaries of the rivers Weser and Ems. In Great Britain, Belgium and Denmark there are smaller moulting grounds which attract 1–2,000 birds. Once on their moulting grounds the birds shed their feathers and become flightless for a period of a few weeks.

The Shelduck is believed to have its origins among the lakes of the steppes of southern Asia. It still occurs extensively in this area and it also occurs in some parts of the Mediterranean, and there is a small Tunisian population.

This species is mainly a salt water bird, but in many places, including southern Sweden, it also lives by inland freshwater lakes and it also occurs on the Hungarian *puzsta*, the great Hungarian plain. The entire western European wintering population has been estimated at 120–130,000 birds, of which an estimated 65–75,000 birds winter in Britain and Ireland.

# Wigeon

## *Anas penelope*

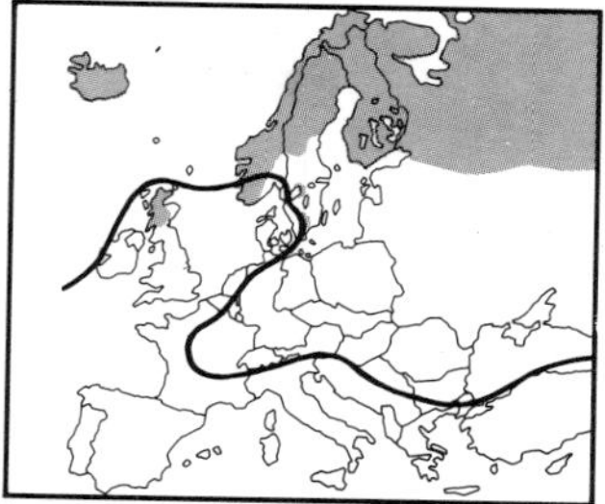

The Latin names for animals and birds often have interesting stories associated with them. Sometimes the name alludes to appearance or to size, or it celebrates the name of the scientist who first discovered the species. He who identifies a new species generally has a fairly free choice in selecting a Latin name for it. When the Swedish naturalist Linnaeus was lecturing his students at Uppsala in the middle of the eighteenth century he called the wigeon *Penelope*, which is a familiar first name for a girl, and by which Carl von Linné sought to honour the wife of the Greek hero, Odysseus. According to Greek mythology, Penelope was "a paragon of domesticity and feminine virtue".

At the beginning of the nineteenth century, naturalists divided the family *Anas* into a number of separate families, with the intention of emphasising that the different species of these various families were not immediately related to one another. And so the Pintail duck ended up in the family *Dafila*, the Shoveller in the family *Spatula*, and the Wigeon was given the family name *Mareca*, and thus its full Latin name became *Mareca penelope*. Latterly the original name has been restored and today the scientific name of the wigeon is *Anas penelope*.

## Characteristics

The wigeon is a compact and plump little duck, with a relatively short neck and body. The bill is comparatively small, the bird's forehead is high and the crown of the head is domed. The mature drake in its second year and older always has a large white patch on the wing, and this is clearly visible when the bird is in flight. The female and immature birds have grey patches on the wings. In eclipse plumage the immature birds and the drakes look very similar to the females. In flight the bird's white belly is very prominent.

The drake is usually slightly larger than the female, and his wingspan falls within the range 239–277 mm, while the females is in the range 230–262 mm. The adult birds weigh up to 1.1 kg.

Wigeon feed chiefly on vegetation, but in the summer part of the diet is composed of small invertebrates and insects. Otherwise the wigeon tends to eat various species of water plants, including pondweed, grass and algae. In winter it feeds readily on various types of sea grasses, especially the *zostera* or eel-grass which plays a big part in its diet. When feeding the wigeon ducks its head but does not dive, and when on the water tends to pick water plants from the surface: more frequently it is seen to graze on dry land. On the water, however, wigeon may be seen in company with flocks of mute swans and coots, and it feeds on the fragments of water plants which are torn up by these other birds and left floating on the surface. Wigeon often feed at dawn and dusk, and on bright moonlight nights they will feed on land during the night. The flights between their feeding areas and resting places take place at dawn and dusk, and on dark and rainy winter days wigeon flocks will also flight by day to their feeding grounds on salt marshes and freshwater marshlands.

The wigeon is a bird which is invariably found in flocks, and from flocks in flight one can hear the whistling calls of the drakes. In the springtime the wigeon drakes whistle especially vigorously, while the females utter a soft *terr* call.

Individual ringed wigeon have been found to live to the age of 18 years.

## Breeding

As with most species of dabbling ducks, the wigeon's monogamous pairings tend to last only for one breeding season. Wigeon are believed to breed as early as their first year, but many do not do so until their second year. The pair bond is broken after the eggs have been laid, when the drakes fly off to moult together.

The female makes her nest in tall vegetation, among low bushes or in osier beds, usually near lakes and slow flowing water-courses, on small islands, and along quiet beaches. The nest may sometimes be situated some considerable distance away from water. The eggs are laid in May–June and a typical clutch will comprise 7–11 eggs, which are yellowish-white in colour. The eggs weigh an average of 45 grams and the dimensions are within the ranges 50–59 mm x 36–41 mm. As with some other duck species, it occasionally happens that two or more females will lay their eggs in one nest. Studies in Finland have revealed that an average of 75% of eggs hatch successfully, and that 34% of the eggs will result in fledged young. The female takes sole responsibility for incubating the eggs and brooding the young, and the ducklings will be fully fledged some 40–45 days after hatching. The whole breeding cycle therefore lasts approximately 2½ months, and the wigeon only produces one brood of young each year.

In the British Isles the wigeon breeds in much of highland Scotland, in the southern uplands and in the northern Pennines. The largest single concentration of breeding wigeon is on Loch Leven, but elsewhere the birds tend to favour upland tarns, small lochs and hill streams. The total British breeding population in the 1970s was estimated at approximately 300–500 pairs.

## Distribution and numbers

As a breeding bird the wigeon is widespread across northern Europe, ranging as far south as northern Britain and the Baltic shores of Germany and Poland. In northern Europe it breeds chiefly in the wooded lakes and marshlands of the taiga zone, with some birds also being found in mixed forest areas and in high mountain regions well above the tree line.

The wigeon is a rare resident in the Baltic Sea area, and most northern European wigeon have their winter quarters in western and south-western Europe. Many spend the winter along the North Sea coasts of Belgium, the Netherlands and Germany, and also in coastal lagoons and on inland waters. More than 95% of northern European wigeon winter in the Netherlands and Belgium. Large numbers of wigeon from Scandinavia and Siberia winter in Britain and Ireland, and the January total of wigeon wintering in the British Isles has been estimated at 300-350,000.

Wigeon migrate mainly by day in the autumn, but the spring return migration in March–May appears to occur mainly at night.

*Males in summer eclipse plumage.*

*Red-breasted merganser pair.*

*Common Merganser males*

# Mergansers

## Red-breasted Merganser — *Mergus serrator*
## Goosander — *Mergus Merganser*

The Mergansers are fish-eating ducks, equipped with long bills which have a hooked tip to the upper mandible and a row of saw-like teeth — hence the name "sawbills". In bodily shape Mergansers are elongated and their legs are located towards the rear of their bodies. In flight they have a fast wingbeat and both species have a large white speculum on each wing, those of the Goosander being the larger and more prominent.

An adult Red-breasted Merganser will weigh 0.8–1.4 kg, and the male's wingspan is in the range 236–265 mm, while that of the female is 204–240 mm. The Goosander weighs 1.2–2.1 kg. and the male's wingspan is in the range 273–295 mm, while that of the female is 253–270 mm.

The female Red-breasted Merganser lays one clutch of 6–12 yellowish-brown to olive-green eggs, and these are laid late in the season and usually in the period May–August. The egg's weight is in the range 65–82 grams, and the dimensions are 59–71 mm x 42–46 mm. The nest is placed in thick waterside bushes or under an overhanging stone, and the young chicks' down is greyish with prominent white markings. The female alone takes care of the incubation, which lasts 29–35 days. The young birds will fledge within approximately two months.

The female Goosander lays only one clutch of eggs each year, and she will make her nest in holes in trees, abandoned buildings close to the water's edge, and in nesting boxes. The 8–12 creamy-yellow eggs are incubated for 32–35 days by the female alone, and the chicks will fledge in about eight weeks. The eggs weigh in the range 69–98 grams, and the dimensions of the eggs are 63–73 mm x 43–48 mm. The nest is usually lined with greyish-white down.

Female Mergansers and Goosanders shepherd their young with great care. When danger threatens the young birds run along the surface of the water, scatter out and then dive.

Both these Merganser species exist in North America as well as in Europe and Asia, and they occur on lakes and along rivers as well as by the coast. Once the young have hatched the males tend to leave the females and young and fly off to their moulting grounds, and several hundred males may flock together in this way. The birds winter in the Baltic and the North Sea, and extensively inland in western Europe. The British wintering total for the Red-breasted Merganser has been estimated at 11,000 individuals, and the total number of Goosanders at approximately 9,000.

Both species are skilled at fishing, although occasionally they are robbed of their catches by parasitic gulls. The Red-breasted Merganser's diet also includes various crustaceans, worms and invertebrates. When fishing these birds swim with their heads stretched out along the surface of the water, looking for what is above and below the water. The Goosander occasionally fishes in large groups which work in unison to drive shoals of fish ahead of them towards a beach or a shallow bay. These birds fish chiefly by day and they undertake their migration movements mainly by night.

## Diving ducks

The Tufted duck *Aythya fuliga* and the Scaup *Aythya marila* are both medium-sized and closely related diving ducks, although the Scaup is somewhat larger. The male Tufted duck has a wingspan of 194–217 mm, while that of the Scaup is 214–230 mm. The females are slightly smaller than the males and are very similar to one another. However, the female Scaup has a broader white ring around the base of the bill; the Tufted duck has a slight tuft of black feathers on the nape, and the sides of the body are lighter. The Scaup drake has a light grey wavy-line pattern on the back, and he lacks the rather conspicuous nape tuft of the drake Tufted duck. In flight a clearly marked white wingband can be seen on the Tufted duck, whereas the Scaup has a lighter colouring on its upper parts and more diffused markings.

Neither the Tufted duck nor the Scaup is particularly vocal. The males make some soft whistling sounds when courtship is in progress. In both species the females make similar *burr*-ing calls, and these are often to be heard when the birds are flying or have been frightened, and they are audible at some distance.

The Tufted duck is confined to the northern parts of Europe and Asia, while the Scaup can also be found in North America. Both species breed in lakes and sheltered ponds in afforested areas, as well as along the sea shore. The Tufted duck has substantial populations in the southern parts of its range, and it often lives in open lakes where there is abundant food. It will also live in the clear northern lakes which are the favourite haunts of the Scaup.

Both species lay up to ten or more eggs in a clutch, and these are coloured greenish- to grey-brown. The females take sole responsibility for incubation and the care of the growing chicks. Both Tufted ducks and Scaups begin to breed in their first year, but a high proportion wait until another year has passed. Like many other ducks they appear to be monogamous, and the pair bond lasts for one season only.

*Greater Scaup male*

The Tufted duck's incubation period lasts for 22–25 days, while the Scaup, being larger, takes several days longer. However this difference in incubation times is offset by the fact that the young Scaup can fly about a week earlier than the young Tufted duck. The male birds often stay with their mates until all the eggs have been laid, and sometimes until the eggs have hatched, although this is more common with the Scaup. The drakes then leave their families to find a suitable place in which to moult. It is not unusual for the female to leave her young before they are quite able to fly, and this is more common with the Tufted duck.

These species' wintering haunts are in western and southern Europe, and many of the Tufted duck which overwinter in the Baltic area have come from remote breeding grounds in northern Russia.

### Waterfowl and their food

Ducks, geese and swans belong to the thriving family of the *Anatidae* of which there are an estimated 151 species world-wide. Their association with water and their dependence upon it have influenced the development of their bills and their legs and feet. The necks of swans and the teal group of ducks — nearly a metre long and barely 10 cm long respectively — are related to their feeding habits, as are their habits of diving or not diving. Competition for the available food in wetland and coastal environments has led to adaptation and specialisation.

All of these birds can dive, use their legs on land, and swim — but in varying degrees. Swans and geese dive only in difficult situations, but then they can stay underwater for quite a few seconds. Diving ducks and mergansers are specialised divers, the Long-tailed duck and the Eider being able to stay submerged for 60–70 seconds. Geese prefer to graze on land, while Whooper swans do so occasionally, but Mute swans only rarely go up onto the shoreline or salt marshes to feed. In shallow water, swans and geese, as well as the dabbling ducks, reach further by stretching their necks downward and tipping over. When large flocks of waterfowl concentrate in limited areas they place a heavy demand on the available resources of accessible food.

Swans eat green plant material that can easily be reached with the bill and the stretched neck. Their body size and flock numbers demand a great deal of easily consumed food, and a Mute swan eats 3–4 kg of plant food every day.

Geese have much the same habits and requirements. They graze on land and require wide marshes and meadows. Certain species eat seaweeds and sea grasses.

Dabbling ducks eat a wide variety of foods. In summer, before breeding and when rearing their young, many very small animals are consumed. The Shoveler skims the water's surface with its curiously shaped bill to catch planktonic creatures, snails, insects and their larvae, and also small shellfish. These make up half of the Pintail's spring and summer diet, the remainder being mainly the seeds of marine and aquatic plants.

Diving ducks feed mainly on snails and mussels from the bottom, and also plant seeds which they find in the shallows. They need to eat almost half their own body weight each day. Pochard normally dive to a depth of about 2.5 metres, while the Goldeneye, the Tufted duck and the Scaup dive twice as deep. Common and Velvet Scoters dive double that depth, while the Long-tailed duck dives deepest of all. Mergansers are fish-eaters, diving to depths of 10 metres and staying submerged for up to 30–40 seconds.

Those which breed in north-west Europe have a shorter autumn migration.

On its wintering grounds the Tufted duck is rather more conspicuous, and it forms large flocks, sometimes of several thousand birds. It has a preference for river estuaries, harbours and ice-free ponds and larger lakes. Here the ducks lie close together, sometimes resting and sleeping all day. At dusk they flight to their feeding places, and they often combine secluded freshwater daytime roosting sites with night-time feeding grounds along the coast. The Scaup also forms flocks of several thousand birds, but these often remain further out to sea and are therefore more difficult to identify and observe.

The Tufted duck is the commoner of these two species over most of their ranges. In Britain and Ireland the breeding population of Tufted ducks has been estimated at approximately 9,000, while the Scaup has only bred in the British Isles on a very few occasions. The north-west Europe winter population of Tufted ducks is estimated at over half a million, of which some 80,000 winter in Britain and Ireland. The north-west European winter population of Scaup has been estimated at 150,000 birds, of which 5–10,000 winter in Britain and Ireland.

## Diving duck shooting

The principal European wintering grounds of migrant waterfowl have been estimated to sustain around 800,000 Tufted duck, 200,000 Velvet Scoters and 1.5 million Common Scoters. These are much smaller than the estimated figures for the long-tailed duck and the Eider, which together dominate the winter scene with winter populations reckoned at 1.5 million and 2 million respectively.

Diving ducks are usually shot by sportsmen in the late autumn and winter, and in Europe the shooting of sea duck species is especially widespread in Denmark. Approximately ten times as many Eider are shot there annually compared with Sweden, and about twice the total numbers of other diving ducks and mergansers. Long-tailed ducks and Goldeneye are next in importance to the Eider as quarry species, but sportsmen have a decided preference for the various species of dabbling ducks. Tufted duck, Greater Scaup, Common and Velvet Scoter are seldom shot by European sportsmen.

*Tufted duck pair*

## Common Scoter — *Melanitta nigra*
## Velvet Scoter — *Melanitta fusca*

The Common Scoter and the Velvet Scoter are also closely related. The Velvet Scoter is the larger of the two, and it can be distinguished from the Common Scoter by the colour pattern of the bill, in the case of the male, and by the light pattern of colouration on the cheeks in the case of the female. The Velvet Scoter also has a white speculum on each wing, whereas the wings of the Common Scoter are completely black. The Velvet Scoter reaches a maximum adult weight of 1.5 kg. while the Common Scoter can weigh up to 2 kg.

Both species occur in northern Europe, and the southernmost breeding range of the common scoter extends to Dalsland in Scandinavia, and also to Scotland and Ireland. The Velvet Scoter does not breed in the British Isles, where it is a winter visitor and the least common of the various wintering sea duck species. Both species breed in lakes, but the Velvet Scoter also breeds along much of the northern coast of the Baltic Sea.

Like the Tufted duck and the Scaup, both these species lay their eggs in nests on fairly open terrain. The females alone carry out the task of incubation, while the males leave on long migration flights to their moulting grounds. Each summer 150–200,000 Common Scoters gather to moult in Danish waters.

Both species lay 6–10 eggs, and like the Eiders and the Mergansers they begin to breed at two or three years old. Both species are holarctic, occurring in the northern parts of Europe, Asia and North America. Because they are adapted to the conditions of northern climes both lay their eggs relatively late, in June.

These ducks are excellent divers: the Velvet Scoter can dive to depths of up to 10 metres, and both species can stay under water for up to one minute. However, like the Tufted duck and the Scaup they usually dive for shorter periods, and all four species normally remain on water which is less than 5 metres deep, and there they search for food along the bottom.

The Scoters eat chiefly small creatures — mussels, molluscs, crustaceans and, in summer, insects and invertebrate larvae. Some of their diet is composed of vegetation and seeds. The Tufted duck and the Scaup have a similar diet, although the Scaup rarely eats any vegetable matter. But it can consume 3–4,000 small mussels in a single day, amounting in total to 75% of the bird's body weight. It can dive to about 6 metres, while the Tufted duck rarely dives deeper than 3 metres. Both species feed principally in shallow waters.

### Pair formation

Most species of duck mate for the entire breeding season, and females often pair up during their first winter. And the longer a female duck survives and breeds, the more of her genes will be passed on to her offspring. Female ducks are very faithful to their breeding places, and mature individuals will often return to the same site for breeding year after year. The young duck readily return to the place where they were born, and there is therefore genetic continuity in particular regions.

Pair formation therefore influences the genetic composition of local breeding flocks of waterfowl. Biological success, however, depends chiefly upon the individual's success in breeding, and the more successful breeding pairs the better. Male ducks have developed rituals for mating and courtship which promote successful breeding.

The gaudy plumage of the mature male duck plays an important role in this. Waterfowl which mate for the breeding season only, such as dabbling and diving ducks, have more striking male plumage than those species which mate for life, such as geese and swans.

# Grey heron

## *Ardea cinerea*

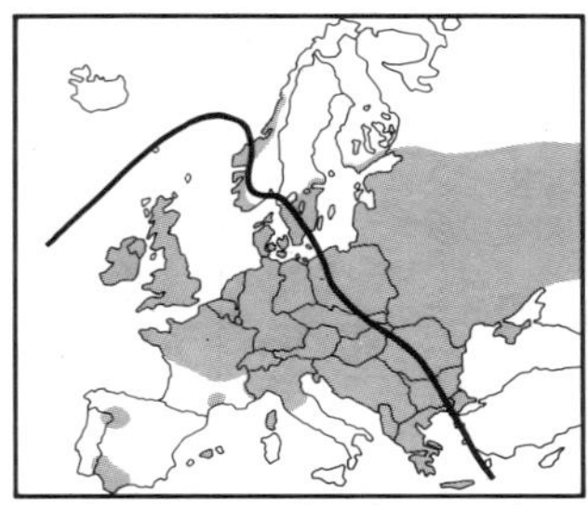

The heron has about sixty related species world-wide, and some of these can be identified as particularly closely related. Among these are the Egret, the Night heron and the Bittern. The familiar Grey heron occurs in North America, Eurasia and Africa. The heron has traditionally been surrounded by a great deal of mystique, legend and folklore, and in former times it played a significant part in man's diet. By the nineteenth century, however, writers expressed no great enthusiasm for the flesh of young herons, although they were said to be edible and decidedly preferable to the tough and indigestible meat of an old heron. The Swedish naturalist Linnaeus remarked that the ancient Romans regarded the entrails of a heron as an especial delicacy.

At one time heron-hawking with trained falcons and flushing dogs was an important field sport among the nobility. The heron's plumes, those long feathers down the bird's back, were very highly regarded as decorations for hats, and the demand was such as to place excessive hunting pressure on several species of the heron family. Other parts of the bird's anatomy also had their uses, and the long leg bones of herons were often used in the eighteenth century as handles for horse whips. In the Sudan the same bone was worn as a decoration hanging from the waistband and was endowed with mysterious protective powers. In Victorian and Edwardian times herons' bills were sometimes used as decorative cigarette holders.

When a heron is fishing it typically stands upright and motionless for long periods, and this patient waiting game is punctuated by the occasional slow forward step and a lightning fast stabbing motion of the bill. At one time naturalists used to believe that the heron's legs gave off some pleasant and alluring smell or taste when it was wading in the water, thereby enticing fish within range of that stabbing bill. Similar beliefs were also attached to the birds' droppings.

### Characteristics

The Grey heron, or the heron as it is more commonly known, cannot easily be mistaken for any other bird except the Crane and the Purple heron. Occasional migrant Purple herons from southern Europe are so rare that there will be few occasions when there is a chance of error. The mature heron has very long black head feathers in addition to the long, thin and pointed light grey back feathers, and the greyish-white flank feathers are also elongated. None of these characteristics is present in the young birds, which are otherwise generally similar in appearance. The sexes are alike, and the bird's eyes are bright yellow, while the legs vary in colour from greyish-green to yellowish-brown. It is interesting to note that the bill of a young heron is yellowish-green on the lower mandible, while the upper mandible is

greyish-green. However, the mature birds have a more uniformly bright yellow bill colour, sometimes with a hint of red.

Herons fly with their necks bent and withdrawn in a S-shaped fashion, and the neck is only extended in flight if they are about to alight or to fly off, or if they are alarmed and chased by birds of prey. By contrast, Cranes and Storks fly with their necks outstretched and with a flatter wing profile than the rather hunched and arched appearance of the heron's wings. herons are slow fliers and are usually seen singly or in twos and threes, although the birds will frequent large collective roosting sites, especially in cold weather.

The mature heron is a large bird, and the sexes are almost the same size. The wingspan of a typical male heron will fall within the range 430–490 mm, while that of the female will be 433–486 mm. The males have noticeably longer bills, necks and legs. Despite their large size a fully grown heron will normally weigh not more than 1–2 kg, a very low body weight when compared to the various duck and goose species with comparably large wingspans.

When they are newly hatched from the egg, young heron chicks weigh only a mere 40 grams, but the young birds' growth is very rapid and fledgelings ready to fly for the first time will be almost the same size and weight as the adult birds. In its first week of life the chick quadruples its body weight; at the age of two weeks it will weigh approximately 560 grams, and this weight will double in another week. By the time the young heron is three months old it will weigh almost 2 kg.

The heron has a large and powerful bill, between 100 mm and 134 mm long. The tip is sharply pointed, but the heron does not normally impale its fish prey with its bill, as used to be thought. Instead it tends to grip tightly with the upper and lower mandibles, and a secure grip is made easier by the fine serrations on the outer edges of the bill.

The legs, and especially the tarsus bones, are very long. When searching for food, the heron often wades in relatively deep water. Its feet are not webbed, but the long and spreading toes enable it to walk about on soft mud and sand. The heron's middle toe is unusually shaped and is used when the bird is preening.

The body feathers of the heron have a dull and coarse surface, and the plumage is covered with small powdery scales. The young herons undergo a partial moult during their first winter, but their first complete moult does not begin until the following spring and early summer, rather earlier than that of the adult birds. By the time they are in their second year, the young birds look like adults, but have shorter feathers on the back and crown.

## Breeding

Herons are capable of breeding when they are one year old, but the majority do not begin breeding until they are two or three years old. herons are tree nesters and breed in colonies, and a heronry may comprise as many as 100 nests and breeding pairs. The size of the heronry often reflects the amount of suitable food available in the vicinity.

The older birds begin nesting before the younger ones, and the nesting site is chosen by the male bird. The older birds take over the bigger, older nests, and the youngsters and late-comers must make do with the poorer nests or build new ones. Typically, the nests are built high in tall trees, and these may be either conifers or deciduous trees. The birds pair up at the heronry, and each pair stays together for the breeding season.

The male heron attracts females to his nest by combination of calls and display signals. He shows off his long chest plumes and the pattern of feathering on his wings by bending his head and neck far back. The courtship display continues with a series of low, soft calls and the male bird finishes his display in a forward, hunched position with his head and neck thrust forwards and downward, with his crown feathers raised. He signals the end of his display routine by a clattering of his bill.

Both male and female participate in the building or refurbishing of the nest, carrying sprigs and small branches in their bills. Occasionally a single pair of herons, or even a handful of pairs, may choose to nest in a tall reed bed by a lake shore, and even in this untypical nesting site the nests are built with twigs and small branches.

A typical heron's clutch will comprise four or five eggs, coloured a uniform blueish-green. Egg-laying can take place at any time between March and July, but April is the peak month. The eggs weigh about 60 grams and the average dimensions are 57–67 mm x 40–45 mm. Incubation is shared by both partners and lasts for 25–30 days, and incubation will often begin when the first egg is laid, which means that the eggs hatch in succession and the chicks vary slightly in age. Young herons are fledged and ready to leave the nest by the time they are 7–8 weeks old, but for some weeks they make only short flights and return regularly to the nest site. The total breeding period from nesting to fledging lasts for about three months, and herons only produce one brood each year. However, if eggs or chicks are lost at an early stage the breeding pair will lay another clutch of eggs.

Near the nest a pair of breeding herons will greet one another by standing in a stiff, upright position. Other birds are driven away by a threat display in which the neck feathers are raised and the bird's head and neck are thrust forward towards the intruder. Often this threat display is accompanied by angry cries, and herons will also chase one another in flight.

## Migration

Outside the breeding colony herons are not very sociable birds. Two or three birds may travel together on longer flights, and perhaps as many as ten may fly together on migration. It is thought that the birds keep together in flight mainly by visual contact, although an occasional call may be uttered by one or two of the birds. The same is true of herons which fly to and fro between the heronry and their feeding grounds. herons seem to call chiefly at dawn and dusk, or when they are frightened in flight. Otherwise herons are relatively silent birds. herons are active throughout the daylight hours although most activity can usually be seen at first and last light.

In most parts of its European range the heron is a migratory bird, but certain populations are sedentary, especially in the British Isles and western Europe. Some herons in central and northern Europe also remain throughout the winter if the weather is mild. A succession of mild winters can cause the numbers of resident birds to increase rapidly, but when winter conditions become severe the birds will either perish or be forced to migrate.

In northern Europe some herons will remain near their breeding grounds if the winter weather is sufficiently mild. Occasionally large numbers of birds may flock together in the vicinity of some especially rich source of food, but the birds are usually more thinly scattered. However, herons are very susceptible to cold weather, and when their feeding grounds become frozen they will weaken and die very rapidly.

Many European herons migrate to winter quarters in southern and western Europe, while others go to northern Africa and also to Africa beyond the Sahara. Young birds disperse in many directions after they leave the nesting colonies, and in summer and early autumn they may be found well to the north of their normal areas of distribution, and this brief dispersal is sometimes referred to as wandering or nomadic migration.

There is a high rate of mortality among young herons. As many as 80% of young birds are destined to die in the first year, and many of those which survive will not live beyond their second year. Only some 6%–7% of birds will survive beyond their second year, and the future of the heron population therefore depends upon their breeding success. Colonies of breeding herons are therefore composed principally of older birds which have succeeded in surviving the rigours of migration and of cold winters.

The heron does not normally encounter any competition from other bird species. However, in some places it will share its nesting sites with cormorants, and this has been seen to occur in Sweden and Denmark. In these instances there appears to be some degree of competition between the species for the best nesting sites.

Herons can be quite long-lived, and some leg-banded birds have been found to live for up to twenty years and more.

In Africa the heron breeds mainly in the south, as far north as the equator; it also occurs in Morocco, Algeria and Tunisia in the north-west, and in Egypt in the north-east. The herons of Europe, Asia and Africa are so alike that they have been considered as belonging to the same species and race, *Ardea c. cinerea*. A race of heron known as *Ardea c. monicae* has been identified from Mauritania, and this is small, light and white-headed. This race or subspecies is thought to be sedentary, but one young bird has been found in Senegal. In Madagascar and some islands in the Indian Ocean, the larger race *Ardea c. firasa* breeds; in east Asia there is the lighter *Ardea c. jouyi*.

## Environment and food

Herons depend upon habitats with wetlands and watercourses where fish occur. They will feed in small streams and brooks, on ponds and larger lakes, and also on estuaries and along the sea shore. For breeding it prefers stands of tall trees not far from suitable feeding areas and preferably on islands and other sites where disturbance is minimal. Occasionally, however, herons will adapt and nest quite close to areas of human activity, including public parks and zoos.

One week after it has hatched, the young heron will eat about 78 grams of food each day. At two weeks old it eats an estimated 195 grams of food, and a mature heron will account for between 330 and 500 grams of food each day. There is one record of a 17-day old heron eating over 300 grams of food in one day. To express it in other terms, the week-old heron's daily food intake is the equivalent of two 15 cm long roach or three 12.5 cm long perch. The daily consumption of a mature adult will be the equivalent of 12–13 roach or two eels each 45 cm long.

The diet of individual herons is largely dictated by the local availability of suitable foods. One group of birds may subsist chiefly on roach, another may feed mainly on eels, while a third may live almost entirely on carp. Large and small sticklebacks are a favourite food, and herons will also eat small rodents, frogs and invertebrates. herons are readily attracted to fish hatcheries and fish farms, where prime food is often available in abundance and easily caught.

Some nesting herons will make long flights to their feeding grounds. Scientific studies in the Netherlands revealed that different adult herons secured food for their young by means of a number of different strategies. Some old birds located individual feeding territories, and defended these against the intrusions of other birds. The younger birds usually had to fly further from their nesting sites to find suitable feeding areas. Studies in Scotland have revealed that some individual herons will persist in feeding regularly at favourite spots by rivers or at estuaries, while others alternate between both types of feeding areas. The latter birds were found to have higher average body weights.

## Distribution and numbers

The heron is found over much of the Old World, in Europe, Asia and Africa. In areas where is does not breed it may be found regularly as a migrant visitor or as a nomadic youngster. The main breeding areas of Europe stretch in a broad band from the British Isles in the west through the countries bordering the North Sea and the Baltic, and eastwards into Russia.

The heron's habit of nesting in colonies has made it reasonably easy to assess numbers and to monitor the increases and decreases in the population. In some areas herons have declined in numbers because of human persecution, but habitat changes and environmental pollution have also played their parts. In general, however, this species appears to be flourishing, and even to be increasing in numbers in certain areas. The European total of breeding birds has been estimated at 50–60,000 pairs, although numbers fluctuate from year to year depending upon the weather conditions in the previous winter. Some sedentary European populations are very badly hit by a severe spell of winter weather. For example, it has been estimated that 95% of herons in Sweden migrate southwards or southwestwards in winter, while the British and Irish populations are almost entirely sedentary.

As mentioned earlier, herons suffer high mortality during difficult winter weather, when the opportunities for successful fishing are limited or disappear altogether. After such winters the breeding colonies inevitably have fewer birds. Owing to the generally high mortality among the younger blrds, and the resulting high average age of the breeding birds, the heronries grow rather slowly after a period of high winter mortality.

However, the size of the heronries is not dictated only by winter mortality. Other limiting factors include the number of suitable nesting trees and the food supply in the vicinity of the heronry. For a specialised fishing bird like the heron, competition for good fishing sites can be a decisive factor in determining the survival of the young chicks even before they leave the nest. Once the young birds begin fishing for themselves, the competition becomes even greater. This must contribute to the tendency of young herons to disperse quite quickly on their so-called nomad migrations. This in turn may take its toll of a proportion of the young birds, but survival rates are still better than if the young birds were to remain in the vicinity of the nesting sites and try to compete in heavily fished waters with older and more skillful birds. In this way the numbers of breeding pairs at a heronry, and the number of heronries within a given area, tend to reach saturation point.

In recent times the British breeding population has been reckoned at 4–5,500 breeding pairs, the Danish population at 2–2,700 pairs and the Netherlands at 9–10,000.

*Wigeon*

*Pintail*

*Teal*

*Mallard*

*Shelduck*

*Goldeneye*

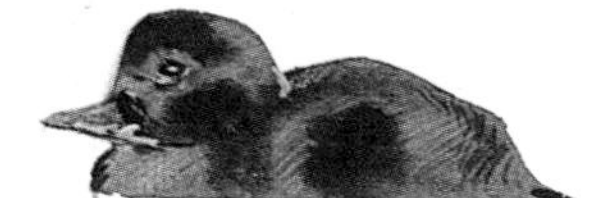

*Tufted duck*

*Greater Scaup*

*Long-tailed duck*

*Pochard*

*Greater Merganser*

*Lesser Merganser*

*Velvet Scoter*

*Common Scoter*

*Eider*

## Young ducks

The chicks of all wild duck species are quite well developed when they hatch. Once all the eggs have hatched the mother leaves the nest with her young. These active young ducks are still covered with down, but they can see, swim and seek their own food. But they remain very dependent upon the care and leadership of the adults. They have a highly developed instinct to follow their parents, and among dabbling and diving ducks the post-hatching duties fall chiefly on the female bird, while in the case of swans, geese and Shelducks both parents normally take care of the young.

Young ducks live in different environments and habitats, each with its own cover and protection, its colours and shades of light and dark. Like the adult female, they can benefit from camouflage plumage. The chicks of one species all look alike and there is no visible difference between the sexes as regards the colouring of the down. This may be uniformly coloured, but it is frequently patterned on the chick's back. The throat and underside are invariably uniformly coloured and paler than the upper parts.

Young cygnets are always grey or greyish-white. Young goslings have a greenish-yellow colouring, while among other species of wild ducks the young are more diversely coloured.

Shelducks, sometimes known as "half-geese" (*Tadornini*), are closely related to geese and swans, but their young are distinctive looking with a richly contrasting greyish-black and white pattern on the down.

European dabbling ducks (*Anatini*) are in many ways a uniform group as regards appearance, body structure and behaviour. This is evident among the adults, for example, by the females' similar plumage throughout the year, and from the males' eclipse plumage. The down of the young chicks is also similar among the different species. All dabbling ducklings have a dark eye-streak with lighter areas above and below; the cheek is normally a light yellowish-brown. They also have a similar pattern on their backs, which are marked with lighter and darker areas in brown and yellowish-brown. The chicks of the Shoveler at first resemble those of other dabbling ducks, but their bills begin to develop the characteristic Shoveler shape at the age of four or five weeks.

The Eider duck is believed to be more closely related to dabbling ducks than to the so-called sea geese, some of which bear a superficial resemblance. Eider chicks are a dark, uniform greyish-black. They have eye streaks, but lack the dark "hat" or crown which is typical of young sea geese.

Diving ducks of the family *Aythya* — i.e. Pochard, Tufted ducks and Scaups — have chicks which closely resemble those of dabbling ducks. But they lack the dark eye-streak and the accompanying pale areas.

The sea geese (*Mergini*) are a varied group, including such different species as Scoters, Mergansers, Goldeneye and Long-tailed ducks. Nevertheless their anatomy and behaviour show that they are related. It is also known that species which are so dissimilar in appearance as Smews and Goldeneye can produce hybrids. All young chicks in this group have greyish-black patterns on a greyish-white or white background, in addition to their dark "hats" and light coloured cheeks.

Mallard adult male
Wigeon adult male
Mallard adult male
Mallard juvenile male
Wigeon juvenile male
Wigeon adult female
Wigeon juvenile female
Mallard adult female
Mallard juvenile female

*Teal: juvenile male* · *juvenile female* · *adult male* · *adult female*

*Goldeneye juvenile male*

*Goldeneye adult female*

*Velvet scoter*

*Pintail*

## Wing profiles

The plumage of swans and geese remains the same all year round, and the sexes are similar. Geese of different ages are also generally similar in appearance. But this is not true of European wild ducks. The male and the female have very distinct and different plumages during the breeding season, and the brightly coloured males subsequently acquire a female-like eclipse plumage. Male Wigeon do not acquire their full plumage until they are two years old, and male Eiders are three years old before they achieve mature plumage.

The males of all species can be distinguished from each other, and from the females, by their wing patterns. This is an important aid to species identification, especially when the males are in eclipse plumage. During the season, shooters must be able to distinguish protected species like the Garganey from legitimate quarry like Mallard and Teal. The wing markings provide a clear-cut means of distinguishing them. Dabbling ducks usually have wing specula of iridescent green and blue, fringed with white, whereas diving ducks normally only have pale coloured areas or long white cross-bands on their wings. Large white wing markings occur, for instance, on mature male Wigeon.

# Waders

Wading birds comprise an exceptionally rich and varied range of species. Over 160 species occur world-wide among the "traditional" wader families, i.e. Old World Oystercatchers, Plovers (*Charadriidae*), Woodcock, Snipe and Sandpipers, Avocets and Phalaropes.

Waders are a distinctive group of birds, in which the length of their legs and the form and length of their bills reveal specialised adaptation to different kinds of wetlands. Phalaropes have webbed feet, can land and take off on water, and spend part of their lives on the open sea. They gather food from the surface of the water and, by rotating their bodies, can churn up food items from beneath the surface. Plovers have short legs and bills, and live along the shoreline and on nearby grasslands. The smallest waders, such as Sandpipers, have long bills but relatively short legs, and are found principally along shallow beaches and on flat meadows and marshland nearby.

The larger waders such as Snipe can wade in somewhat deeper water and frequent rather longer vegetation on marshes and wet meadows. Common Sandpipers, however, will lay their eggs on the ground in pine forests, often at some distance from the water. Curlews, Oystercatchers and Avocets are highly specialised wader species with exceptionally long bills and legs.

Several wader species occur in the Arctic and large numbers of them overwinter in western Europe. There are also species such as the Lapwing and the Dunlin which have recently declined in numbers, owing to environmental deterioration and habitat destruction, among other causes.

Waders are especially well adapted to life on marshy land, damp meadows and long, shallow beaches, and in many parts of Europe these environments are under threat. Land-filling, drainage, dredging, cultivation, peat cutting and pollution are just some of the principal threats to waders and their habitats.

At one time, almost all the wader species mentioned here were hunted by man, and a number of wader species are still shot annually by sportsmen in Britain and Europe. In many countries the principal sporting species of waders are the Common Snipe and the Woodcock, but Sandpipers and Golden Plover are also shot in some countries as part of their sporting tradition.

During the last century the Great Snipe (*Gallinago media*) was a highly prized sporting bird in mainland Europe, although it was always a rarity in Britain and Ireland. There has been a very significant decline in this bird's European population, due largely to the drainage and reclamation of its traditional northern nesting sites, and shooting pressures may also have been excessive in some areas. There are historical records from Scandinavia showing how hundreds

*Dunlin*

of Great Snipe were killed in a single day's sport. In certain countries where it was formerly a common breeding bird, the numbers of breeding pairs have been reduced to fewer than ten pairs, and Norway and Sweden are now the species' principal European stronghold, with approximately 1,000 breeding pairs. The Great Snipe's principal winter quarters are in equatorial East Africa.

The Golden plover (*Pluvialis apricaria*) is now much less common in certain areas of Europe than formerly. Shooting pressure has been heavy in some countries, especially in the Netherlands. The Golden plover is about the same size as the Great Snipe, and has a body weight of about 200 grams.

In Denmark, the Netherlands, Belgium and Germany there are now no breeding Golden plover, and the population in southern Sweden has also declined. The British and Irish breeding populations are together estimated at approximately 30,000 pairs.

The Golden plovers of the southern areas of this bird's breeding range are considered to comprise a separate subspecies, *Pluvialis a. apricaria*, while the more northerly Golden plovers belong to a different race, *Pluvialis a. albifrons*. The Golden plover occurs as a breeding bird in a somewhat limited range which extends in a band across northern Europe and adjacent parts of north-western Siberia.

*Purple Sandpiper*

Where numbers have declined this is chiefly attributed to the drainage and cutting of peat bogs.

However, in northern Europe many hundreds of thousands of pairs of Golden plover still breed. They migrate southwards and southwestwards to winter in western Europe and in the Mediterranean lands, and November counts in the Netherlands have revealed up to 400,000 birds. The total wintering population in Britain and Ireland has been estimated at 500–600,000, many of which come from Iceland.

The Curlew (*Numenius tarquata*) is the largest European wader, with a body weight of up to 1 kg. It occasionally occurs in very large flocks; however, Curlews usually migrate in small groups of about ten birds, and there may be up to 100 at any one time resting together on migration.

Traditionally the Curlew has been a favourite sporting quarry, because they make good eating, their flight is fast and sporting, and they can be decoyed and called with whistles within reach of the fowler's gun. In Great Britain the Curlew was granted full protection in 1981.

Curlew numbers have declined in recent years in many areas, and this has been largely attributed to loss of habitat and other environmental changes. However, the Curlew has succeeded in adapting to breeding in arable and other cultivated areas, and the species' principal wintering grounds in western Europe are in the British Isles.

The oldest known Curlew, which was leg-banded in Sweden, was shot when it was no less than 32 years old!

*Whimbrel*

# Lakes and waterways

Some natural habitats can be changed and manipulated to satisfy the interests and wishes of mankind, while others are not so easily changed. Low-lying wetlands, areas of open water and the natural lagoons and ponds on marshland can all be altered and controlled by man. As a result of widespread damming, drainage and land reclamation, much of Europe has lost large areas of wetlands, which have traditionally been some of their richest and most productive environments.

In the past there were many natural wetland refuges for birds and animals, and habitats for many plants and insects, which have subsequently been drained and reclaimed. As a result of this, there are many areas where flash floods, heavy rain and snow-melt no longer constitute part of the natural rhythm, and they no longer provide suitable habitats for the rich animal and plant life which they formerly supported. However, in those areas where thriving wetlands have been preserved, there is clear evidence of the great importance of this type of environment for many forms of natural life.

Brooks, streams, rivers, lakes and marshlands all impart life to their surroundings. The combined powers of water and gravity have carved and shaped the land and helped to mould the appearance of the landscape. Meandering streams wander across level plains to rich and muddy lowland lakes, just as bubbling rivers have cut their ways through rocky landscapes and upland forests to reach clear, clean and acidic lakes in the uplands.

Industrial waterways and artificial reservoirs also provide evidence that many forms of plants, fish and animals are ready to adapt and accept artificially created environments, regardless of aesthetic considerations or the permanence of such new wetlands. For new life to colonise such artificial waters, the principal considerations are security and a source of nutrition.

Man has been forced to recognise the importance and vulnerability of water and wetlands, for mankind is a consumer of water and has exploited it for natural and industrial uses, for fishing and for recreation.

Many low-lying areas which have been drained and reclaimed have suffered from a shortage of water, and the result can be dusty soil swirling over a dessicated plain. Agrochemicals used in farming and forestry have an insidious way of creating over-enrichment of certain waters, and even of poisoning them, and animal life is affected slowly but surely. In recent times conservationists and game managers have sought to co-operate in the protection of our remaining wetlands, and also in the restoration of neglected waterways, as well as the creation and management of special areas of conservation wetlands. Internationally there are moves to develop a general system of wetland environments which will constitute a safe chain of resting places, breeding grounds and wintering haunts for many species of migrant water birds. Efforts are also being made to co-ordinate the legislation relating to the shooting of migrant species.

There is an international responsibility to improve and maintain healthy wetlands, for rain and snow and meltwater recognise no manmade national or international borders. As waste products and pollutants are carried along by flowing water, acidification and enrichment and poisoning can be carried along too. Drainage in one area may mean a death sentence for wetlands elsewhere.

# Beaver

## *Castor fiber*

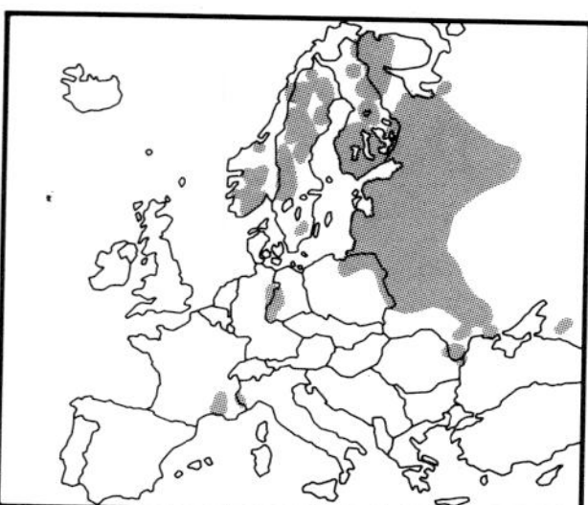

The beaver has a good claim to be called the "royal game of Europe's waterways", for it has been hunted in every part of its natural range, often to the point of extinction. The beaver's pelt, and also its meat, have been eagerly sought after. Castor (*castoreum*) is a secretion excreted by glands near the animal's anal opening, and this substance used to play an important part in medicine, and up to the early part of this century it could be bought in pharmacies, where it was recommended as a remedy for certain nervous complaints. One chronicler recommended it, and said that "half a spoonful in a strong glass of spirits was the very elixir of life".

In medieval times monks made use of the beaver's meat during Lent, for they pointed to its scaly and fish-like tail and maintained that this form of meat was permissible during times of fasting.

The last native beavers in Finland and Sweden disappeared towards the end of the nineteenth century, owing to a combination of excessive hunting and habitat loss, but the species retained an important presence in Norway, although there its numbers dropped to fewer than one hundred animals in the early part of this century.

Over much of Europe the same decline occurred, and the beaver disappeared from the British Isles, the Netherlands, Belgium and Austria, although small populations continued to exist in parts of France and Germany until quite recent times. The last beavers disappeared from most of eastern Europe around 1900.

However, since the 1920s beavers have been successfully re-introduced to a number of areas, where there has been full protection and careful habitat management.

## Characteristics

The beaver is the largest European rodent, and the second largest of all the world's rodent species, which number some 1,650 species. Only the South American capybara is larger, weighing some 60 kg. The beaver can be distinguished from its relatives by the presence of a broad, flat, "scaly" tail. At first sight the beaver's fur appears to be coarse and straggling, and this impression is especially strong if the animal has just emerged from the water. In reality, the fur is very soft indeed. The underfur is very thick, especially on the animal's belly, and it is overlaid with longer and stiffer bristles or guard-hairs which are quite flexible and elastic, being rounded at the base and flattened towards the ends. There is no hair on the beaver's tail or feet. What is often described as scaling on the beaver's tail is actually a chequered patterning of the skin. The beaver's outer ears are small and are quite hidden in the animal's skin and fur. The muzzle and nose are hairy, and the beaver's whiskers are black to dark brown in colour. The beaver's eyes are rather small and brown in colour. The body is almost cylindrical in shape, and the tail looks flattened, like the blade of an oar.

As an adaptation to aquatic life the beaver has a well developed webbing between the toes of its hind feet, and there is also a small degree of webbing between the toes of the front feet, which makes the beaver a very strong swimmer. The fur is water repellent, and air can be trapped beneath the long bristles. The nose and ear openings can be closed when the beaver dives beneath the water.

The beaver weighs on average 30 kg, and there are some recorded instances of beavers weighing up to 35 kg, but the normal body weight of a Scandinavian beaver seldom exceeds 20 kg. The length of the body seldom exceeds one metre, but individuals have been reported measuring up to 120 cm, and the tail adds a further 25–30 cm to the overall length. The tail is 12–15 cm broad, and the sexes look alike.

At birth the young beaver weighs approximately 500 grams, and they are born with their eyes open and with a well developed covering of fur. The rate of growth is rapid in young beavers, and by the time it is three months old it will weigh about 4 kg, eight times its birth weight. Thereafter the rate of growth slows down, and at the age of eighteen months the adolescent beaver will weigh about 10 kg. The weight continues to increase over the first three years of the animal's life, and rather longer in the case of an exceptionally large and heavy individual.

Typically, a beaver can live to 25 years of age, but the majority are thought not to live beyond 10–15 years, and a Russian study of beaver ages carried out by examining the animals' tooth wear indicated that only about 2% of beavers were more than 15 years old.

Like other rodents, the beaver has two different types of teeth. The incisors are large, reddish-brown and have a distinctive chisel shape. They are specially developed for gnawing and are separated from the molars by a gap. The incisors grow continuously, an adaptation to the extreme wear and tear caused by their gnawing activities, and only the incisors are enamelled. The molars often wear down with age, but the rate of wear is also influenced by the animal's diet, and the surest way to age a beaver from its teeth is by counting the annual layers of enamel.

## Reproduction

The beaver mates once a year, in winter. The female is only capable of being fertilised for a period of half a day, but if fertilisation does not occur at first she will come in season again within a few weeks. The mating period can therefore last for up to a month or more. The mating behaviour is rather specialised, since both the sexes' reproductive organs lie within the cloaca and the broad tail is an obstacle to mating in the conventional way. Mating is carried out in the water, when the animals turn side to side, belly to belly. Beavers in Scandinavia usually mate in late winter, in February, and the gestation period lasts for 16 weeks, which means the young are usually born in early June. Since mating can occur both earlier and later, litters can be born in May or in late June, and there is only one litter of young each year.

The female beaver usually reaches sexual maturity at the age of 2½ years. The young beavers usually stay with their parents until the third year of life, when they have reached the age of two years. By the time they leave the family group their youngest half-brothers and sisters will be a few months old. In ideal conditions they are now ready to undertake parenthood themselves.

Beavers are believed to be monogamous. What are sometimes referred to as beaver colonies are actually stable family groups comprising the parents and two or sometimes three different litters. A clear "pecking order" gives stability to the community and minimises the risk of conflicts. The long period of adolescence gives the young beavers plenty of experience of life, and their comparatively long lifespan gives them 10–12 years of reproductive life. This reduces the

*The beaver has a total of only 20 teeth.*

need for large litters, and a typical annual litter will comprise three young, although the number can vary from one to six in individual cases.

## Senses and communication

The beaver is primarily a water animal: if frightened on land its first reaction is to retreat to the water for safety. It is most active during the hours of darkness, but if it is undisturbed it can also be very active during the daytime. A large part of its life is spent "indoors" in the protection and darkness of the beaver's lodge.

The need for mutual contact, and alertness to danger or attention to attractive food sources nearby, are all maintained using the animals' various senses. The beaver's brain is considered to be the largest of any rodent in proportion to its body size, and its mental abilities seem to be proportionately well developed. Despite its small outer ears, the beaver's hearing is very acute. When in danger beavers signal to one another by striking their flat tails against the surface of the water. On hearing this warning signal the whole family will dive for safety, as well as the individual beaver who gave the warning signal.

Beavers maintain contact with one another by grunting and hissing, and the young are said to hiss also. Within the territory, however, the most important contact is achieved by scent markings. All members of the family mark and indicate their presence by using the secretion from their castor glands, and the males are especially energetic in doing this.

By means of these scent markings, trespassing beavers can learn about the size of the family's territory, the number of animals in the colony, and similar information.

## Adaptations and behaviour

The beaver is superbly adapted to life in the water. Its anatomical characteristics include good swimming ability, assisted by the webbed hind feet and the powerful tail, the lack of external sexual organs and the ability to shut their nose and ear openings. They can dive for long periods of time, and can remain under water for up to 15–20 minutes, although most dives do not last so long.

The beaver's forefeet have moveable "fingers" and a prehensile thumb, and this makes them excellent tools for digging, gathering building materials or holding food. On the second toe of the hind foot the claw is split, which is useful for grooming the fur. Social grooming is common, in which the various members of the family tend to one another's fur.

The young beavers do not usually go into the water during their first month of life. However, there is no risk of drowning as the youngsters are lighter than water and float buoyantly. If they wander off the mother carries them back with a firm grip of her teeth across the back or the scruff of the neck.

Beaver dams or lodges are a well-known phenomenon. When building dams beavers will use their strong teeth to grip large branches and drag them across the ground, as well as through the water. They do the same with the large branches and slender boughs which comprise their reserves of food. Many environmentalists believe that beaver dams helped to create and shape the lowland topography of the older wilderness areas of Europe, Asia and North America.

When beavers have chosen a stream for their future home they mark the boundaries of their territory with scent as they set about building the dam. It is important to create the correct depth of water in the area of the new lodge, and for this reason the beaver must have some degree of foresight. The dam serves as a refuge from enemies but it also provides a suitable avenue for transporting trees and branches. In addition, the water must be deep enough to allow the animals to pass easily under the winter ice.

The beaver's ability to cut down large trees has given rise to a great deal of discussion and some fantastic descriptions of what they can do. In fact, however, beavers mainly fell small trees, up to 10 cm in diameter, and they can do this in one night. They are quite capable of felling much larger trees, up to 50 cm in diameter, but to fell such a large tree would take several nights' work.

Young beaver pairs build their lodges and dams in the autumn, and it takes 6–8 weeks to build a lodge. Carrying building materials and branches and trunks for food is a time-consuming activity, but the food stores must not be neglected. Everything must be ready before the ice of winter sets in, because in northern regions this restricts the beaver to an indoor life throughout the winter.

The lodge contains a bed-chamber and a food store which also serves as a multi-purpose room. From the food chamber there is a direct exit to the open water. Sometimes there is an extra food supply consisting of sections of tree trunks and branches anchored to the bottom outside the lodge, and in winter the beaver sometimes swims out underneath the ice.

In summer the beaver has other building structures to maintain. These are the canals that provide the animals with an open passage through the rushes and the rich vegetation of the marshy meadows.

## A year with the beaver

**Winter.** The beavers spend most of their time in the lodges. The female is in season and the beavers mate (January–February), mating occurring mainly in the water, but sometimes in the lodge. The beavers' diet consists of bark from branches and twigs which were stored in the "pantry" of the lodge, or fastened to the bottom outside the lodge.

**Spring.** As soon as the ice cracks the beavers leave their confined winter quarters. Trees are cut down near the lodge, the beavers are less cautious than usual, and they eat the fresh foliage of the newly felled trees with relish.

**Summer.** The beaver changes its diet to plants which are found among the aquatic vegetation, but also eats bark and small branches. The two-year-olds leave the family group. The occupants of the lodge comprise the parents, the new litter and the one-year-old previous year's litter.

**Autumn.** The 2½-year-old females become sexually mature and can form pairs. Newly formed pairs seek out suitable nesting places and build their first lodges. There is intense tree felling to secure building materials for the dams and to ensure that there are adequate food supplies for the coming winter. The diet consists of more bark and twigs.

## Population structure

Beaver populations comprise a number of related families that have their lodges and dams along stretches of the same watercourse. In densely populated areas one dam or family territory may be directly adjacent to the next one. The individual families allow other beavers free passage along the waterways. It is also not unusual for beaver families to share food gathering areas. When the young beavers leave the family colony they go wandering to find territories of their own. Distances of 20–30 km are considered normal, but they sometimes wander as far as 100–200 km away. The young males tend to wander furthest.

## Distribution and diet

Beavers are found principally in lowland areas, especially in well wooded valleys. The essential combination is plenty of deciduous trees close to a waterway. The watercourse must be sufficiently deep to ensure that it does not freeze all the way to the bottom in winter, and a flowing current with areas of open water is best.

Beavers can live along rivers and streams in coniferous woodlands with a sprinkling of deciduous trees, but they much prefer water with dense stands of aspen and sallow, and they like waterlogged alders and birches along the river banks.

Like many other rodents, including rabbits, the beaver eats its own excrement, a form of behaviour known as refection. By doing this the animal makes the most of the available protein and vitamins.

The mother suckles her newborn young, providing them with a diet rich in fats and protein. When the young are a few weeks old they are given a taste of tender parts of plants, but they are not fully weaned until they are three months old. By this time they can already gnaw quite hard bark and tender twigs.

During the summer and especially towards the autumn the beavers fell relatively large trees to get at the tender branches, twigs and foliage, and they gnaw the bark from the felled trunks. In summer their diet comprises many species of herbs and grasses, and includes both aquatic plants and species from marshy meadows. Beavers eat water lillies, reeds and bullrushes, meadowsweet and other plants, including the twigs of young sallow and osier bushes. Twigs, branches and slender trunks are transported to the lodges and kept at winter supplies, some inside the lodge and some stuck down into the muddy bottom outside. The aspen is perhaps the beaver's most important food.

It is doubtful if the beaver will ever regain its former popularity as a source of food, but it is still eaten by hunters in various parts of Scandinavia. Beaver tail, which is very fatty, used to be regarded as a great delicacy. In addition, castor — the secretion from the two anal glands — was a most important commercial by-product of the beaver. Castor now has little use in medicine, but it was in regular use up to the 1930s. In the eighteenth century Linnaeus stated that "castor . . . has a powerful effect on the nerves . . . " and his pupil, Carl Peter Tunberg, later a professor of both medicine and botany, wrote in 1798 that "beaver glands . . . especially when dissolved in alcohol and distilled to a tincture, is the most powerful remedy for the nerves".

## Hunting and predation

Beavers were formerly hunted and trapped by various methods. Before the widespread availability of firearms, trapping was the most important method, and because it is relatively easy to trap beavers this method remained in use even after shooting became widespread. In the northern parts of its range hunters used to pursue beavers on skis, using their ski poles as lances.

Beavers used to be hunted extensively in Scandinavia, and the kings of Sweden maintained stores of beaver skins whose numbers were carefully recorded. There was also an export trade in beaver pelts. The records show that in 1574, for example, some 3,381 beaver pelts were exported from Stockholm.

## Distribution and numbers

The beaver used to be common in large areas of Europe, Asia and North America, but persecution has led to a sharp reduction in beaver numbers in large parts of these areas.

In the 1930s the beaver remained in only three places in western Europe in the Camargue region of the Rhone delta in France; in the middle reaches of the river Elbe; and in Telemarken and around the river Mandal in south-eastern Norway.

In Sweden the beaver was reintroduced in 1922, when some animals were placed in the River Bjur in northern Jutland, in Tarna in Lappland in 1924, and in Varmland in 1925. These introduced animals were brought from the Telemarken area of Norway. In Finland both the European and American beavers were reintroduced.

A sustained programme of reintroductions at various sites, combined with total protection of the species throughout Scandinavia resulted in a population explosion in beaver numbers. The Norwegian population increased from around 90 animals in the 1890s to some 10,000 animals in the early 1980s. At this time the Swedish beaver population was about 45,000 animals and the Finnish population about 5,000. Most of the latter were descended from stock of American origin, and in south-west Finland there are probably not more than 200 European beavers.

Likewise, the present day population of beavers in Poland is of American origin. This stemmed from animals which had escaped from fur farms and established themselves on the Pasleka river. In Russia, beavers were released in the Karelia area. Recent counts have estimated there are several hundred beavers in Poland, about 500 in West Germany, 1,000 in France and 10–20 in Switzerland.

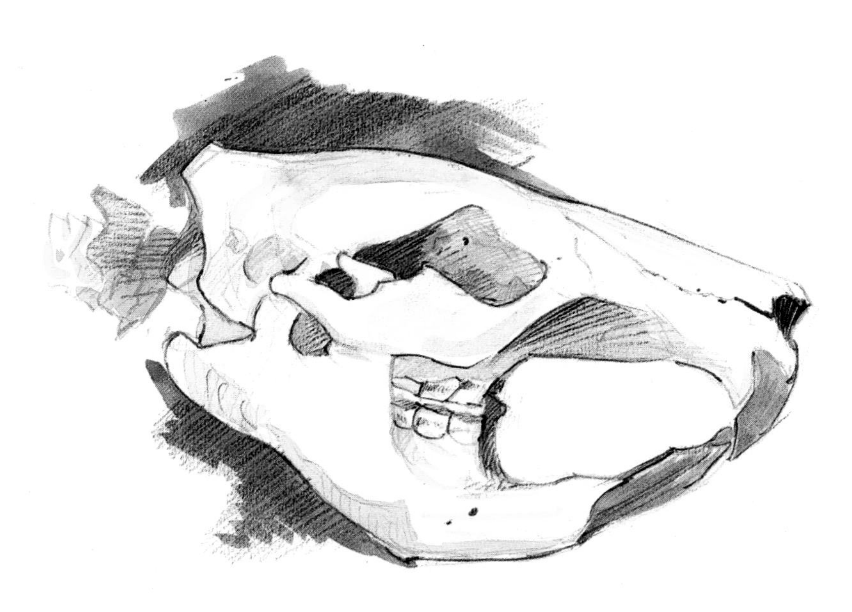

Beaver pelts are still an important commodity, especially in Canada and the Soviet Union, and there may eventually be a return to the controlled commercial harvesting of beavers in some parts of Scandinavia. In Canada the beaver population is estimated to yield some 200,000 pelts annually. Beaver skin hats are common, and coats also. At one time there was also a demand for beaver meat for the table.

Today, beavers are hunted mainly by the method known as "still-hunting", in which the hunter places himself carefully and quietly close to the beaver lodge and waits in silence for the beaver to appear. This method calls for extreme vigilance and stealth, for the slightest sound or movement will frighten this wary quarry away.

Historically, man's persistent hunting and persecution have been the single most serious threat to beaver populations. Among the beaver's natural enemies only the wolf poses a really serious threat, but in most of Europe the two species' range does not overlap. However, predation by foxes can occur in areas such as Trysil in Norway, Varmland-Dalarna in Sweden and in Karelen in Finland. There are a few instances on record of a fox succeeding in overpowering a beaver.

Apart from human pressures and natural predation, the beaver's worst enemy is the species itself. Where certain areas are over-populated this can lead to the spread of disease and fighting and general competition for food and territories. A serious problem for the European beaver is the matter of competition for territories with the introduced North American beaver (*Castor canadensis*). This latter species already accounts for about 90% of the beaver population in Finland. If the North American beaver is introduced elsewhere or is allowed to spread naturally from areas where it has become established in the wild, there is a serious risk that it will spread at the expense of the indigenous European beaver.

There is little information available about the various causes of beaver mortality. From studies of one particular German population along the river Elbe it is known that one-third of the 33 beavers found dead had been killed by poachers. Eight deaths were indirectly attributable to man (entanglement in fish nets, poisoning etc.), and six beavers had died of disease and old age.

*The beaver's thumb grip.*

Some naturalists regard the European and American beaver as one species, *Castor fiber*. The European beaver is then referrable to the subspecies *Castor f. fiber*, while the American beaver belongs to the subspecies *Castor f. canadensis*. Other scientists disagree and regard the two animals as belonging to two different species, and they give the name *Castor canadensis* to the American beaver. One point of distinction between the two animals is the shape of the skull, but the principal difference lies in the numbers of chromosomes, the European beaver having 48 chromosomes and the American beaver having 40. From this it seems clear that the two should be regarded as quite distinct species, and this is borne out by the fact that no hybridisation has been observed, even in those parts of Finland where the range of the two animals overlaps.

Yet another view held by some researchers is that the European beaver can be distinguished from more easterly populations in Eurasia by the structure of the skull, and they therefore designate the European animal *Castor albicus*. Other local populations have also been described as if they were independent subspecies, but the distinguishing features are often barely perceptible.

# Otter

## *Lutra lutra*

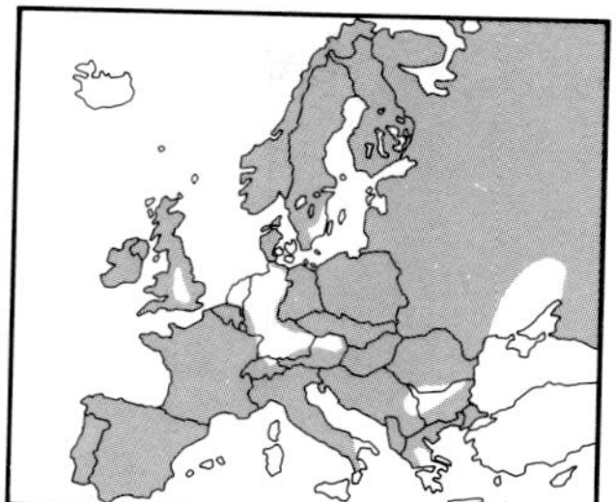

The otter used to be found throughout all of Europe, but it has now vanished from many parts of its former range. Human persecution and environmental changes are responsible for much of this decrease in numbers and distribution. In the fifteenth century large numbers of otter pelts were exported to western and central Europe from Scandinavia. In addition to the trade in otter skins, there was a demand for otter meat. An old account from the early 1800s states that "in France people prefer to eat the fish-otter's meat, and Catholics in general take advantage of this as a Lenten dish, since (curiously) they regard the otter as a species of fish".

In many parts of Europe fisheries management included widespread destruction of otters, and a bounty was often payable for animals killed. In 1653 Isaac Walton wrote in his celebrated book, *The Compleat Angler*: "I am, Sir, a brother of the Angle, and therefore an enemy to the Otter: for you are to note that we Anglers all love one another, and therefore do I hate the Otter, both for my own and their sakes who are of my brotherhood".

At that time and until the 1960s special packs of otter hounds were kept specifically to hunt the otter, but these packs now concentrate on hunting feral mink. Today, man's principal threat to the otter is in the form of the pollution of rivers and the destruction of the environment.

The otter has an elongated body which is well adapted for a life spent in the water. The head is small and neat, the neck short and very muscular. The tail is thick at the base and very narrow at the end. The legs are short and muscular, and the feet have toes with claws, with webbing between the toes on all four feet. The fur is shiny, and the underfur is dense and yellowish-white but gives a mottled appearance when the fur is stroked the wrong way. The guard-hairs of the coat are brown. The fur on the underside of the neck varies in colour from grey to shiny silver. The ears are very small and rounded, the otter's whiskers are numerous and bristling, and the muzzle is dark and flattened in shape. Both sexes look alike, but the male otter is larger and heavier.

The otter has few natural enemies, although young otters may be vulnerable to predation early in life. Hunting took a heavy toll in the past, but the otter now enjoys complete protection throughout most of Europe.

### Size and growth

The total length of the body of an otter, from its nose to the tip of its tail, can be up to one metre, about half of which is accounted for by the length of its tail. Otters in western and northern Europe can weigh up to 14 or 15 kg, but normal weights are around 10–12 kg. On average a mature male otter will weigh 30% more than the female. At birth the cubs weigh 150 grams, but some have been recorded as heavy as 300 grams. At the age of two months the cubs will weigh about 1 kg, and they are considered fully grown at the age of two years.

### Reproduction

The European otter generally mates twice a year, mainly during the winter months, but probably also again in the summer. The female otter attracts sexually mature males by emitting special scents from her anal glands, and also by her shrill whistling voice. After mating the pair split up and go their separate ways. The female will already have created her lair or burrow, known as a holt, and this usually has one entrance above the water in thick cover, and another below the level of the water. The holt is lined with grasses, reeds, moss and similar vegetation.

The gestation period varies, but it is normally reckoned to last for two months. With the delayed implantation of the foetus it therefore takes a total of 9–10 months, and the young are usually born in the months from April to June. However, newborn young have been seen at all times of the year. In central Europe the winter mating takes place earlier and the young are born during the winter months. The cubs are born blind and helpless, and have no teeth. Their eyes open after about 4–5 weeks and they begin to swim after a further month. The female takes sole care of the young, which are normally in litters of 2–4 cubs. However, some litters of up to six cubs have been recorded. The young suckle until they are three months old, but they are taught to fish before they are weaned. The female brings injured fish for the young to practice catching. The family usually remains together until the time of the winter mating, and they occasionally remain together until early spring. It seems that not all females have cubs every year. Female otters become sexually mature when they are almost two years old, and the males at a slightly older stage.

### Senses and communication

The otter is endowed with all the typical mammal senses, but the sense of smell is especially important because the scent markings (considered to smell rather like violets) distinguish territories and also attract the males to females in season. Under water the sense of smell is probably unimportant because the otter's narrow nostrils are tightly shut. However, the long and bristling whiskers play an important role as feelers in muddy water and in the dark.

The otter's eyesight is well developed both on land and under water. The eye muscles, which affect the shape and efficiency of the eye lenses, are very powerful, and it is important for the otter to be able to see well under water. Otters also make a variety of vocal sounds. Most of these are to establish and maintain contact between females and their cubs, or between sexually mature animals at mating time. Otter "slides", created by otters sliding down the steep banks of streams and along icy or snowy slopes, are also thought to be a part of the system of marking out territories and may play some role in courtship and mating behaviour.

### Adaptations and behaviour

The otter's closest relatives are land mammals, but otters are highly developed for life in an aquatic environment and are totally dependent upon the availability of suitable streams and watercourses. The otter retains all the distinguishing features of carnivorous land mammals but is also superbly adapted to life in water. The webbing between the toes is perhaps the most obvious example of this. The cylindrical shaped body and the short legs are also adaptations for life in the water and for moving in narrow passages and cavities, and the same is also true of the otter's small ears and eyes. The fur is also adapted to life in the water, for the relatively short but numerous and thick guard-hairs prevent water penetrating to the otter's dense underfur.

This is therefore able to maintain an insulating layer of air, and the animal's skin does not become wet and chilled. The otter is able to stay under water for up to seven or eight minutes, and it can dive to depths of up to ten metres. Usually, however, it dives for shorter periods and in relatively shallow water. The nasal openings are shut when diving, and the ears are closed by three folds in the skin.

The otter has a reputation as a playful animal, and this is true not only of the female with her cubs but also of adult otters playing together. This play serves to teach the young otters to be quick and agile, and it also acts as an important form of social contact. Otters will often swim on their backs, especially when transporting food they have caught. The forelegs and paws are dextrous and prehensile, well adapted for holding prey such as fish. Otters will sometimes drive shoals of fish into shallow bays and towards beaches where they can be caught more easily, and when hunting otters can move surprisingly fast. Occasionally otters will hunt together. Small fish are generally eaten in the water, while larger ones are taken onto dry land. Where fish are abundant otters will build up caches of food.

## Population density and age structure

Both male and female otters are territorial, although they do not share the same territory, and usually are only together during the mating season. As a rule the males maintain larger territories than the females, and they will range far and wide, especially at night. The females remain closer to home, especially when they have young cubs.

The territory of a female otter is roughly 7 km in diameter, but as the otter's territory tends to lie along beaches, rivers and streams the actual shape of the territory will tend to be elongated. A male otter's territory may have an average diameter of 15 km, but the territory could actually be 40–50 km in length. The male otter's range may also extend into the territories of several different females. In coastal areas with numerous islands the territorial arrangements may be very different.

In ideal conditions, such as on lakes well stocked with fish, densities of one otter per 2–3 km of shore have been recorded. This is about half the usual length of river which an otter requires. There have been few studies of the age structure in otter populations, but family groups are thought to break up when the young are about one year old. Female otters become sexually mature before the males, and probably establish their breeding territories at the age of 2–3 years. As is the case with many other mammals, the young male otter will travel much further away from his birthplace than the female. Otters in the wild may live from 11 up to 20 years old.

*The otter has a very flat and elongated skull, like other weasel-like animals but unlike the domed skulls of cats. Otherwise, both types of animals have short, blunt noses. In Europe, the former type have 34–38 teeth, but the latter have only 28–30. The otter's 36 teeth are structured as in the formula*

$$\frac{3\quad 1\quad 4\quad 1}{3\quad 1\quad 3\quad 2}$$

## Habitat and diet

Otters live in the immediate vicinity of either fresh or salt water, in marshy wetlands and along rocky shorelines. In lakes where there is lush marginal vegetation and along streams with dense bankside growth otters will live very shy and inconspicuous lives. Otters will also frequent rivers in upland areas, and in Europe otters have been found at altitudes of up to 2,800 metres above sea level. Otters tend to be most active at night, but where they are not disturbed they may be quite active during the daytime. Most of the otters movements and activities are in search of food, and they have been known to eat waterfowl, including their eggs and young. They will eat small mammals including fieldmice and and young rabbits, and their diet may also include frogs, crawfish, crabs and earthworms. However, their staple diet is fish, and although individual otters' diets vary from place to place, an estimated 90% of its food intake will consist of fish of some kind. The otter's daily food requirements correspond to about 10%–15% of its body weight.

Otters prefer lakes which are rich in fish and plant life, and especially those which form part of linked river-and-lake systems. Such conditions provide species of fish such as roach, bream and perch which often predominate in the diet of individual otters. When frogs go into hibernation in late autumn and winter the otter has less choice of prey, and in winter the otter may eat only fish.

## Distribution, populations and subspecies

The otter is found in most parts of Europe, Asia and north-west Africa. Its range does not extend into the northern tundra regions where winter conditions are too cold for it to survive. Over much of its wide range otter numbers have been severely reduced.

Twelve different species of otter have been noted world-wide. Half of these belong to the same *Lutra* genus as the European otter. These otters occur in Europe, Asia and in the East Indies, in Africa, and in North, Central and South America. Therefore otters may be regarded as having established themselves as ecologically successful mammals.

The European otter (*Lutra lutra*) is widely distributed, having colonised suitable habitats around lakes and along watercourses, and also on many islands. *Lutra lutra* is also found in Java, Sumatra and Japan, as well as in Europe and the British Isles. Curiously, otters appear never to have colonised many of the Mediterranean islands.

The tendency of otters to move about far and wide has tended to keep the species uniform, and to avoid the evolution of local races or subspecies. However, Irish otters are regarded as sufficiently different from those which occur elsewhere in Europe that they have been assigned to a distinct subspecies, *Lutra l. roensis*.

There are very few accurate records to show how otter populations have changed over long periods of time, and most information is informal and anecdotal. Some deductions can be made from records of the sales of otter pelts, and in the 1890s it was estimated that around 20,000 otter skins were sold each year on the Swedish market. By the 1930s this had dropped to not more than 1,500 pelts per year. Throughout Europe there is evidence that all otter populations have been depleted to a tiny fraction of their former size.

*The smaller weasel-like animals — weasels, polecats and otters — often survey their surroundings by sitting up on their hind legs, using their tail as a support.*

# Swans

## *Cygnus* spp.

Swans have always played an important role in mankind's view of the natural world. They have always been admired for their beauty, and have been kept as ornamental birds in parks and wildfowl collections. Historically, they have also been the focus for a great deal of legend and mythology, and there was a time when the swan was prized for its flesh.

In many countries these large white birds have been accorded the status of royal game. Traditionally, swans on the Thames belong to the English sovereign; and in Sweden an edict of the early seventeenth century made it an offence punishable by death to shoot or otherwise kill a swan. In Denmark, however, King Christian V organised a celebrated swan hunt in 1692, which resulted in a total bag of 420 Mute swans in one day.

In Britain swans have a long tradition of being reserved as birds for royalty and the nobility. Ownership of the Royal swans on the river Thames was supervised by the Keeper of the Royal Swans, and the Royal Swanmaster registered the annual crop of young birds. The swans were individually marked on the bill, and adult swans were caught in late summer to be checked, and the new season's cygnets were recorded and marked for the first time. In country areas swan-herds were appointed to keep records of the swans kept by noblemen and other specially licensed persons. This careful concern for swans is fully documented from the fourteenth century onwards, and in recent times swans from the Thames have been caught up and marked. Royal ownership of swans has been repeatedly proclaimed in numerous statutes and byelaws, but the death penalty no longer applies. However, there are records of imprisonment and fines being imposed upon those who have killed swans or taken their eggs or young.

Elsewhere in Europe some celebrated local swan populations were established and maintained. In Germany the swans on the Alster lakes near Hamburg are famous, and like the English swans they are still looked after by a "Schwane-warten" or "swan warden". The population is kept at an optimum size, and this is usually achieved by removing some eggs after incubation has begun, so as to leave each swan pair with only two eggs. Injured or sickly birds are culled, and the Mute swans of Hamburg are kept as a sort of elite breeding nucleus. There are some scarcely credible accounts of some of these swans attaining weights of up to 25 kg!

The swans of Hamburg have ended up on dinner tables in recent times, though not at the tables of royalty. During the First World War the population was reduced from some 400 individuals to only 16 by the time the war ended. By the outbreak of the Second World War the population had recovered to its former levels, but by 1945 only two swans remained in the Hamburg flock!

For as long as records have been kept, man has played his part in the destiny of these birds. The result is that Mute swans now exist in every continent, and successive British monarchs have played their part in this, for swans from the Thames have been given as royal gifts to Brazil, South Africa, Australia and Canada.

In central Europe the Mute swans in the rivers and wetlands around Berlin and Potsdam were also historically the property of successive kings and emperors, and these populations are believed to have been established in the sixteenth century. In the nineteenth century their numbers totalled many thousands. The swans were caught up in summer to have their down plucked for use in pillows and quilts, and there was a second catch-up at the beginning of winter. Of these once-huge populations only about 20 birds survived the First World War, and the numbers were restored by bringing eggs from the lake of Luknajno in Lithuania. The swans of Havel and Spree are therefore descended from Baltic stock.

The Whooper swan has also been the object of a great deal of interest over the centuries, but largely as a hunting quarry among country people. Whooper swans have been caught by stalking up on them, and also by using small boats to move the birds within range of shooters waiting in ambush. One tactic used in Finland was to drive a long pole into the bottom below the water and attach a heavy stone to it. A cord was tied to the stone and at its other end was attached a hook, baited with some attractive form of food. The trap was then set for the Whooper swan, which is a shy and wary bird. When the swan took the baited hook the stone was pulled from the pole and the swan was dragged down into the water where it drowned.

The Whooper swan has not been so important as an ornamental bird, largely because it was not considered to be as elegant and beautiful as the Mute swan. It was also aggressive and difficult to breed in semi-captivity. Yet the Whooper swan appears to figure more prominently in literature than the Mute swan. In particular its ringing call and bugling cries have given rise to the myth of the "swan-song". The ancient Greeks and Romans were divided in their opinions of the Whooper swan. Plato, Socrates and Cicero considered that the bird's song expressed a "presentiment of the bliss and peace of death", but the Roman naturalist Pliny considered this idea to be false.

The Greek gods, Adonis and Aphrodite, are depicted in classical art as drawn in a chariot pulled by two swans. Like Apollo they were brought by the swans to Hyperborea, the land of paradise. This myth arose from the migration of the Whooper swans between their northern breeding places and their southern wintering haunts. The conspicuous migrations and the stirring calls of the Whooper swans made them more conspicuous than the Mute swans. Yet the two birds are so closely related that interbreeding occasionally takes place.

One of the best-known tales of magical metamorphosis is Hans Christian Andersen's fable of the ugly duckling which was transformed into a beautiful swan. Every swan chick develops from a greyish little cygnet into a dazzlingly white adult.

In the "Dialogue of Creatures Moralised", published in Latin in 1483, one can read the following:

> "The swan is an entirely white bird; the raven on the other hand is totally black. Now the raven was envious of the swan because of its whiteness and purity . . . One night when the swan was asleep the evil raven sneaked up on the swan's nest and blackened the swan, making it as black as coal. In the morning the swan awoke and saw that he was besmirched with blackness, and he bathed and rinsed his plumage until he was totally white once again, whereupon he said 'He who would preserve his happiness must remain immaculate and unblemished'." In this allegory the raven is intended to represent the devil.

The swans are the largest waterfowl, and are found in many parts of the world. Australia has its own species of Black swan (*Cygnus aratus*), and in the far south of South America there is the Black-necked swan (*Cygnus melanocoryphus*). In Europe and Asia the Mute swan (*Cygnus olor*) has a wide natural range, while the Whooper swan (*Cygnus cygnus*) occurs across Europe, Asia and North America. The Bewick's swan (*Cygnus bewickii*) has a wide range across Europe and Asia. The largest species of swan is the North American Trumpeter swan (*Cygnus buccinator*), which weighs up to

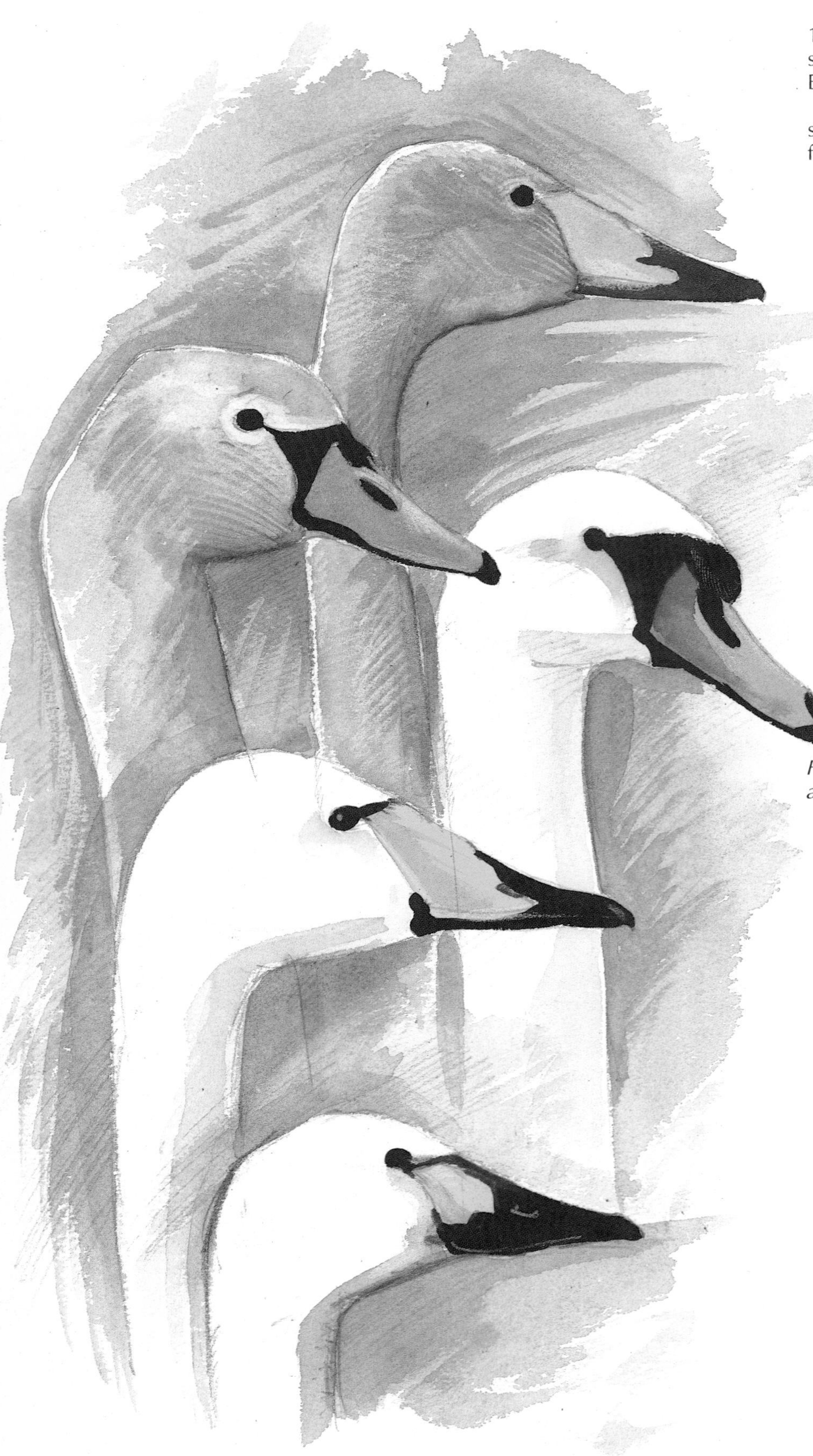

15 kg. This swan is sometimes considered as a distinct species and sometimes regarded as a subspecies of the Whooper swan and the European Mute swan.

In all swan species the male is larger than the female and weighs some 20%–30% more. The smallest of the Cygnus swans is the female Bewick's swan, which weighs about 5 kg as an adult.

*From top: young Whooper swan, young Mute swan, adult Mute swan, adult Whooper swan, Whistling swan.*

# Mute swan

## *Cygnus olor*

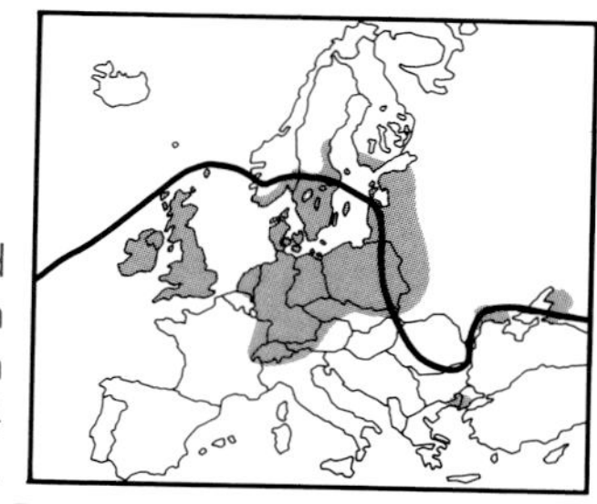

### Characteristics

The Mute swan and the Great Bustard (*Otis tarda*) are the two largest European birds. The male Mute swan is larger than the female and his wingspan is 2.3–2.4 metres, with average weights of 7–15 kg. The average length of the male's wing is 560–630 mm, while that of the female is 540–622 mm. The length of the neck, measured from the top of the breastbone, is 82.5 cm for an adult male and 75.5 cm for an adult female. The neck is usually carried in an S-shape and the bird's tail is pointed and slightly cocked upwards. The bill is a reddish-orange in the adults and greyish-pink in the case of young birds, apart from the black point of the bill and a black triangular marking on the sides of the bill in front of the eyes. Their feet are greyish-black or black, and the iris of the eye is brown or sometimes greyish. The downy chicks are grey and the young bird is a greyish-brown in the first year of its life, but its plumage becomes white in the second year, apart from the brown colouring at the base of the tail. (Some chicks, occasionally referred to as "Polish swans", are born all white with beige-brown legs.)

The Mute swan, as its name suggests, is silent apart from occasional grunts and snuffles, and in flight its wings make a powerful swishing sound. When alighting, and also when laying claim to their territory, they make a clattering sound with the soles of their feet. When in a resting position on the ground they keep one leg lifted and tucked up. The Mute swan requires a long distance in which to take wing or to land.

The Mute swan is active by day, and it is a plant eater which primarily derives its food from the water: this may include sea-grass, reeds, Water Milfoil and other aquatic vegetation. Adults will eat about 3–4 kg (wet weight) of vegetation each day. The oldest recorded Mute swan was leg-banded in Sweden and attained the age of 25 years.

Mute swans, Coots and Wigeon often form groups for the purpose of finding food. All three species are especially fond of sea-grass (*Zostera*) which occurs in bays and inlets along the coast. The coot is the only one which dives for its food, and it often tears up more leaves and roots than it will eat. The Mute swans and Wigeon therefore swim along with the coots and feed eagerly on the floating

leftovers. On the other hand, the Coots will occasionally feed on the remains of what the dabbling swans have torn up from the bottom. The Wigeon, which can neither dive nor dabble at any great depth, are the only ones which always benefit from the feeding activities of the others. In some coastal bays hundreds of swans and thousands of coots and Wigeon can be found feeding together off the underwater meadows of sea-grass.

*Hybrid of Mute and Whooper swan.*

## Reproduction

The Mute swan is sexually mature at two years old, but it does not usually breed until it is 3–4 years old. Many do not breed for the first time until they are eight or nine years of age. But they can go on breeding up to the age of 20 years. The Mute swan is monogamous, but it is not unusual for the pair to split up, and new pairings take place quite quickly after such partings. Occasionally one male will be found to have mated with two females.

Mute swans produce one clutch of eggs annually, but they will lay again if the first clutch is lost. There may be from 2–12 olive-green eggs, each weighing about 330 grams, up to a maximum of 410 grams. The average dimensions of the egg are 99–124 mm x 68–80 mm. The breeding pair builds a large nest of reeds on freshwater lakes and ponds, and on coastal islands they will construct a nest from seaweed or twigs. Mute swans can be found to breed not only in their original inland, freshwater habitats but also along the coast. Usually they have large territories, but in some places where the birds are numerous they form breeding colonies. The species' breeding success can vary, and on average only 40% of breeding attempts will be successful.

The newly hatched chick weighs about 200 grams, but by the time the young are fledged they will weigh 6–8 kg. The female does most of the incubation, occasionally relieved by the male, and incubation lasts for approximately 35 days. The breeding season from egg-laying to the young fledging takes approximately 22–23 weeks. The Whooper swan has adapted to the shorter breeding season in its northerly haunts, and completes the season in about 18–19 weeks. The chicks leave their parents shortly after they have fledged, but the family group normally remains together until the following January–February. The breeding pair usually return year after year without their young to the same breeding place.

## Distribution and numbers

The Mute swan occurs in two distinct populations, one in Europe and the other in Asia, and at some considerable distance apart. The population in countries bordering the North Sea and the Baltic Sea has increased significantly since the 1950s, and the European winter population probably consists of 120–130,000 individuals. In the 1920s the total Danish/Swedish population had been reduced to a mere four or five breeding pairs, after much relentless hunting, but the breeding population has since increased to 6–7,000 pairs.

# Whooper swan

## *Cygnus cygnus*

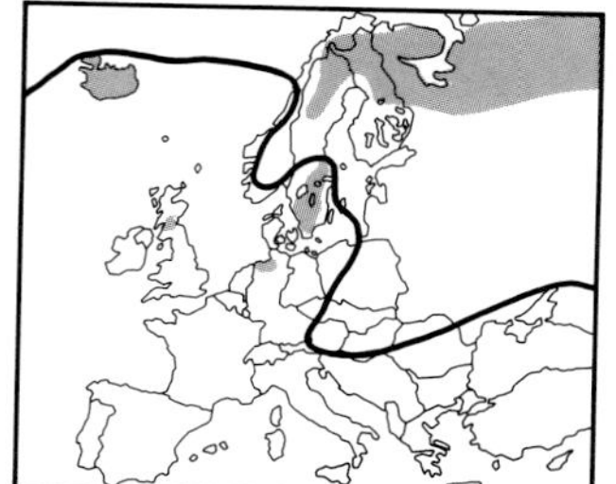

The Whooper swan is smaller than the Mute swan and rather larger than the Bewick's swan. In its second year the greyish youngster acquires its white adult plumage. It therefore requires close observation to distinguish the older birds from the younger ones during the second winter: the birds which are 18 months old have a few greyish crown feathers and greyish wing coverts. The young bird is lighter and even more grey than the Mute swan chick, and it has a greyish, flesh coloured bill. The adult birds have a yellow bill with a black tip, on which the yellow forms a wedge shape. The tail is cut off squarely, and this is especially obvious in flight. There is no whistling from the pinions in flight, but there are loud and resonant trumpetings. The bird stands or swims with its neck erect, and the sexes are similar in appearance, although the male is somewhat larger. The feet are black and the tarsus is longer than that of the Mute swan. The iris of the eye is brown or greyish. The length of the adult male's wing is in the range 577 mm–660 mm, and that of the female 562 mm–635 mm. The oldest wild Whooper swan, leg-banded in Denmark, survived to the age of 23 years. The closely-related and similar-looking Bewick's swan is also shown here: it has a shorter neck and is rather more goose-like.

## Reproduction

The Whooper swan is normally sexually mature in its fourth year, although there are some recorded instances of third-year birds breeding. Nests are located by lakes and in marshes in the swamps of the northern coniferous and birch forest regions, and during recent years this species has started to breed on some of the southern lowland lakes, as well as in marshes and on lakes which are not so nutritionally rich. Although the southern breeding birds lay their eggs in April, the northern breeders will not lay their eggs until June. A typical clutch will contain 4–8 oval, yellowish-white eggs measuring 105-126 mm x 69-76 mm, and weighing about 320 grams each. The female incubates the eggs for about one month, and the little greyish-white chicks will fledge after about 13 weeks. After hatching both parents take care of the brood of chicks. The large nest, which is at least a metre in diameter and 30–40 cm high, is placed in the smaller marshy spots or on the margins of a pond. After the eggs hatch the family wanders off to larger lakes nearby. The Whooper swan is usually monogamous with long-lasting pairings, although there are some recorded instances of one male mating with two females. The family stays together until the adults return again to their nesting site.

## Distribution and numbers

The Whooper swan belongs to the bird life of the palaearctic, which means it occurs both in Europe and the northern parts of Asia. North America has the closely-related trumpeter swan, which is sometimes regarded as a subspecies of the Whooper swan. The Scandinavian population was in serious decline, almost to the point of extinction in some places, but in recent years it has recovered noticeably. Today, many Whooper swans breed well to the south of the species' former southern limit, in West Germany, Poland, southern Sweden, the countries of the south-eastern Baltic, and in Scotland.

In the Baltic and North Sea areas about 25,000 Whooper swans over-winter, and there is a similar number on the Black Sea. In summer there are large congregations of moulting birds around the coasts of Iceland. The Whooper swan moults in the same way as the Mute swan. Outside the breeding area they stay in coastal bays, flooded marshes, along rivers and on lakes, and they will graze on land. They migrate in a traditional pattern, along the same flight lines year after year.

*Mute and Whooper swans are unlike in several ways. The former has a more blunt tail than the latter, as seen when flying. The Mute swan also swishes its wings loudly. It keeps contact with its partner by gentle turns of its head, while the Whooper swan does so by trumpeting and beating its wings. Mute swans seldom graze on land, but Whoopers do so readily.*

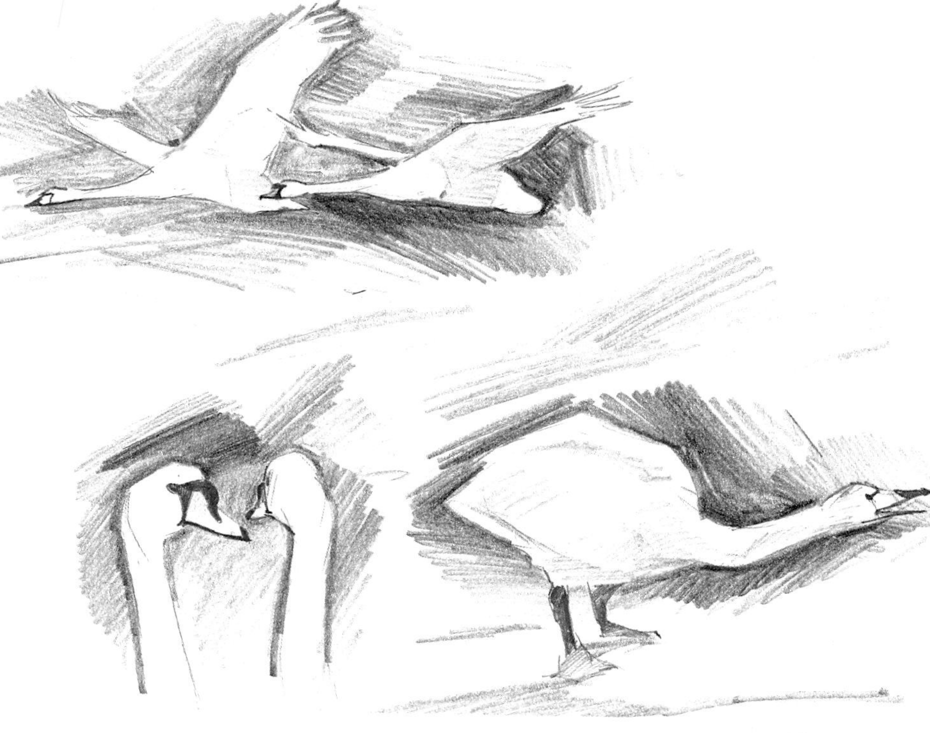

# Bean Goose

## *Anser fabalis*

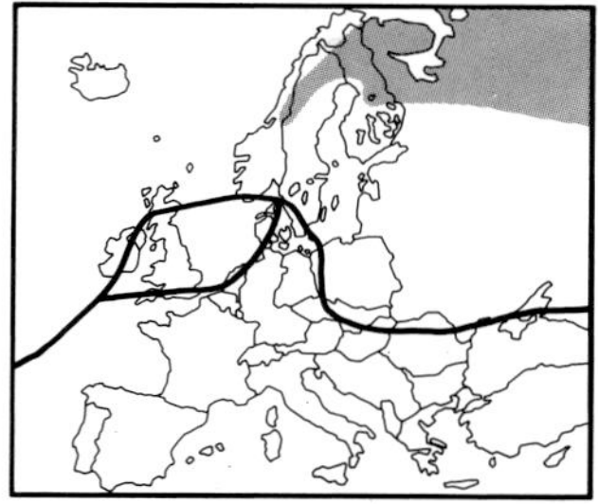

The Bean goose is principally a daytime migrant, but at dawn and dusk they flight between the lakes, where they roost and sleep, and their grazing areas. These movements often take place in total darkness. In various parts of Europe these geese with their wild calls became woven into local fears and superstitions, and were sometimes referred to as "the hounds of Odin" or "the devil's hounds". Bean geese maintain close family bonds, and when flighting they will maintain contact with harsh calls and higher pitched cries as the older birds call and are answered by the younger geese.

## Characteristics

The Bean goose is a large and relatively long-legged bird with a prominent bill. The plumage is much darker than that of the familiar greylag goose, and the rump is white. In flight the white band across the base of the tail is visible. The bill is orange with a dark base and a brownish-black tip. The legs are a uniformly orange colour. The sexes are similar in appearance, and the adult birds often have white feathers around the base of the bill. The Bean goose has brown eyes; and the young goslings are coloured olive-green on their backs and yellow on their underparts.

The body weight of the Bean goose is between 3.2 and 4.1 kg, and the wing of a mature male will measure 445–498 mm long, with that of the female falling within the range 410–450 mm. From this it will be seen that the female is decidedly smaller than the male. The newly hatched goslings weigh about 80–95 grams, and their early growth is rapid. By the time they have fledged at the age of two months they will weigh about 3 kg.

The tundra Bean goose (*Anser f. rossicus*) differs from the Bean goose in having a shorter bill which is mostly blackish-brown, often with little more than a yellowish or yellowish-orange band before the tip. Its shorter bill gives the appearance of being set higher, and the head appears more rounded, like that of the Pink-footed goose (see below). Moreover, its body is smaller and plumper, with an apparently shorter neck.

Bean geese form long-lasting pair bonds, and they are ready to breed as soon as they return in spring to their breeding grounds. Usually they return while there is still a covering of snow on the breeding territories, and the birds wait in meadows and river estuaries. As soon as the snow has disappeared the pairs spread out, and the nests are placed on hummocks, large tufts of vegetation and other raised areas to keep them clear of the boggy ground and the melted water. Nests are often built beneath the cover of osier bushes and they are made of grasses lined with light grey down.

The clutch consists of 4–6 off-white eggs which are laid at the end of May or in early June. The eggs weigh an average of 145–150 grams and their dimensions are 80–90 mm x 56–60 mm. The females incubates her eggs for 27–29 days, and after hatching both parents care for the young goslings. The parents take the young goslings away from the nest site after they have hatched, and they often wander for long distances to find suitable lakes which will be their homes while the young geese are growing and fledging. The period of time which elapses from egg-laying to fledging is about 14 weeks. In areas where the pairs of geese nest close together and wander to the same lakes, flocks of adults and goslings from several families are formed. The family remains together until the following breeding season.

## Adaptations and behaviour

During the winter both the Bean goose and the tundra Bean goose are found in Europe. Their relationship is explained below. The tundra Bean geese leave their breeding haunts earlier than the Bean geese, and they appear to be more sensitive to low temperatures and snowfalls. The two types differ somewhat in their food preferences and in their daily routines in winter and at their roosting sites.

The Bean goose is normally active during the day, and roosts on the open water of lakes by night. From its roosting lakes the geese move in flocks to their feeding grounds, and this may involve a flight of up to 25 km. In the middle of the day they often return to the lakes to drink, rest or bathe, and in the afternoon they graze again until dusk. Normally these geese move to and fro at dawn and dusk, but in the late autumn and winter they prolong their feeding time so that their return flight is made in the dark, and they tend also to leave the lake early in the morning, before dawn. In this way they increase the grazing time available to them in the short, dark days of the winter.

When winter conditions become too severe, and the roosting grounds and feeding areas are covered with ice and snow, many of the geese move on. This winter migration means that the geese move in their tens of thousands to the south and the south-west. Some may remain, however, and these geese can survive for up to a month by making use of their reserves of body fat.

## Food

The Bean goose is entirely vegetarian, and grazes on various types of greens, including grassy fields, sprouting cereals and fields of clover. They will also pick over the stubbles after harvest and are attracted to potato fields. In their northern breeding areas they will also eat various types of berries. It has been observed that these birds can distinguish between foods of differing quality, and well fertilised clover and sprouting corn are preferred to any more meagre and less nutritious form of food.

## Distribution, migration and numbers

The Bean goose is part of the palaearctic fauna, and its natural range extends in a broad zone across northern Europe and Asia. In the taiga regions of the north it is largely confined to swampy land, marshes and lakes among the coniferous forests, but it can also be found out on the tundra.

The Bean goose migrates to its winter quarters after the breeding season and the fledging of the young birds. Ringing of Bean geese and subsequent studies of the birds' movements has indicated that there is a dividing line over the area of west Taymyr, which divides the winter populations of the Bean goose in two — one of which migrates to the south-west and the other to the south-east, with winter quarters in Europe and Kazakhstan/China respectively. Bean geese which breed around the Taymyr area usually move southwards to winter quarters in Turkestan and Iran. Bean geese from Scandinavia migrate south-westwards through Denmark and western Sweden on their way to their final wintering grounds in Scotland and north-eastern England.

Tundra
bean goose
Bean goose
Old bean goose
Bean goose
Asian bean goose
Pink-footed goose
Snow goose

Large numbers of Bean geese migrate to their moulting grounds in northern Norway, and leg-banded individuals indicate that these birds have come from Scandinavia, Finland and north-western parts of Russia.

In recent decades the Bean geese which winter in Europe have significantly changed their migration movements. At first a decrease in numbers was noticed, and this was attributed to excessive shooting pressure, egg collecting and the killing of the geese at moulting time. But at the same time as this decrease was apparent in western Europe, numbers of Bean geese were observed to be increasing in Skane and East Germany. This change was brought about by milder winter conditions in the latter areas. At the same time the geese which wintered in central Europe, from Hungary down to Italy, also began to move to these more northerly winter haunts. The Bean geese which formerly wintered in northern Africa, especially Tunisia and Morocco, have now disappeared.

In the nineteenth century two different species of Bean goose were identified, *Anser arvensis* and *Anser fabalis*. It now seems that these denoted the Bean goose and the tundra Bean goose. The former is the Scandinavian Bean goose (which also occurs in Finland and north-west Russia) and this is the nominate form, *Anser f. fabalis* which gave the species its first scientific description. The tundra Bean goose, *Anser f. rossicus*, breeds on the tundras along the Arctic Sea coast of western Siberia. A question arises as to whether other eastern subspecies such as *A. f. johanseni* and *A. f. serrirostris* also occur in Europe. There are transitional forms of this species between the different subspecies, usually encountered along the boundaries of the various subspecies' ranges.

Today there is an estimated winter total of about 600,000 Bean geese in Europe, of which 100,000 belong to the nominate form, and thus constitute the entire world population. Some 60,000 of these winter in southern Sweden, with good numbers along the Baltic shores of northern Poland and Germany, and also a few in the British Isles.

The tundra Bean goose is found to the south of the area favoured by the nominate form, with a total of some 400,000 wintering in East and West Germany, Poland, the Netherlands and Belgium, and a further 200,000 are found in Czechoslovakia and Hungary. Thousands of these birds used to winter in Spain and Italy. From the available figures it would appear that the nominate form of the Bean goose has declined very seriously to about 25% of its former numbers, while the tundra Bean geese have simply shifted their wintering grounds.

## Relatives of the Bean goose

Ducks, swans and geese together comprise the family *Anatidae*, which consists of 146 different species. Only nine of these species are accounted for by the Bean goose and its closest relatives in the *Anser* family. Europe is home to three of four relatives of the Bean goose: the Greylag goose (*Anser anser)*, the White-fronted goose (*Anser albifrons)*, the Lesser White-fronted goose (*Anser erythropus*) and the Pink-footed goose (*Anser brachyrhynchus*), of which the latter is sometimes considered to be a subspecies of the Bean goose. In addition there are the Snow geese (*Anser caerulescens*), a rare but fairly regular visitor from across the Atlantic, and also a south-eastern Asiatic species, *Anser indicus*, which is an escapee from parks and waterfowl collections.

All these geese form flocks, especially outside the breeding season, and some also tend to breed in colonies, for example the White-fronted goose and the Snow goose. Occasionally Greylag and Pink-footed geese nest so close together that one could describe them as nesting in colonies.

These geese are sexually mature in their third or fourth year, and occasionally in their second year, and this gives rise to large reservoirs of non-breeding young birds. The females incubate alone, but both adults take care of the brood once the eggs have hatched. When the young goslings are growing the families often group together to form flocks. The smaller, northerly species of geese have the shortest periods for egg-laying, incubation and fledging of the young, which is an adaptation to the shorter summer breeding period available to them in the far north.

The *Anser*-type geese have many features in common. They like to graze on open meadows and fields, and only feed occasionally on or near water, although the breeding greylag is an exception to this. These species' migration routes and choice of wintering grounds are occasionally subject to abrupt changes. The non-breeding birds move along specific migration routes to the same moulting areas year after year, and here they undergo a complete moult. Likewise the birds' choice of roosting lakes and feeding grounds remains the same year after year. These geese follow well defined traditional patterns of movement, despite their occasional changes of migration routes.

These geese are believed to be very long-lived, and the oldest known leg-banded wild geese have lived for up to 17–21 years, although the majority die sooner. The annual rate of mortality among the old birds is estimated at 25%–35%, and the rate of mortality is slightly higher among the young birds. Rates of mortality vary from one species to another, and depends among other things upon local conditions and shooting pressure.

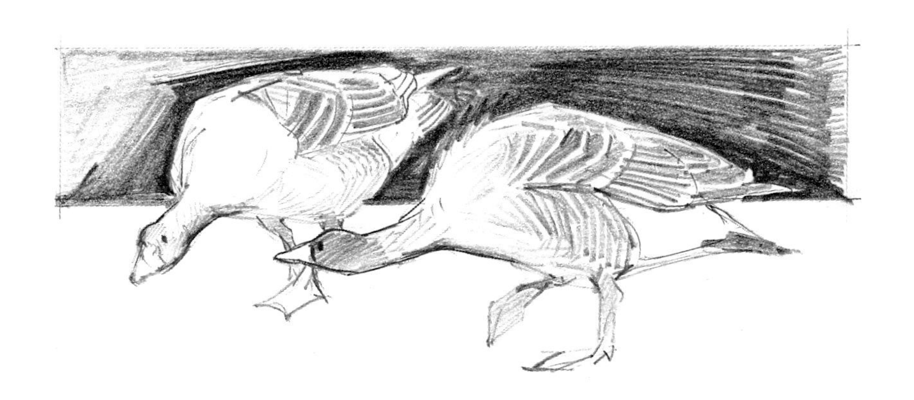

## How to identify the *Anser*-type geese

In the field, the Bean goose and its relatives seem to be coloured a uniform greyish-brown, and there are no striking patterns in the plumage, such as can be seen in the Canada goose and its relatives among the *Branta*-type geese. It should be noted that the Greylag goose, the White-fronted goose and the Lesser White-fronted goose all have a pale tip to their bills, and have pale-coloured toes, while those of the other species are blackish-brown. However, young White-fronted geese have a dark tip to their bills.

The Greylag goose is the largest of the European geese, and adults usually weigh between 2.4 and 5 kg. The male is usually larger than the female, but the sexes are otherwise indistinguishable. Greylag geese can be distinguished from other closely-related geese by the light, silvery-grey colour of the front of the wing. The bill is a uniform orange colour, inclining to pink, in the northern European subspecies. The Asian breeding subspecies *rubirostris* has a bill coloured reddish-pink with a white tip. Both subspecies have pinkish, flesh-coloured feet. The plumage of *rubirostris* is somewhat lighter, and it has black spots on the belly.

The Greylag goose has been released and introduced into many parts of Europe. Its breeds freely along the coast of Norway, by the shores of lakes in southern Sweden, in Denmark and in south-west Finland. Elsewhere it breeds on the coasts of the central and northern Baltic Sea, and also in Estonia where there is plenty of suitable habitat in the form of islands and small bays. In recent years the Greylag goose has also colonised the islands off the west coast of Sweden.

The Greylag goose breeds in separate pairs and also in more densely packed colony-like groups. In Estonia densities of up to 15 breeding pairs per 100 hectares have been noted, especially along the reed beds, and this is the highest known density.

The clutch comprises 4–8 off-white eggs, and these are incubated by the female for 28–29 days. The male stays with his mate and guards the nesting territory. Both adults rear the young goslings until they have fledged, which takes place about ten weeks after hatching, and thus the breeding period lasts in all for about three months. Only one brood of young is raised each year, and the family stays together during the autumn and winter. By the age of two years the young are themselves capable of mating and breeding. The greatest age known to have been attained by a Greylag goose in captivity is about 25 years. Among the wild geese the adult mortality account for about 25% per annum.

The Greylag goose is chiefly a vegetarian feeder. It likes to eat reeds in lowland lakes and will graze on grass along shoreline meadows. In winter it is often found grazing on sprouting cereal crops.

The Icelandic breeding birds' winter quarters are in Scotland, while the Danish and southern Swedish birds migrate to Spain. The geese which breed in eastern Sweden, Finland and eastern Europe winter in Hungary, Yugoslavia and Greece. There is a so-called "flight divider" across the Baltic Sea, which divides south-westerly migrating Greylags from those which migrate in a south-easterly direction. Non-breeding Greylags gather together in small flocks, and moult together on lakes close to the birds' breeding grounds.

The European wintering population of Greylag geese has been estimated at around 120,000 individuals, and the majority of these winter in the British Isles, in Spain, the Balkans and Turkey. The highest breeding numbers occur in Iceland, Denmark, East Germany and Poland.

The White-fronted goose is an arctic species, and is considerably smaller than the Greylag goose and a little smaller than the Bean goose. A typical adult bird weighs between 1.9 kg and 3.5 kg. The young birds and the females are smaller than the mature males.

*The sexes look alike, but male geese are usually bigger than females. One can identify the sex of any swan, goose, or duck by turning its cloaca inside out. The males have a prominent penis.*

A century ago the White-fronted goose was rather uncommon in Europe, and tended to be seen either singly or as individual specimens mixed in with very large flocks of Bean geese. At that time the White-fronted goose was found chiefly in its winter haunts in southern Europe, from Hungary down to Greece. At the beginning of this century the White-fronted goose began to appear in larger numbers, and by the 1930s and 1940s this had become the

### Two goose mysteries

At one time the so-called Suschkin goose was one of the great enigmas of the bird world. It closely resembled the Bean goose, and occurred among flocks of Bean geese in several parts of Europe during the late nineteenth and early twentieth centuries. But its feet and bill were *pink*. A Russian ornithologist named Suschkin gave it the scientific name *Anser neglectus*. Sightings of this goose were especially numerous in western Siberia; between 1910 and 1920 flocks of thousands were seen in the province of Ufa. It was also found in the winter flocks (then very large) of Bean geese in Hungary. No-one ever managed to explain it, whether as a mutation, or perhaps as a race connected with some undiscovered breeding location. Suddenly the birds disappeared, and only a few museum specimens remain. Rumours that families of Suschkin geese have been seen in recent years in the Netherlands have not been confirmed.

A curious note was made by one of Scandinavia's old historians, Olaus Magnus, who was bishop of Uppsala in the 1500s, and author of *A History of the Nordic Peoples* (1555). He referred to the Roman historian Pliny, who had written in the first century A.D. that white geese with black wing-tips migrated over Scandinavia. The Snow goose does migrate from breeding places on Wrangel Island in eastern Siberia to its winter quarters in California, but could Snow geese from, say, Baffin Island or western Greenland have reached Europe in large numbers in Pliny's time? It is not impossible, and individual Snow geese have often appeared in England and western Europe, either as escapees from waterfowl collections or as passing vagrants from the other side of the Atlantic. In the Netherlands, however, a flock of eighteen was seen in 1980, one of which had been leg-banded as a fledgling in Manitoba, Canada.

*The Lesser White-fronted goose breeds mainly on forest tundra and nearby swamps.*

*The White-fronted goose breeds on the Arctic tundra, which is characterized by mosses, lichens and low bushes.*

commonest goose species in south-eastern Europe. At the same time it was seen more and more around the Baltic shores, and in the Netherlands and Belgium.

Around the middle of this century the White-fronted goose population declined from many hundreds of thousands to a few tens of thousands in central and south-eastern Europe, but at that time it was becoming more common in Germany, the Netherlands and Sweden. There in the 1950s the winter population was estimated at almost 20,000 birds, and by the 1960s this had increased to 75,000. This was followed with peak winter populations of between 180,000 and 250,000 in the early 1980s.

The western Greenland subspecies *Anser albifrons flavirostris* normally winters in Ireland and in some parts of Scotland and England. This subspecies can be distinguished by its yellow bill and its darker plumage, but the population has recently declined to less than half of its former numbers, which used to be around 15,000.

Like the Bean goose, the White-fronted goose likes to feed and rest on meadows and open fields, and its diet consists of various types of grasses. Also like the Bean goose, it moves to and fro between the lakes where it roosts and the fields where it feeds. The European population is estimated between 500,000 and 600,000 individuals, and it is considered the most numerous species of goose in Europe and Asia. It has a circumpolar breeding range in the arctic.

The Lesser White-fronted goose is found as a breeding bird in an area to the south of the White-fronted goose's breeding range, and principally in the swampy areas between the tundra and the timber-line in Norway, Sweden, Finland and Russia. This species is the smallest representative of its family, and in many respects it is similar to the White-fronted goose. But the white patch on the front of the head extends rather higher up towards the crown, and the bill is a little shorter in proportion, which gives this bird a more round-headed appearance. The average body weights are between 1.8 kg and 2 kg, with adult males being the largest and heaviest.

Around the 1940s–50s there began a marked decline in the numbers of this goose breeding in Scandinavia, and many traditional breeding grounds were abandoned. In the 1970s the Swedish breeding population was estimated at 100 pairs, but by the early 1980s the total Swedish, Norwegian and Finnish breeding population was believed to be not more than 90 breeding pairs, with a non-breeding reservoir of not more than 100 individuals.

The Scandinavian lesser White-front usually migrates, and leg-banding studies have shown that the birds move in a south-easterly direction towards winter quarters in south-eastern Europe and the areas around the Caspian and Black seas.

Despite this considerable decline in the numbers of lesser White-fronted geese breeding in Scandinavia, this species was still considered in the 1960s to be a very common winter visitor in eastern Hungary. Three to four thousand birds winter there even now. However, little is known of the exact size of the winter population, and the old data which gave accounts of tens and hundreds of thousands of birds wintering in Hungary and southern Russia are now outdated and inaccurate. The most pessimistic estimates are of only a few thousand birds, and there has been some concern regarding the survival of the species. Only 200–300 winter now by the Black Sea and perhaps 3–5,000 by the Caspian Sea.

The Pink-footed goose has its nesting territories on the islands of Spitzbergen, but the species also breeds in eastern Greenland and in Iceland. The body weight of a typical adult is between 1.4 kg and 3.9 kg, and the females weigh less than the males. The Pink-footed goose is therefore quite close in size to the Bean goose, even closer in size to the White-fronted goose, and noticeably smaller than the Greylag goose.

Pink-footed geese from Greenland have their winter quarters in Scotland and England, while those which breed in Spitzbergen migrate to the coasts of Denmark, Germany, the Netherlands, Belgium and France, along the shores of the North Sea and the English Channel. The migration flight from the breeding grounds in Greenland and Iceland is direct and non-stop across the north Atlantic. Geese coming from Spitzbergen also seem to fly non-stop to their first landfall in Denmark, and are thought to fly over Norway at very great heights.

Pink-footed geese have shown a significant recent increase in numbers. In the British Isles a total of 95,000 of these geese were counted in November 1981, and on the mainland of Europe a similar number were seen each November from 1961 to 1964 along the North Sea coast. Improved breeding conditions, a milder climate and better availability of good food are thought to have contributed to this great increase.

In Sweden, Norway and Finland the Pink-footed goose only occurs as a migrant visitor, and in small numbers, usually mixed in among the flocks of Bean geese, and the same is also true of central and southern Europe.

# Mallard

## *Anas platyrhynchos*

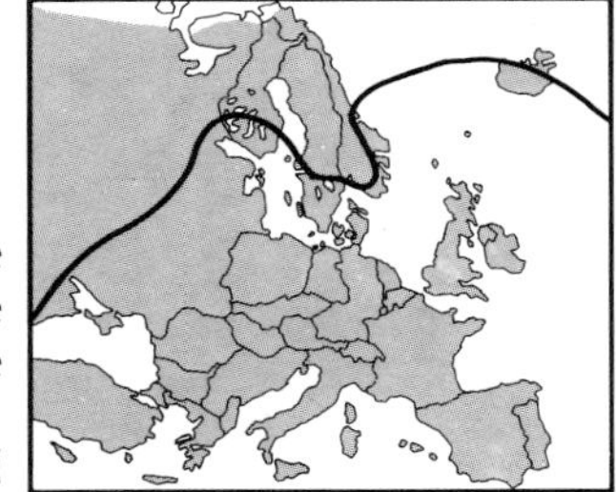

In wildlife and sporting books of the eighteenth and nineteenth centuries one can discover numerous references to the hunting and shooting of young Mallard, barely fledged, in the months of July and early August. So-called "flapper shooting" often resulted in very large bags of immature birds, which can surely have presented little or no sporting challenge, although they were doubtless pleasant to eat because of the birds' habit of feeding on fields of barley and oats at harvest time.

Apart from the Mallard shooting and catching of late summer, there was also the catches made by European hunters from their "duck huts". These were found along the coasts of the North Sea, especially in the Netherlands and on the North Frisian Isles, where Mallard were trapped in their thousands in winter. One single duck hut in the year 1784 accounted for no fewer than 67,000 Mallard, all taken in nets and snares. The same hut took an annual average of 11,750 Mallard in the years between 1839 and 1851. It is not clear what impact, if any, this catching of Mallard had upon the species' total numbers, but there can be no doubt that this catch represented only a very small proportion of the total population.

An estimated 750,000 Mallard are shot annually in Sweden, Denmark, Norway and Finland, and of this total three-quarters are shot in Denmark. In light of this large annual harvest it may seem curious that the Mallard has been able to maintain its position as the commonest duck species in Europe right up to the present day.

## Characteristics

During part of the summer both ducks and drakes have a very similar plumage, but the male can always be distinguished by the colour patterning on his wings, and its bill always has an overall greenish hue, while that of the female is greyish-green with orange-beige edges and dark spotting. The upper parts of the male Mallard are always darker and more evenly coloured than those of the female. The male assumes his full plumage in late summer and early autumn, and keeps it until the following June or July.

Young birds look similar to the female, which has a brown-flecked plumage throughout the year, mottled with dark — often black — spots and markings. The young birds have orange feet, while those of the adults are of a deeper reddish colour. In autumn the different ages can be distinguished by, for example, the shape of the principal tail feathers. An old method used by sportsmen for telling the age of a shot duck is to hold the bird by the lower mandible and give it a sudden jerk: if the mandible breaks easily, the bird was young.

Mallard are active both by day and night, often feeding and migrating by night, but always conducting their courtship during the daylight hours. Much of the daytime is simply used for resting. When taking flight the Mallard springs vertically upwards, and this is believed to be an adaptation to life around rivers and lakes where there is lush, high vegetation, and where the birds feed in thick cover. Here they can find both protection and nutrition. In recent times many captive-bred Mallard have been released into the wild,

*From top: Mallard adults and young with normal markings; a hybrid and colour variant; and a Pintail.*

and these can tend to be of a rather different type from the truly wild birds, with relatively shorter bills and wings and weighing considerably more. Fears have been expressed that these different characteristics may have a negative effect upon the wild Mallard population, in matters such as choice of food, flying ability and successful chick production. The "tame ducks" can be more competitive than the wild ones, pushing them aside, although the latter are naturally better adapted for a life in the wild. By interbreeding with the wild populations, these reared birds may change the migration behaviour and the mortality rate of the wild Mallard. Mallard are also known to hybridise freely with other species of dabbling ducks.

Outside the breeding season the Mallard tends to live in flocks. As soon as the female has laid her eggs and begun to incubate, the males will gather together and flock at certain moulting sites. Like most other ducks they lose their ability to fly at moulting time, which lasts for about four weeks in July–August. Females with young moult their plumage at the nesting site, but this does not begin until the young are a few weeks old. At the autumn gathering places and in their wintering areas thousands of Mallard of both sexes and all ages will mix together.

The Mallard's sense of sight is its principal means of protection when danger threatens, but it has an extensive repertoire of calls to spread information among the flock and to maintain contact. The Mallard's various calls have many functions, including warning and aggression, but they are also caused by fear or uncertainty, and during contact between the sexes.

The Mallard has various kinds of ritualistic behaviour which have different functions and meanings, and these are part of an elaborate visual signalling system. For example, this includes the pretence of cleaning and preening under the wing, which has the effect of stretching and exposing the wing speculum.

## Size

The Mallard is Europe's largest dabbling duck, and typical body weights for mature adults lie between 1.0 and 1.7 kg. The male is larger than the female, and his wing length is within the range 263–304 mm, while that of the female is within the range 247–281 mm. The male also weighs on average about one-eighth more than the female. During the moutling period, which takes place principally in July, weight losses of up to 20% have occurred, and in extremely severe winter weather Mallard may lose up to half of their body weight.

Mallard body size varies with different local populations. The Greenland Mallard are regarded as the largest wild ones, but Mallard which have been reared in captivity and released into the wild may weigh up to 25%–30% more than wild birds. Newly hatched ducklings weigh about 35 grams, but they normally lose several grams weight in the first few days of life, perhaps as much as 25%–30% of their original weight. Recent studies in the USA suggest that the growth rate of Mallard chicks differs between separate populations and is caused by genetic factors.

Where the feeding is good, young Mallard grow quickly and by the time they are one week old their body weight will have tripled. At the age of three weeks they weigh, on average, 290 grams, at six weeks 610 grams, and at nine weeks 875 grams, after which their speed of weight gain drops off. Birds of all age groups reach one of their two annual peaks of body weight in early winter, but the young birds are normally lighter than the older ones. After a decrease in weight as winter gives way to spring, Mallard become heavier again in late spring and early summer. The females undergo some weight loss during the laying and incubation of the eggs.

As winter approaches Mallard build up their stores of fat under the skin and around the internal organs. They can live on this fat for long periods if food is not available, for up to two weeks under normal conditions. In cases like this the mortality is less among the females than the males, but artificial feeding soon restores the birds' weight. Apart from a shortage of food, and in addition to it, temperature plays a part, and in conditions of extreme cold the birds lose weight quickly as their reserves of fat are burned up. In western Sweden the average body weight of Mallard is some 8%–17% higher in December than in January, and the females lose more weight than the males. Mortality caused by extreme starvation and emaciation occurs when body weights drop to around 5–600 grams.

## Reproduction

Mallard pair up in the winter months, and the male and female remain together until all the eggs have been laid. Pairs therefore rarely last for more than one breeding season, although the females return again and again to the same nesting spots each year. Breeding Mallard usually arrive in pairs at the nesting place.

The female makes her choice of a nest site and creates a nest hole or scrape which, as with other ducks, is lined with a combination of local vegetation and feathers and down from the female's body. As a rule female Mallard are ready to breed at the age of one year.

Mallard nesting places can be found in some curious spots! These may include flower boxes on balconies in towns, in the flower beds of busy urban parks, and in old crows' nests in trees. More typically, however, they will nest in grassy vegetation, often beneath the overhanging shade of bushes, among reed beds and rushy areas, in high vegetation, on heathland and so on. But whatever the chosen nesting place it should be associated with water in some form — a pond, marsh, stream, river or lake. However, the nest is often some distance from the water, and in extreme cases the female may have to lead her brood through the busy traffic of a city or, more often, across fields and through woodlands, where predators like foxes and mink may occur.

Mallard eggs are usually laid between March and June. In westernmost Europe the eggs may be laid as early as February, and the breeding season may be extended as late as September–October. The clutch typically consists of 7–11 greenish eggs, and the female incubates for a period of 24–28 days. Each egg weighs 50–55 grams and measures 51–62mm x 38–44 mm. The size of the clutch depends upon the age and the physical condition of the female, the time of year and the environmental circumstances. Second clutches tend to be smaller, and younger females lay fewer eggs and do so later in the spring than older breeding females.

The female lays one egg each day, and incubation begins as soon as the clutch of eggs is complete. The female is solely responsible for the incubation and the care of the brood, for the male leaves once all the eggs have been laid, and goes off to begin moulting. Mallard chicks hatch simultaneously, and they leave the nest as soon as they have hatched and are dry. They are attracted to the first moving thing they see, and since this is usually the mother bird they will follow her instinctively. It takes about 7–8 weeks for the ducklings to hatch, and the full breeding season lasts for approximately 3½ months.

Many Mallard broods are lost during brooding, and mortality is also very high among newly hatched chicks. If the first clutch or brood is lost, a second clutch will usually be laid. But losses are high, even when the birds breed in captivity and are secure from predation and bad weather. About 30% of all adult female Mallard fail to breed, and only half of the birds that breed actually succeed in rearing their young to the fledging season.

Clutches which hatch successfully can comprise up to 7.5–10.5 chicks, based on studies carried out in various parts of Europe, and this indicates that Mallard have a high reproductive capacity. The largest number of successfully fledged broods, and the largest clutches of eggs, tend to come from the early nestings. In general, the average size of broods which fledge successfully is 3–6.

The female bird leads her brood, but the chicks may wander off to a distance of several metres away from her in a quest for food, especially when the ducklings are well grown. When danger threatens the chicks bunch together and sneak away into the cover of vegetation, led by the mother. Sudden and extreme danger can cause the family to scatter, with the chicks rushing off in all directions, and the mother bird may make a display of being injured, trying to divert the source of danger by fluttering away and often calling shrilly. Both the chicks and their mother may also dive when danger threatens.

## Mating behaviour

Mating usually takes place while the birds are on their wintering grounds, and much of the mating behaviour and courtship takes place when the birds are on the water. A Swedish study of a large flock of 3,600 Mallard wintering on a lake near Gothenburg showed that between 73% and 91% of the birds which remained on the open, ice-free water were males, all displaying and courting the females, and moving in a closely bunched mass. In display the male Mallard were noted to utter shrill, high-pitched whistling calls. In this study, even the birds on the open water close to the ice were principally males, between 54% and 67%. Further out on the open water were the established pairs, dispersed and resting calmly, and therefore the ratio of the sexes was more equal. Among the pairs, the male birds chased away potential rivals who tried to come close to them as they walked on the ice.

The prenuptial display of Mallards has been studied thoroughly by many ornithologists, and it consists chiefly of numerous postures and movements. Many of these serve to show off the males' breeding plumage. The female attracts and excites the male by stroking her bill along the side of her body, and also uttering a staccato call. The male responds by swimming alongside the female, stretching his head and neck forward and upwards while shaking the bill back and forth.

Having paired, the male and female excite one another before mating by bobbing their heads up and down vigorously. The male mounts and mates with the female on the water, gripping the feathers of the nape of her neck in his bill. After mating he will stretch himself upwards on the water and bob his head. Mating may be followed by both birds bathing and preening vigorously.

## Population

It is difficult to give an accurate figure for the total number of Mallard in Europe. Those countries which have carried out a detailed census of breeding Mallard have combined to give a total of 1.5 million breeding pairs, but this is based on counts covering barely one-fifth of Europe. If one extrapolates from this figure, a possible total of 7.8 million breeding pairs seems possible. International winter counts have revealed a winter population of around 3 million Mallard in Europe.

Recoveries of leg-banded Mallard which have bred in Europe reveal that the majority of these birds winter in western Europe, mainly in the countries around the North Sea, and also around the Mediterranean.

## Distribution and Food

The Mallard occurs in Asia and North America, as well as in Europe. It is therefore part of the holarctic fauna, breeding in the Old and New Worlds. The Mallards' winter quarters lie well to the south of their breeding zone, but the southern and western populations of Mallard are sedentary and resident. Even the Mallard which breed as far north as Greenland, Iceland and the Faeroe Islands often remain there through the winter.

In summer Mallard prefer habitats like shallow lowland lakes, small ponds and similar areas of open water, especially where abundant vegetation affords both cover and food. Mallard also frequent open shorelines and small islands where they can roost and rest without disturbance.

In winter Mallard have their favourite places where they rest by day. These may be ponds in public parks as well as more secluded lakes, marshes and streams, and quiet areas along the coast. At dusk they flight from these resting places to their feeding grounds. The distances covered vary and depend upon local conditions, but Mallard will fly for several kilometres, and will make the return flight early the following morning.

The winter diet of the Mallard is chiefly vegetarian, and there is a much higher proportion of animal food in the birds' diet during the breeding season. Ducklings in particular depend upon the availability of insects and other invertebrates as an important source of protein when they are growing, and the breeding female likewise depends upon these rich sources of protein at nesting time, when she uses up a great deal of energy.

In autumn and winter Mallard feed on a diet which is mainly composed of vegetable matter, including seeds and the soft parts of water plants. The seeds of rushes are eaten eagerly, as are acorns and beechmast, and these birds are especially attracted to cereal foods such as the spilled and scattered grains of barley and wheat.

## Migration and Subspecies

Many Mallard are migratory, but some populations are sedentary and others spend the winter in northerly areas where they do not wholly escape the effects of the winter. Once Mallard have reached their winter quarters they tend to remain there. In the northernmost parts of its range the birds often frequent ponds and other areas of open water close to human habitation, where they can be sure of finding some open water at all times, and where they may find good feeding, directly or indirectly, from the human population.

Mallard from different breeding populations will mix together during the winter. Research in Scandinavia showed that a single Swedish wintering flock of 3–4,000 birds was composed of birds from three main areas. Approximately one-third were found to be local birds, while a further third were migrants from central and northern Sweden, and a further third had migrated from Finland. A very few birds were found to have migrated from the White Sea area of Russia. These different groups arrived and departed at different times, but it was noted that the females tended to remain loyal to their wintering grounds and returned year after year.

## Relatives of the Mallard

Dabbling ducks have so many features in common that they are all assigned to the same scientific family, *Anas*. Mallard, Wigeon and Teal are highly prized by sportsmen everywhere, while the Pintail (*A. acuta*) and the Shoveler (*A. clypeata*) are of more local sporting interest. The Garganey (*A. querquedula*) is a rather unfamiliar species, not unlike the Teal. The two are almost identical in size and appearance, except in the breeding season. The Gadwall (*A. strepera*) is rather rare in most of Europe.

Male dabbling ducks are in their full breeding plumage during winter and spring, and this plumage plays an important part in their courtship displays and mating rituals. Later in the year a protective and more camouflaged plumage, not dissimilar to that of the female, can be seen, and in this eclipse plumage the male and the female can be distinguished by the speculum on the former's wing.

Dabbling ducks depend upon shallow ponds, lakes and marshes which are rich in their preferred foods. They also like open, calm water for mating. Their nests are often to be found along the margins of ponds and lakes, or in grassy vegetation not far away from the water. In autumn and winter dabbling ducks frequent shallow bays along the sea shore, marshlands and swampy areas, river estuaries and lake shores, and in ice-free areas they will frequent small ponds and shallow still-waters.

In order to improve the wild populations of dabbling ducks, to enhance their breeding success and to provide them with good wintering areas, a great deal of time and effort has been invested in wetland habitat management, by sportsmen and conservationists alike. These measures are often beneficial to a large number of other species also. Food and shelter are the two principal factors in making

Mallard throughout Europe, Asia and North America closely resemble one another, and are therefore assigned to the same race, *Anas p. platyrhynchos*. But the Greenland Mallard is paler and more greyish in colour, and the male birds' breeding plumage has black flecking on the chest. This form belongs to the subspecies *Anas p. comboschas*. In addition to this, a number of unusual subspecies have been found in the coastal areas around the Gulf of Mexico (*A. p. fulvigula, maculosa, diazi*) and also in Hawaii and the Lysan Islands (*A. p. wyvilliana, laysanensis*).

an area attractive for duck, and it is best if the margins of lakes and ponds can be rich in suitable forms of vegetation, including seed-producing plants, soft leaves and stems, and plenty of marginal and floating vegetation, which encourages insects and many forms of invertebrate life. The creation of small artificial islands can give resting and breeding birds security and also provide undisturbed breeding and "loafing" areas.

The foods which dabbling ducks eat vary from season to seaon, rather than from one species to another. All species eat more insect and invertebrate life in spring and summer than in autumn and winter. Just before egg-laying the females depend on suitable sources of protein-rich foods, and the newly hatched chicks also need a high protein diet if they are to survive and thrive. Insect and larval life provide a suitably protein-rich source of food.

All species are to some extent migratory, and the autumn migration takes place between August and November. Garganey migrate by September; Pintail, Shoveler and Gadwall migrate before October; while Mallard, Teal and Wigeon are still on the move in November. Migrants return to their breeding haunts in the period March–May, and most migration takes place by night, although Wigeon will migrate by day also.

## Special Characteristics and behaviour

Dabbling ducks are similar in many ways, and one of these common factors is the way they feed. They find much of their food on the surface of the water, and they extend their scope by dibbing their heads and bills under the water. They will often tip right up to reach food material on the bottom of shallow water. They rarely dive except when danger threatens, or occasionally in search of food.

Dabbling ducks take flight from the water in a direct upward jump, unlike diving ducks which take flight by running along the surface of the water. This is an adaptation to life on wetlands where there is high surrounding vegetation. Their tails are never flat along the surface of the water, but are cocked upwards at an angle. Their necks are relatively long, and these physical features give the birds a distinctive silhouette. Except during the breeding season these birds are found in flocks, and courtship, pairing and mating begin in late winter and early spring. Courtship and mating take place on the water, and most birds arrive at their nesting sites in pairs.

Sometimes during the winter pairing and mating, a drake from a local, sedentary population will mate with a migratory female, perhaps from a more northerly breeding population. The females are faithful to their nesting sites and to the places where they themselves were bred, and they will lead the males back there with them, and this is known as abmigration. It can prevent the formation of local races and subspecies by spreading the genes among different populations of duck.

## Moulting and moulting migrations

The males of the dabbling duck species leave the females after the eggs have been laid and fly off to quiet, food-rich wetlands to undergo the annual moult. This usually takes place in June–July, and some very large gatherings of moulting males can occur. On the delta of the river Volga half a million or more dabbling duck drakes will gather annually, some having flown up to 1,500 kilometres to get there. Pintail, Teal and Garganey are common, and large numbers of Gadwall also occur. The majority of these ducks are believed to come from northern Russia, but the moulting grounds of the ducks of western Europe are smaller and more scattered, and therefore less dramatic than that which takes place on the Volga.

*Gadwall (above) and Garganey, males in breeding plumage.*

## Territorial limits and guarding of females

Dabbling ducks are not considered as territorial as geese or swans, but they nevertheless have some decidedly territorial characteristics. Females with young ducklings remain within a relatively small area, especially when the chicks are very young. The females return annually to the same nesting places and use the same territories for rearing their young, but these territories are not defended. The size and the boundaries of a female's chick rearing territory are determined largely by the availability of suitable foods. Territorial behaviour can vary from one species to another, and the contrasts are most easily seen by studying sedentary populations.

Mallard breed early in the year, and they often choose a nesting site which is quite far away from the water. This may be an adaptation to avoid areas where the water levels are subject to fluctuation, perhaps by flood water or by snowmelt. Shoveler, Garganey and Teal breed later in the season and tend to place their nests closer to the water, often breeding along the margins of small lakes and ponds.

The females are watched and guarded by their mates, although the success of this can be limited by the distance the female may have to fly from her nest site to her feeding area. The males try to ward off "rape attacks" by other males, by following their partners.

## Competition for food between and within species

The diet of dabbling ducks comprises green vegetation, seeds, parts of plant roots and the stems of both freshwater and salt water plants. They will also eat various small creatures, including insects, larvae, molluscs, worms and small crawfish. The basic diet of these species is so similar that there must be some element of competition between them for the available foods. Individual ducks' diets tend to become more specialised as a result of their territoriality, and the availability of food within the territory. The distinctively shaped bill of the Shoveler is a permanent adaptation by the species to feeding on plankton and certain small invertebrates. The Wigeon is a specialised plant eater, and it differs from other species by its regular tendency to feed on marshland and shoreline fields, as geese do. The Wigeon's bill is short and set high. Mallard will flight to arable fields after harvest, to eat spilled grain, and will also frequent oak woodlands to feed on fallen acorns.

The ability of dabbling ducks to tip up in the water has been said to give different feeding opportunities to the various species, depending on the size of their bodies and the length of their necks. Teal, Garganey and Shoveler tend to feed in the top ten centimetres of the water, while Mallard will reach down almost 50 centimetres. In addition, there are differences in the structure of the various duck

*Shoveler pair*

species' bills, and this combines with the birds' reach under water to create individual feeding adaptations for each species.

A study in Finland has shown that there is a relationship between the size of a Teal flock and the depth to which the birds will reach for their food. Larger flocks did not cover larger feeding areas, nor did aggression develop. Instead, individual birds with lower status in the flock sought their food in the less nutritious areas of the feeding grounds, and the entire flock dabbled deeper under the water's surface for food.

*Pintail pair*

# Teal

## *Anas crecca*

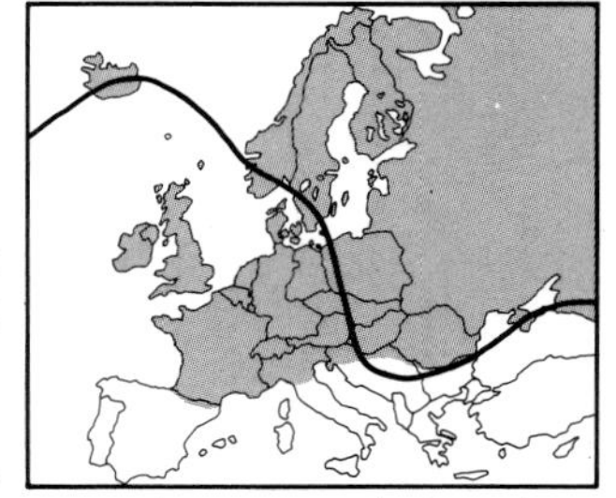

The Common Teal is our smallest duck and, next to the Mallard, the most widespread. But from the ecologists' point of view it is the biomass which counts. Numerically the Teal attains approximately 30% of the Mallard's European winter population, but in biomass they represent less than 9%. This can be interpreted to indicate that the Mallard requires 11 times more nutrition in the winter for their survival.

Large creatures require more food than small ones, but in winter a small body demands proportionally more food than a large one; the body surface is greater, and thus so is the loss of heat in proportion to the volume. Perhaps heat economy and competition for food work together, for the larger, stronger Mallard over-winter in colder environments than do the Teal. There is one Teal per ten Mallard in western Europe, while in the Mediterranean there is one Teal for every two Mallard.

Likewise, in areas where the Mallard dominates, the Teal is a popular sporting bird which, despite its small size, has always been shot extensively. Teal have suffered heavily from shooting and netting at duck-huts around the North Sea, much more so than Mallard and other dabbling ducks. One such hut on the island of Sylt, in the north Frisian Islands, accounted for a total of 141,034 Teal during the period 1832–81, amounting to some 88% of the entire waterfowl catch at that one site.

### Characteristics

A pair of Teal in breeding plumage are illustrated here. The drakes lose this plumage in June–July when the annual moult begins, and in their eclipse plumage they resemble the females until the moult is complete by late summer, by which time the sexes can easily be distinguished once again by their different plumage. However, in his summer eclipse plumage the Teal drake can be told apart from the female by the coloured plumage of his wings, and the male also has a non-spotted belly and fewer spots on the breast feathering than the female. His upper parts are a more uniform dark colour.

The male Teal is larger than the female, with a wing length in the range 182–199 mm, while that of the female is in the range 171–189 mm. The adult body weight is in the range 260–500 grams, and males weigh about 10% more than females. The birds achieve their highest body weights in December, having stored substantial amounts of fat in November. However, body weights during the winter can vary considerably, owing to the effects of cold weather. Newly hatched chicks weigh 20–25 grams, but by August the fledged ducklings will weigh 260–300 grams, and the males will be heavier on average than the females.

Because of its small size the Teal cannot readily be mistaken for any other species of duck, except perhaps the Garganey. However, the Teal is altogether darker in colour. The upper side of the Garganey drake's wing is a blueish-grey, while that of the female and juveniles is greyish-brown. The wing speculum of the Garganey is a paler green colour than that of the Teal.

Difficulties can arise in distinguishing the females of these two species, and also the juveniles. The head of the Teal appears more rounded than the somewhat elongated head of the Garganey, and the Teal's bill is slightly shorter. The Teal has a light band above the eye, while the Garganey has a distinct light band both above and below the eye.

The male Garganey makes a distinctive "creaking" call, which is quite unlike that of the Teal. However, the females of both these species make very similar quacking and croaking sounds.

*Teal plumage: from above, female, young male, male in eclipse plumage, and male in breeding plumage.*

## Reproduction

Pair formation takes place while birds are on their wintering grounds, and pairings usually last for one season only, since the male Teal, like the males of other dabbling duck species, leaves the female after the eggs have been laid. It follows that the chances of a pair becoming reunited are very slim.

Teal breed as early as two years old, and occasionally as one year olds. Only one clutch of eggs is laid annually, although the female will lay a replacement clutch if the first eggs are lost. A typical clutch comprises 8–10 rounded eggs, which are greyish-green to yellowish-grey in colour, and the females lays these at the rate of one per day. The eggs weigh 26–30 grams, and the dimensions are 41–51 mm x 31–35 mm. The nest is lined with straw, dry grasses, some down and some breast feathers, and it is usually located on the ground in dense vegetation. The female chooses the nest site and builds the nest, which is sometimes located at quite a distance from the nearest water. More usually, however, it is built among shoreline and bank-side vegetation, or on small islands in wet marshes and reed beds.

The female takes sole charge of the incubation of the eggs and continues to look after the chicks until they are fledged. Incubation last for 21–23 days, and the chicks will fledge after a further 44 days. Thus the full breeding cycle lasts for 12 weeks, from the first stages of nest site selection to the fledging of the young ducks.

The chicks learn to follow the female during their first hours after they have hatched, but even while they are still in the egg and some hours before hatching, the chicks utter weak, high-pitched calls which establishes the earliest contact between the young and their mother.

Because pair formation takes place in winter, when other closely-related species may also occur, the Teal's plumage characteristics are very well defined and conspicuous, especially the males'. In addition, the Teal's small size may also prevent confusion and reduce the likelihood of hybridization occurring. (The closely-related and equally small Garganey migrates to Africa and tropical winter quarters.)

During courtship, the male utters a high *"krick"* call, which he also uses as a danger signal. The courtship display involves exposing the distinctive markings on the drake's head and tail. The male also draws himself upwards and plunges his head downwards, tossing water into the air with his bill. The female stimulates the male before mating by making a series of jerky head movements. Courtship is primarily a matter of ritualised visual signals.

## Behaviour and adaptation

Teal are active by day and night, and usually the birds flight to their feeding grounds at dusk and return to their daytime resting places at first light. Migratory flights, however, are only undertaken in the dark. If one is in the right place at the right time, one can hear the noctural calls of migrating Teal.

Except when the birds are paired on their breeding grounds, Teal flock together, and a flock may comprise between 20 and 100 birds. The flock formation is compact, and their flight is very fast and with a rapid wingbeat. If Teal are frightened and disturbed at their daytime resting place, the flock will fly quickly in wide turns, and may often come into land again near the place from which they were flushed. Teal usually only flock together, but occasionally a Garganey or a Shoveler may be found among a flock of Teal as they rest by day.

Although Teal migrate by flying relatively high — perhaps from 50 to 100 metres high — their daily flights are much lower, and they are inclined to fly low across the water, more so than most other dabbling ducks.

Like many other dabbling ducks, the Teal has a soft and sensitive bill, which plays an important part in the bird's search for food, because of its sensitivity, rather like a human's taste buds. Apart from this, the bird's eyesight and hearing are its most acute senses. For this reason, acoustic and visual signals play an important part in the contact between the birds, whether as nesting pairs, as members of a flock, or between the adult female and her brood of young.

## Population structure and relationships between species

The Teal is, as mentioned, the second most common dabbling duck in Europe, in both summer and winter. However, the population density varies widely. In the Scandinavian countries the population of Teal is much higher in the northern parts of this region than in the south. In West Germany the Teal is only the fourth most common dabbling duck, for Mallard, Garganey and Shoveler are all more common. In Scandinavia the Mallard is the commonest, with the Teal taking second place; then come the Wigeon and the Pintail. In continental Europe it appears that the Teal encounters more competition from southern species, especially the Garganey and the Shoveler, whose ecology is closer to that of the Teal than the Wigeon or the Pintail. Mallard must also be considered a competing species.

## Enemies, disease and mortality

The principal causes of mortality among post-fledging Teal are considered to be shooting and starvation in winter. In some areas they also suffer from lead poisoning, environmental pollution and outbreaks of botulism.

Europe's Teal population has been adversely affected by the drainage and reclamation of wetlands, especially in the birds' wintering areas. Various parts of Scandinavia and northern Europe have

revealed signs of a decline in the numbers of breeding Teal, which is generally attributed to changed forms of agriculture and land use.

The greatest lifespan for Teal has been estimated at 13–17 years, based on data taken from leg-banded birds. However, over 50% of fledged young die within the first year, and only 10% survive to their third year. Since Teal can breed at one year old, the Teal breeding population is dominated by younger age groups.

## Occurrence and diet

When breeding, the Teal is found close to small wetlands, ponds rich in vegetation, meres, marshes and swampy boglands, especially inland but sometimes also near the coast. They often breed by shallow lakes in wooded areas, but will also be found in more open country beside stillwaters, estuaries and water meadows.

The Teal is not a common species around large lakes fringed with high, reedy vegetation, for it prefers other types of marginal growth. In continental Europe the Teal breeds in higher latitudes than the Garganey and the Shoveler, often on forest lakes. The nest is usually close to the water, although there are occasional exceptions. The nesting site is usually among tall vegetation, shrubby growth, grasses or in the fringes of woodlands close to open water.

Outside the breeding season, Teal flock together and are to be found on open, food-rich waters, inland and along the coast. Like most other dabbling ducks that move to and fro between their feeding and resting areas, they like an undisturbed and settled existence in and around ponds and lakes which afford both cover and food, and where the water is up to ten centimetres deep. They walk about on the margins of the water, stretching their necks and quacking; sometimes they duck their heads under the water. In deeper waters they tip up and stretch their necks towards the bottom, with their tails high in the air. However, their reach is limited by their short necks, and they rarely dive for food.

## Distribution and migration

The Teal breeds throughout most of Europe, northern and central Asia, as well as in North America. Along the southern fringe of its breeding range, however, it often occurs only sporadically. Teal numbers are highest in the far north. Teal also breed in Iceland and right across Asia as far as the Pacific coast.

The corresponding American species (*Anas crecca carolinensis*) is found in Alaska. It generally resembles the Eurasian Teal in its habits, behaviour and migration. Occasionally it occurs in western Europe, just as our Teal does in North America. Several hundred have been imported into the United Kingdom and have bred there.

At the end of the winter shooting season the European population of Teal is approximately 2.5 million. The species' winter quarters lie almost directly to the south of the breeding zone, although some Teal are semi-sedentary and winter in the southern parts of their breeding areas, for example in the British Isles and in France. However, there they represent only 7% of the total European winter population. Only a small proportion of birds go to Africa south of the Sahara.

Based on counts made of breeding pairs of Teal in Europe, the density can be estimated at 0.4 pairs per km$^2$ in Europe, north of latitude 60°N. The breeding population is therefore twice as numerous in that area as in the more southerly parts of Europe, around the Mediterranean and the Black Sea.

The best data on Teal migration comes from leg-banding results from France, the British Isles and Russia. Results from these sources show that of the 150,000 Teal estimated to winter in western Europe, especially in countries around the North Sea, many are summer breeders in Scandinavia and in north-west Russia, but that the British Isles Teal population is mainly sedentary. The Teal of the Mediterranean area in winter, numbering around 750,000 in the Camargue area of the Rhone delta, come from summer breeding grounds in Finland and north-western Russia. Some Teal which have been leg-banded in the Camargue and around the North Sea, have been found in winter in the Iberian peninsula and North Africa, while others have been recovered south of the Sahara in Mali.

The Teal which fly to the moulting grounds on the Volga delta spend the winter mainly around the Caspian Sea, but they also extend as far as the Balkans, the Near East, Egypt and Syria. Like some of the Camargue Teal they breed in the area east of Moscow-Leningrad-Archangel. There must also be a high proportion of Teal which breed in north-east Europe in the winter population of 1.5 million found around the southern Caspian Sea. Many Teal which breed in Eastern Europe migrate via Egypt to the Sudan, where they winter in large numbers.

## Pochard

### *Aythya ferina*

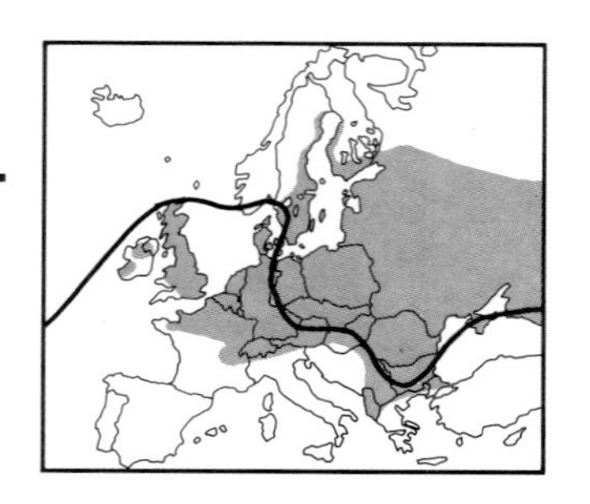

The pochard belongs to a group of wetland birds which are especially associated with lowland lakes, and which spread westwards during the nineteenth and early twentieth century from their eastern haunts. Other species which also extended their range at the same time include the Reed warbler, the Black-headed gull, the Tufted duck, the Black-necked grebe and the Slavonian grebe.

### Characteristics

The pochard is a diving duck, and it is characterised by a plump body shape and a comparatively large, long head and neck. The line of the concave upper bill leads back to a relatively flat forehead and a rounded crown. The male has an unmistakeable breeding plumage. However, the female is rather similar to the female Tufted duck and the female Scaup. But she lacks the attempted "tuft" of the Tufted duck female, and she is also rather larger. Both the Tufted duck female and the female Scaup have a conspicuous white area around the base of the bill, which the female pochard lacks, and they also have yellow eyes, in contrast to the pochard's reddish-brown eyes (both male and female). The young birds resemble the female. The downy chicks are yellowish-brown, in contrast to the blackish-brown chicks of the Tufted duck and the Scaup, and they also lack the dark band through the eye.

In his breeding plumage, the male pochard in flight gives the impression of a much lighter coloured bird than other species of diving ducks, but he lacks any marked white bands. Like other species of diving ducks, the pochard takes flight by running a short distance along the surface of the water; dabbling ducks make a direct upwards take-off. The pochard presses its tail against the surface of the water while swimming, and it flies rather heavily but fast.

The pochard's length is 46–47 cm, and the adult male's wing length lies within the range 207–223 mm, while that of the female is in the range 198–221 mm. Mature wild pochard have a body weight between 0.5 and 1.3 kg, and the female is normally a little smaller than the male. The newly hatched chicks weigh about 35 grams, but they grow rapidly and by the time they have fledged they will have achieved body weights at the lower levels of the adults' weight range. Pochards achieve their highest body weights in September–October. Pochard in captivity have been known to live to 20 years of age, but wild birds which have been leg-banded rarely exceed a maximum of 10–11 years.

## Adaptations and behaviour

The pochard is active both by day and night, although migratory flights are always nocturnal. The autumn migration takes place in September–October, and occasionally begins in late August. The return passage takes place in March–April, and sometimes lasts into May. To moult, the males gather in special moulting areas, where they are joined by non-breeding females. The period of the moult lasts from July–September, and the birds are flightless for a period of 3–4 weeks. By the time pochard return to their breeding grounds the birds will already be paired. There will also be a scattering of single males, and this species, like many ducks, has a surplus of males. Breeding females which have failed in their first breeding attempt and have been abandoned by their mates can form new pairs with these surplus males. It is possible to see courtship displays as late as June, and the display is carried out on the water. The behaviour of the male is similar to that of most diving ducks: he swims with his body pressed low against the water with the head and neck thrown backwards, and he raises and lowers his neck. He also utters sudden polysyllabic "whistling" calls as part of his display. Unpaired female pochard may also be chased in aerial flights by a number of male birds.

The pochard breeds when it is about one year old, although some individuals wait until their second year. The female lays a single clutch of eggs, which may number 3–13, although an average of 6–9 eggs is more usual. The eggs are broad and oval, and are a uniform greyish-green to yellowish-grey. They are comparatively large, and measure 58–66 mm x 41–45 mm, and weigh 66–70 grams. The female chooses the nesting site and builds the nest close to water. Sometimes nests float among reeds or lie on dry tussocks or in shoreline vegetation. The down lining of the nest is black with white flecks. The female takes sole charge of the nest, incubation and rearing the young. Incubation begins when the last egg has been laid, and lasts for 23–28 days; in a further eight weeks the young ducklings will have fledged. Eggs are laid in May–June and the chicks fledge in August. Pochard form monogamous pairs, but the male leaves after egg-laying and goes off to moult. The female and her brood rarely venture out onto open water.

## Distribution and numbers

The pochard is a palaearctic species, i.e. the birds' breeding range is limited to Europe and Asia. It frequents food-rich lakes, large and small, and it also breeds in coastal bays which are rich in vegetation. In winter most of the pochard of the Baltic Sea countries migrate to western Europe and southwards to the Mediterranean. In Sweden only a handful of pochard remain in the winter, while nearby Denmark is host to thousands. Around the North Sea and in the Bay of Biscay there are usually up to 250,000 wintering pochard, while the Mediterranean winter population is thought to be three times as large. Some birds migrate to sub-Saharan Africa, mostly to river deltas in Senegal and Niger.

The numbers of breeding pochard in the countries bordering the North Sea and the Baltic probably comprise 15–20,000 breeding pairs. Young pochard ducklings which have been leg-banded at breeding sites in this area have been found chiefly in the North Sea area during the following winter. If one estimates a chick survival rate of 2–4 chicks per breeding pair, the total population falls very far short of the 250,000 individuals estimated to winter in this area. The numbers are swelled by large influxes of pochard from Russia, and British wintering pochard have been found to come from beyond the Ural mountains. Pochard have a uniform appearance in all the species' main breeding and wintering haunts, and these birds are found from western Europe as far eastwards as Lake Baikal. It is therefore not possible to identify the eastern winter migrants from their physical characteristics. Two similar and closely-related American duck species, the American pochard (*Aythya americana*) and the Canvasback (*Aythya valisneria*) have occasionally occurred in western Europe.

## Food

The pochard enjoys a mixed diet. However, plant foods are more important for pochard than for the other large northern European diving ducks, such as the Tufted duck, the Scaup and the Goldeneye. Pochard will readily eat seeds, the soft parts of plants, roots of freshwater vegetation, as well as different species of water-weed, including water milfoil, hornwort, duckweed etc, and it will also eat sea grasses. The pochard will also feed on small freshwater animals such as insects, larvae, worms, small crawfish, invertebrates and molluscs. Pochard prefer shallow water and rarely dive for food at depths greater than 5 metres, and the bulk of its food is found at depths of not more than 2 metres. Pochard will feed on the surface of the water, by dabbling and by diving, and in winter the daily food requirements amount to almost one kilo of small clams, which may mean up to 3–4,000 small clams.

# Coot

## *Fulica atra*

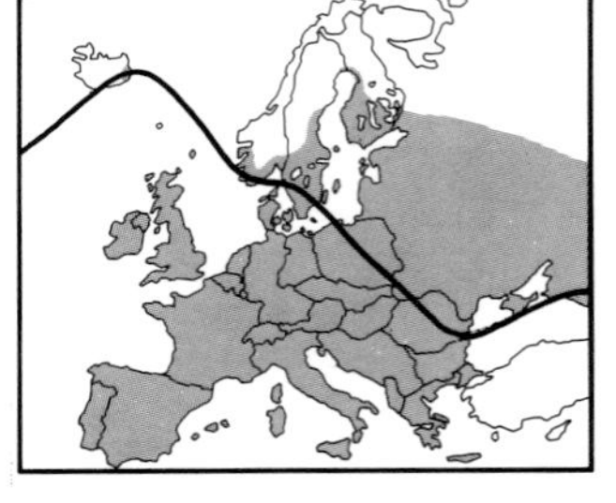

In many parts of Europe the coot is a desirable sporting bird. A Swedish sportsman-naturalist of the mid-nineteenth century wrote that "the fried flesh of young coots is very good, but the older birds taste like engine oil if the skin and all the fat are not removed completely"! Records from Germany dating from the 1920s indicate that large numbers of coot eggs were offered for sale in the market places of large towns and cities every spring. Coot eggs were said to have the same consistency and flavour as chickens' eggs.

### Characteristics

The coot has a round, plump and compact body. Its neck is comparatively long, but it is usually kept retracted so that the small head seems to be attached directly to the bird's body. The tarses are long, as are the toes, and the latter have slight webbing. The colour of the legs is greyish-green. The short and conical bill is white, like the skin which forms a prominent patch on the coot's forehead, and this patch is largest in the spring. The adult birds' eyes are red, while those of the juveniles are brown. In addition, the forehead patch is smaller, or entirely absent, on young birds. The young bird's bill is yellowish-white and the body is coloured a matt brownish-black, with a pale throat and a greyish-white chest. The downy chicks are black, except for the head, which is reddish-orange with red and blue frontal markings. The base of the bill is red with a white tip, and males and females are similar in appearance.

Coots swim with jerky movements, and the coot takes flight by a splashing run along the surface of the water, usually flying off low over the water with rapid wingbeats. In flight the body looks surprisingly long and narrow; the long legs stretch out far behind the bird's short tail. Coots like to dive, taking a short running jump, and they dive and emerge in the same place. Coots can stay submerged for a period of 15–30 seconds.

The male coot's wing measures in the range 193–246 mm, the female's 176–233 mm. Adult birds weigh between 0.5 and 1.3 kg, and the females and young are usually smaller and lighter than the adult males. On hatching a coot chick weighs about 25 grams, but it grows quickly, doubling its weight in the first week, and by the age of three weeks it weighs about 245 grams. By one month it will weigh half a kilo.

Mortality during the first two years is very high, up to 70%–90%, and freezing winters cause heavy losses. After hard winters the numbers of coots seen to be breeding may have fallen very considerably. The oldest recorded coots have been between 13 and 18 years of age.

The spring migration takes place mainly in March–April, and the autumn migration in August–November. Some coots overwinter in the northernmost areas of the breeding grounds, but most northern and eastern birds migrate for the winter to western Europe and the Mediterranean.

The coot is sexually mature as early as one year old. The male takes a territory and lays the foundations of the large floating nest, which is later completed by the breeding pair. The nest is usually found in reeds and shoreline rushes, but occasionally the nest will be out in the open among the floating vegetation. The clutch of eggs averages between 6 and 12 eggs, but 8–9 is more typical. The eggs are oval and yellowish-white, richly flecked with small blackish-brown spots. Larger clutches are invariably caused by two or more females laying their eggs in the same nest. The eggs are comparatively large — 45–49 mm x 34–39 mm — and weigh about 40 grams. Clutches which are laid late in the season — second clutches or re-laid clutches — usually have fewer eggs. Both male and female take turns in incubating the eggs, which takes a total of about 22–26 days. The chicks will fledge in a further eight weeks.

The chicks are active and mobile as soon as they hatch, but the adults and young will usually remain in the area of the nest for a day or two after hatching. The nest has a small "shelf" or step on which the birds can easily climb up or down, for the level of the nest is sometimes placed as much as 20–30 cm above the water. The eggs are covered when the birds leave the nest, but egg losses to crows etc. can be very high. The clutch can be relaid two or three times; and a typical brood of fledgeling coots will comprise 2–4 birds.

### Occurrence and food

The coot is a cosmopolitan bird, found around the world except in Central and South America. This wide distribution has given rise to several subspecies, the only European one being the entirely black *Fulica a. atra*. It lives by fertile lakes and on coastal waters, and when breeding it prefers water surrounded by vegetation and with clumps of reeds and rushes. It spends the winter on open, ice-free water such as large lakes and sea coves and bays.

Coots eat both vegetable and animal material. Among British coots plant matter makes up 84% of the birds' diet, but this can vary with the area and the time of the year. In summer coots will eat insects, larvae, snails and mussels, and the coot's body weight varies considerably depending upon the kind of food available.

The European winter population of coots is estimated to be almost three million birds, which indicates a breeding population of 500–600,000 pairs.

## Snipe

### *Gallinago gallinago*

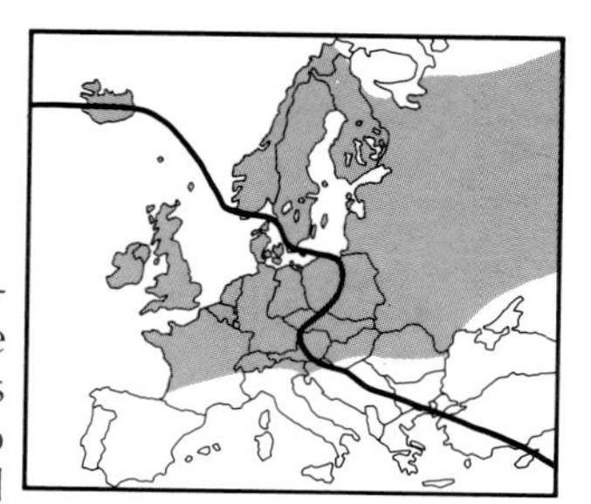

### Characteristics

The Common Snipe is a relatively long-legged wading bird with a distinctive long, straight bill in proportion to its small body size, which is comparable to that of a starling. It has a white belly and its flanks are striped. The back is coloured brown with elongated rust-coloured and black markings framed in yellowish-gold, which make bands along the back. A broad, dark band runs through the comparatively large, dark eye. The bird's crown is also dark brown. The male and female look alike and cannot be distinguished by external characteristics. The chicks are chestnut coloured, with a pattern of black and pale spots on their backs.

Males and females are the same size, and average body weights fall within the range 95–144 grams, depending upon the bird's age, the time of the year, and the bird's physical condition. The newly hatched chicks weigh 10–11 grams, but within a week they will have trebled their weight, and this will double again after a further week. By the time they are fledged the young birds will weigh 70–80 grams. The Common Snipe, the Woodcock and the Golden Plover are the only members of the wader family which may legally be shot in Great Britain. The shooting season begins on 12th August and ends on 31st January. Occasional snipe are shot on duck-shooting days, but the majority of snipe winter on wet meadows, on marshland and on raised heathery bogland.

Closely related to the Common Snipe is the Great Snipe (*Gallinago media*), which is a larger bird with a white belly with bands across it. The Great Snipe's most distinctive field characteristics are its size and the distinctive white of the bird's tail feathers as it flies away. The Common Snipe has 14 tail feathers, while the Great Snipe has 16. Another related species is the Jack Snipe (*Lymnocryptes minimus*), which is smaller than the Common Snipe and has a relatively shorter bill. It has 12 principal tail feathers, and the plumage on its back has an attractive and subtle greenish iridescence. It also has two conspicuous yellowish-gold bands along its back, and these are visible as the birds flies off. Its flight is weak, rather slow and the bird usually drops into cover after a short distance.

The Common Snipe rarely occurs in large flocks; however, it is commonplace to see them in small groups or "wisps" of up to 10–12 birds. Common Snipe are usually flushed singly, when they will fling themselves into flight uttering a harsh *scaape-scaape* call. The birds rise into the wind and then swing off downwind in a very rapid, zig-zag flight, often climbing quite high as they fly off. Often suitable habitat will be found to contain a number of snipe, which can be flushed singly. By contrast, the Great Snipe rises and flies off in a straight and unhurried flight, while the Jack Snipe tends to lie tight and only to rise when the approaching human or dog is almost on top of it. Its flight is straight and silent.

The Common Snipe has the ability to flare out the two outermost feathers of its tail. This action plays an important role in the birds' courtship and the maintenance of a breeding territory by a pair of nesting snipe. The territorial flight displays involve the birds climbing to a moderate height and then diving downwards at a steep angle. The protruding outer wing feathers vibrate rapidly and produce a most distinctive drumming or "bleating" sound, lasting for two or three seconds.

The snipe's bill is strikingly long. The tip is somewhat flattened and rather soft, and is equipped with very sensitive nerve endings to assist the bird in identifying food items when it probes in mud and soft earth. The end of the upper mandible is also flexible and can be raised upwards, giving the bird a forceps-like ability to grasp food items. This makes the snipe well adapted for feeding by probing for worms and other invertebrate foods in soft ground.

The snipe's tarsi are rather short compared with those of other wading birds of similar size, but the toes are relatively long and enable the snipe to walk on soft mud.

## Reproduction

The Common Snipe breeds from its second year onwards, and one clutch of eggs is laid each year. The male snipe claims a territory in suitable breeding habitat, and the female selects the nest site within this territory. There are usually four eggs in the clutch, and incubation begins as soon as the last egg has been laid. The female incubates for 19–21 days, and the eggs weigh 15–16 grams and measure 36–43 mm x 26–30 mm. The weight of each egg is therefore about 15% of the female's total body weight, and the clutch represents the equivalent of 60% of her total weight. The eggs do not hatch simultaneously, and the first one or two chicks to hatch are usually cared for by the male bird, while the female takes care of the later ones to hatch. The young chicks are active and on the move as soon as they are dry, and fledging takes place after 19–25 days.

The nest is little more than a simple depression, lined with dead grasses and placed among low vegetation such as heather and reeds. In damp areas the nest will be placed on a drier raised hummock or tussock.

The annual mortality for the Common Snipe is estimated at 50%, but it is probably true to say that mortality is higher than this among young birds, and somewhat lower among older birds. The oldest known leg-banded snipe have lived for up to 11–15 years.

## Occurrence and distribution

The Common Snipe nests in marshy ground, on coastal salt-marshes, by rivers and lakes and on heathery blanket bog and upland moors.

During migration snipe will be found in similar types of habitat, but they will occasionally be found on wet meadows and shallow flooded farmland. Throughout the year the Common Snipe frequents areas of land which have been flooded to a very shallow depth, where they will feed around the margins and wade in water up to 4–5 cm deep. They feed with rapid movements, thrusting their long bills repeatedly into soft mud and ooze, and the sensitive bill locates worms and other invertebrates below the surface.

The Common Snipe occurs widely across western and central Europe, Asia and North America. In also winters in South America, Africa and southern Asia. In South America and Africa there occur races of the Common Snipe which are considered to be closely related to *Gallinago gallinago*, while others regard them as distinct species. These southern populations of snipe are geographically distinct and isolated from the northern populations of the Common Snipe.

The Common Snipe of the Faeroe Islands was distinguished from the typical European Common Snipe as early as 1831, and its plumage is rather different from that of the nominate species. It is distinguished by the scientific name *Gallinago g. faeroensis*.

The European winter quarters of the Common Snipe are in the west and south-west, although some birds migrate as far as North Africa. Others, such as the British and Irish breeding populations, are mainly sedentary, but they are joined in autumn and winter by large numbers of snipe from Scandinavia and northern Europe. The principal times for migration are in September–November and in March–April.

# Open countryside

In Europe the managed woodlands and the open landscape of fields and plains are features of the lowlands: more than 60% of Europe's total land area lies less than 200 metres above sea level. In northern Europe the lowlands appeared on the scene before the development of the forests. As the inland ice retreated to the north, areas of former lakes and areas of the sea bottom were exposed, as were areas of the Arctic and sub-Arctic tundra. Later forests began to encroach upon the areas of open land. But low temperatures and a short growing season inhibited the spread of forests in the far north of Europe, while elsewhere poor soils and inadequate rainfall have prevented woodlands from developing. When human communities turned to farming they cleared areas of forest and their agricultural activities produced what are sometimes referred to as the "cultivated steppes" of northern and central Europe.

In the past much of the countryside was extensively cultivated and grazed, and thereby kept free of the growth of shrub species and trees. This was especially true of heathlands and moorlands, where livestock were grazed. Cultivated fields maintained the open quality of the landscape, and even the marshy lands were exploited, with reeds and coarse grasses being cut to feed and bed cattle in winter. In these and various other ways man's activities have established or maintained open countryside in many places, where his domestic animals live alongside wild creatures, and sometimes compete with them for the available habitats.

Parts of the open countryside are relatively dry, especially in summer, and in western Europe the heathlands are especially important examples of this.

Many parts of the world have grasslands that provide food and living space for large herds of herbivorous animals, and in eastern Europe the saiga antelope of the south Russian steppes is an example of this. However, the open lowlands of Europe are of particular interest because of the many species of birds which are to be found there. These range from the great bustard to the red grouse of the open moorlands of Scotland and northern England, although large numbers of wild red deer have also adapted to life on the open hills in Scotland. The open moorlands and plains of southern Europe are home for species such as sand grouse and red-legged partridge.

In intensively cultivated area of lowland there is often abundant food for wild creatures, and this makes farmland attractive to species like the hare and the pheasant. Wood-pigeons have also adapted well to this environment, and they feed avidly on many arable crops such as rape, peas, clover and cereals. Some important species of large herbivores also occur, such as the roe deer, the fallow deer and the wild boar.

# Brown hare

## *Lepus europaeus*

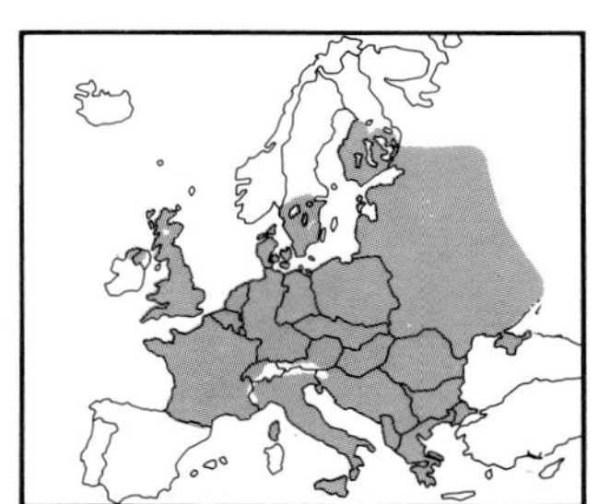

The landscape and topography of much of northern Europe has changed very considerably over a long period of time, and this has affected the way in which various species of animals are distributed in Europe. When the ice sheets of the last Ice Age receded northwards, many animals species colonised the land which was revealed. At first arctic conditions followed the retreat of the ice, and this had a direct effect on the fauna, and as the climate became milder and the ice moved further away to the north, conditions were created for a new and very rich range of wildlife species. The Brown hare is one such mammal which colonised much of western and northern Europe.

## Characteristics

The basic colour of the Brown hare is a light sandy brown. This is most conspicuous on the animal's breast, and its legs are similarly coloured. The fur of the hare's back is darker, and this species retains much the same body colour throughout the year, although some individuals appear a little paler during the winter months, especially along the flanks. The hares of Britain and Europe grow a rather reddish coat during the late summer. The upper side of the hare's tail is black and the underside white. The ears are especially prominent and long. Brown hares are larger than the Arctic hare (*L. timidus*), which is another important species in much of northern Europe and in northern parts of Great Britain. (In Europe the Brown hare is occasionally referred to as the "German hare".)

A mature adult hare can weigh as much as 6–7 kg, although a more typical adult weight is 3½ kg. The females are rather larger than the males, and the newborn leverets weigh about 100–110 grams.

The special distinguishing characteristics of the Brown hare in comparison to the Arctic hare and the rabbit are its warm yellowish-brown coat and its extremely long ears. The Arctic hare's tail is completely white, and shorter than that of the Brown hare. The ears are also very much shorter, although longer than those of the rabbit. The foot pads of the Brown hare are hairy, although they do not have the dense fur of the feet of the Arctic hare in winter. Both species of hare have amber-coloured eyes, while those of the rabbit are dark brown. The Brown hare has longer legs than the Arctic hare.

The tooth structure is the same for all three species:

$$\frac{2\ 0\ 3\ 3}{1\ 0\ 2\ 3}$$

The incisors grow continuously, usually in relation to the speed at which they are worn down by eating. Injuries can occasionally cause unusually long, deformed and twisted teeth in all three species.

## Adaptations

Hares are adapted in various ways, one of which, as we have just noted, is the way in which the incisor teeth grow throughout the animal's life.

As regards their feeding habits, many animals share the hare's characteristic of eating their own excrement, a behaviour known as refection. It is thought that they eat only the soft excrement produced during the day, and that the final stage of digestion involves a fermentation process.

The Brown hare moves with a hopping gait; the hind legs are long and powerful but the forelegs are relatively much shorter and weaker. When they fight the males stand upright on their hind legs and flail at each other with their forepaws. When danger threatens all hares will press themselves close against the ground, and adult hares have regular "forms" where they will crouch down and rest, out of sight and in shelter.

The hare's senses of hearing, touch and sight are very acute, although they often fail to see people who stand quite still, especially if the hare is running directly towards them. The male hares make important use of their sense of smell to identify females in season. These animals' vision is good, the eyes being set well back on the sides of the head, and protruding to give an almost all-round field of view.

Brown hares are very fast, and individuals have been clocked at speeds up to 70 km per hour. When chased by, for example, a dog the Brown hare takes off on longer and straighter runs than the Arctic hare, although both can turn and twist with great dexterity. When hard pressed they will take to the water and swim.

Brown hares are mainly nocturnal in their activities. They become active at twilight, and in the spring when mating time occurs they can often be seen to be very active during the daytime. They have a preference for open areas, and gatherings of up to 10 or more hares can be seen. In other respects, however, the Brown hare is not a very sociable animal and tends to lead a solitary and sedentary life.

Hares use secretions from their anal, sexual and nasal glands as a means of marking the boundaries of their individual ranges.

## Reproduction

The Brown hare is polygamous, and breeding can take place throughout the year. Pregnant females have been found in Britain and Europe in every month from February to October. When she is in season, the female Brown hare emits secretions from her genitalia and from her anal glands. This attracts the attention of the males. The number of young in each litter is usually between two and five leverets, although there have been exceptional instances of six leverets in a litter. The first litter of the spring and the final litter in the autumn are often smaller than litters born in late spring and summer. In Scandinavia each adult female is reckoned to produce 2–3 litters each year, and in mainland Europe there is an average of four litters in the year. The gestation period lasts for 42–45 days.

The young are born in a hollow or form in the open, and sometimes in the cover of dense vegetation. The young leverets are born with fur and with their eyes open, and within a day or so of their birth they are moved by their mother away from their birthplace to individual forms. Thereafter she visits them in turn to suckle them. Suckling takes place mainly at night and lasts for only 5–6 minutes. The hare's milk is extremely nutritious with a fat content many times greater than that of cows' milk. After less than three weeks the young are quite independent, and they reach adult size after about eight months, by which time they are also considered to be sexually mature.

## Distribution and subspecies

The Brown hare is found throughout most of Europe except for Norway, northern Finland, Spain and Portugal, and it does not occur in south-western Asia. As a result of extensive introductions the Brown hare now exists in all continents except Africa.

The Brown hare spends virtually all its life in fairly open countryside. Its origins lie in the extensive steppes and grasslands of eastern and northern Europe, and from this area it spread through western Europe after the last Ice Age. In Europe the hare is fond of cultivated farmland, but it is also found in some woodlands, especially where the trees are a mixture of deciduous species, and it can also occur in the public parks and private gardens of urban and suburban areas. In Europe it occurs at elevations up to 1,500–2,000 metres, but it is not found in the Alps.

We have already discussed something of the Brown hare's relationship with the Alpine or Arctic hare. Some authorities believe that the Brown hare is a subspecies of the Cape hare (*Lepus capensis*), which is widespread in Africa and in large areas of Asia as far east as China. One argument in favour of this conclusion is the similarity of the structure of their skeletons; another is that the Cape hare inhabits Spain and Portugal, where the Brown hare is absent, and hybrids are found in some areas where the two distribution zones border one another.

The Arctic hare is also closely related to the Brown hare, and it is believed that hybrids between these two species can occur very occasionally. The Arctic hare and the Cape hare are the only two European hare species which are closely related to the Brown hare. However, there are a number of local populations which have been described as subspecies, and these particularly relate to the isolated populations on islands such as Rhodes, Corsica and Crete. Some eastern populations have also been given the status of subspecies, although opinions are divided on this.

## The hare and the environment

The hare depends upon a cultivated rural environment, and this places it in a position which is both privileged and vulnerable. Hares have little competition from other wild herbivores, especially since rabbit numbers declined due to myxomatosis, but changes in land use and farming methods can seriously affect hare populations. Where the conditions are favourable hare numbers can build up to remarkable densities. In parts of Poland the population of hares has reached a density of 50 hares per 100 hectares. In such conditions each hare's range overlaps with others, and individual hares may range up to 1–2 km during the night. Studies of hare populations in southern Sweden show that hares prefer those fields which are well fertilised and rich in their preferred foods. Stubble fields, especially where they have been undersown with clover, provide excellent feeding areas before they are ploughed in during the autumn, and in stubble and clover fields the hares have no competition from grazing cattle, such as they find on pastures.

Winter can be a difficult time for hares on contemporary farmland. In arable areas there are few or no winter pastures, and the landscape lies ploughed, barren and black. Hares are unlikely to be plentiful in areas where there is intensive monoculture farming of cereal crops. Under such farming regimes the hare will best be conserved by leaving a proportion of the land unploughed and managed for wildlife. Rough uncultivated corners of fields are attractive to many form of wildlife, and hares may also benefit from the provision of some supplementary winter feeding. A mixture of beet and hay is highly suitable for this.

## Nutrition and diet

Hares are herbivorous. In summer they eat herbs and green plants of various different kinds, including grass, rape and clover. In winter their diet is augmented by bark and dried grass, as well as cereals and the winter stalks of clover and rape. In autumn hares will also frequent sugar beet fields. They also frequent orchards, nurseries and gardens where they will eat the stems of young trees and destroy young plants.

## Enemies, diseases and mortality

The fox is the principal predator of the hare; man is probably the second. Stoats, polecate and mink will also take the occasional young animal, and Golden eagles will also take brown hares.

A significant cause of death among young hares is coccidiosis, a disease caused by small, one-celled organisms. It is a parasitic disease which emerges when the young hares are in poor physical condition. It attacks the digestive tract, the liver and the kidneys, and the symptoms are paralysis, emaciation and diarrhoea. Another disease, also caused by a single-celled parasite, is toxoplasmosis, which is usually indicated by frothy blood around the animal's nostrils.

Shooting takes a heavy toll of hare numbers in many European countries. Shooting in Britain is chiefly carried out as a control measure, when hare populations have increased. In mainland Europe, however, the hare is more prized as a sporting quarry, and it is estimated that one million hares are shot annually in West Germany. The annual bag for Denmark has been reckoned at 190,000, and in Sweden it is estimated at 100,000 per annum.

## The hare as a sporting animal

The Brown hare (sometimes referred to as the "German hare") is a creature of grasslands and open agricultural land. In most parts of Europe the hare is especially common on farms, and although its activities are mainly nocturnal, the hare may often be seen in the early mornings or late in the evenings grazing in the fields. Hares often make their forms right out in the open, in a furrow, in a stubble field or at the edge of a ditch or drain. While it frequents small woods and the woodland edge, the hare will rarely venture more than a few hundred metres into extensive forests.

The hare's choice of habitat has influenced the tactics which are used by sportsmen when they are shooting. It often happens that a solitary Gun — or perhaps a a line of Guns — will range up and down across flat fields, hoping to flush hares from their forms. In large fields or on open plains a line of beaters is often deployed to drive hares towards standing Guns. Hare shooting is not considered a difficult test of marksmanship, and those who pursue the Arctic or Alpine hare are often condescending towards the shooting of Brown hares on farmland.

While most sporting shooters will use shotguns, hares are also a good quarry for a skilled rifleman using a small calibre rifle. Hares can provide an exciting test of the shooter's skill and fieldcraft when they are stalked in the open.

Brown hares can also be hunted, both by greyhounds and other "long-dogs" and by scenting hounds such as harriers or beagles, and both kinds of hunting are exciting and demanding of skill and stamina on the part of the dogs and hounds.

# Red grouse

## *Lagopus lagopus scoticus*

The Red grouse is one of sixteen different local subspecies of the Willow ptarmigan (Lagopus lagopus), which occurs widely across northern Europe and Asia. But the Red grouse subspecies is the one which occurs on the heathery hills and moors of Scotland, northern England, Ireland and parts of Wales. Sometimes referred to simply as "the grouse", this particular subspecies of the Willow ptarmigan has always enjoyed very high esteem as a sporting bird.

During the latter part of the nineteenth century many attempts were made to introduce the Red grouse to various parts of continental Europe, including Germany, Belgium, Poland, Sweden, Norway and Denmark. Unfortunately all these many attempts were unsuccessful and the released birds died out after quite a short time. Those which were introduced into Germany, and into the Ardennes region along the border of Germany and Belgium, were perhaps the most successful, but the various experiments are believed to have failed chiefly because the habitat was not suitable for this bird, which is specially adapted to the heather uplands of Britain and Ireland.

## Characteristics

Like other birds of the grouse family, the Red grouse has feathered tarsi and toes. Their summer and winter plumages are very similar, in contrast to the northern European Willow ptarmigan which have a white winter plumage. In fact, *Lagopus l. scoticus* is the only subspecies of the Willow ptarmigan which never turns white on its upper parts in winter. Grouse feathering is a rich russet-red, a combination of red and brown and blackish colours which makes it extremely well camouflaged in its typical habitat of ling heather. The well-grown young birds resemble the mature females, and up to their first moulting at about one year old the young birds may be distinguished by the fact that their two outermost primary feathers are pointed, and not rounded like the adults'. The chicks are yellow-brown with dark brown camouflage patterning on their backs.

The mature male Red grouse is slightly larger than the female, with a wing length of 200–214 mm, compared to the female's 190–200 mm. The maximum body weight is 750 grams (mature male), with the female weighing about 100 grams less. The young chicks develop rapidly; at two weeks old their quills are sufficiently developed that they are capable of short flights.

## Reproduction

The Red grouse is monogamous, although in exceptional cases one male may have two mates. As early as October a pair will take possession of a breeding territory, which they will defend and maintain until the end of the following breeding season. The grouse is a ground-nester and the eggs are laid in a depression or scrape in the ground which is scratched out by the female. The nest is almost

always placed among heather, but nests are occasionally found in grassy tufts and tussocks not far from heather. The Red grouse lays one clutch of eggs per year, and from 6–12 eggs are laid. The eggs are a cream colour heavily flecked and spotted with dark brown and brownish-black spots. Eggs are laid at the rate of one per day and measure 41.5–50.1 mm x 34.4–29.8 mm, and weigh about 24 grams. The eggs are laid in late April or May, although clutches can be found both earlier and later than this. Incubation begins as soon as the last egg has been laid, and the female takes sole charge of the business of incubation. The eggs hatch after a period of 20–24 days and the chicks may be fledged as soon as 12 days after they hatch, although they are not fully grown until late August or September. The chicks are active as soon as they hatch, and follow the mother bird, but the male also stays nearby. (If a clutch of eggs is destroyed the bird will lay a replacement clutch.) When immediate danger threatens a female with young, she will often pretend she is wounded, to lure the predator away from her chicks. At night and in cold or wet weather the adult bird will brood the young chicks under her wings. Hatching success can vary, and in good weather almost 100% of the eggs will hatch. In adverse weather, however, the hatching rate can be 75% or less. The family group remains together until about September, by which time the adult birds will once again be preparing to take up their breeding territories.

## Occurrence and diet

As mentioned previously, the Red grouse is a subspecies of the Willow ptarmigan, which is widely distributed in Europe, Asia and North America. However, the Red grouse only occurs in the British Isles, and British ornithologists and sportsmen have long been proud of their "own" gamebird. In the eighteenth century naturalists like Latham referred to the Red grouse as *Tetrao scoticus*, and there are innumerable later references to *Lagopus scoticus*. However, this unique subspecies is more correctly designated *Lagopus l. scoticus* — the British Isles race or subspecies of the Willow ptarmigan.

The Red grouse is to be found over large parts of Great Britain and Ireland. It occurs on heather-dominated moorland and raised peat bogs from sea level in Ireland and the Hebrides to altitudes of up to 800–900 metres in the Scottish highlands. Above that altitude its place is taken by the ptarmigan *Lagopus mutus* which is a grouse of the high hills and rocky tops of mountains.

The Red grouse is a sedentary species, and the mature male red grouse claims and defends his individual territory for much of the year. However, in extreme weather conditions the birds will pack together into quite large flocks and move about in search of snow-free areas of heather on which to feed. When the family coveys split up in September the young birds do not travel far, and studies of marked birds show that most do not travel more than a few kilometres at the very most. The Red grouse is capable of achieving astonishingly high densities when the habitat is suitable and when there has been a good breeding season. Breeding densities of up to 50–60 pairs per square kilometre have been found on the most productive moors of northern England and Scotland.

The Red grouse has been the subject of a great deal of thorough research. In particular scientists have studied the factors which determine grouse population density. It has been shown that the availability of ling heather is of the utmost importance, for the shoots of young heather constitute over 90% of the diet of adult grouse. With the approach of autumn the family coveys split up, and the males begin to take possession of their territories, in preparation for the coming breeding season. It is almost vital for the survival of grouse through the winter that they have their individual territories, and research has shown that birds without territories tend to drift off to the less suitable parts of the moor, and have a very poor chance of surviving the winter and breeding the following spring.

Grouse are also susceptible to a number of diseases, including the virus "louping ill" which can be transmitted by sheep and hill hares, and most importantly the parasitic disease known as strongylosis. This is caused by the infestation of the birds' digestive tracts by many thousands of small threadworms (*Trichostrongylus tenuis*), which has the effect of making the bird debilitated, vulnerable to death by predation and bad weather, and less able to produce a good clutch of fertile eggs. The larvae of this parasite live on heather shoots and are individually eaten by the grouse as it feeds. They build up in numbers in the caecum or blind guts of the bird's digestive tract and a burden of several thousand worms will result in considerable weakening of the affected bird. Strongylosis is a disease which is directly related to the density of the grouse population, and where grouse densities are very high there may be ideal conditions for the spread of the disease. Various management techniques may be used to minimise the chances of a severe outbreak of strongylosis, including the shooting of sufficient birds to leave only sufficient winter stocks of grouse to allow good breeding the following spring. Moors which are under-shot and over-populated may be at serious risk from this density-related disease.

Experiments have also been carried out, which involve the catching-up and dosing of individual hen grouse with a vermicide to rid the bird of the strongyle threadworms. In addition, a "grit pill" has been developed, whereby the birds will dose themselves against the threadworms by ingesting pills treated with a vermicide along with the fragments of grit which the grouse ingest to help them macerate and digest their fibrous diet of heather shoots.

The object of research into strongylosis and the management techniques which have been developed from it is to reduce the dramatic ups and downs of grouse numbers which have characterised most moors in the past. If grouse can be kept at relatively high densities and free of disease, grouse moor management as a form of economic land use becomes a much more attractive option for the landowner.

Grouse numbers can rise and fall dramatically owing to strongylosis, but other factors have caused a steady long-term decline in grouse numbers in many areas where the birds were formerly plentiful. Good heather management is essential to promote suitable habitats for grouse to feed, to nest and to find cover. Heather is best managed by a programme of rotational burning in small patches or in long strips not more than 30 metres wide. Old woody heather can be burned off to promote the vigorous growth of fresh young heather shoots which grouse require for food, while middle aged and older heather provide the birds with cover and with suitable nesting habitat. Ideally, heathers of all ages and lengths ought to be available to the grouse within their breeding area.

Also important are insects, for it has been proven that grouse chicks cannot survive and thrive unless their diet in the days after hatching comprises a high proportion of protein-rich material such as insects. The hen bird will lead her brood to wet places on the moor, where insect life abounds and where the young chicks can pick insects like craneflies off the heather and mosses.

The chief predator of the Red grouse is the fox, and efficient grousemoor management involves keeping foxes and also hooded crows under control. Predator control is especially important in winter and spring, when the birds and their young are at their most vulnerable.

Various changes in land use have militated against the grouse, especially in western parts of its range in Britain and Ireland. Excessive grazing by hill sheep and cattle can result in the progressive loss of heather, which is replaced by rough hill grasses, which are unattractive and unsuitable for grouse. The spread of bracken is another factor which has caused grouse populations to dwindle, both because of the loss of heather and because bracken provides an ideal habitat for the proliferation of the sheep tick, which can spread virus diseases like louping ill to the grouse.

Very large areas of heather upland have been ploughed and planted with even-aged stands of non-native coniferous trees. This quickly transforms the habitat and makes conditions quite unsuitable for grouse, and for the heather upon which they depend. Conifer plantations are also known to harbour large numbers of predators, and these affect the breeding success of Red grouse and other upland birds.

# Red-legged partridge

## *Alectoris rufa*

# Rock partridge

## *Alectoria graeca*

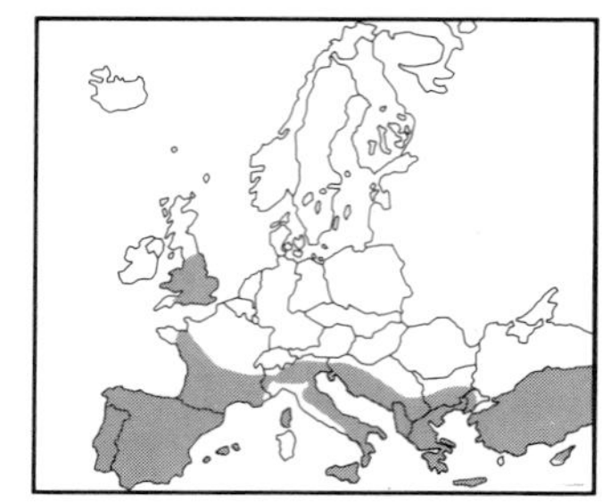

Many countries around the Mediterranean are characterised by landscapes which are dry and dusty, with sandy, light soils and broken up by areas of cultivation, vineyards, and the growth of *maquis* scurb. This habitat is the home of the Red-legged partridge and the Rock partridge, both of which originate from Mediterranean lands. The Red-legged partridge is common in south-western Europe, while the Rock partridge is a common gamebird in south-eastern Europe.

Both species are highly prized as sporting birds, and consequently there have been numerous attempts to introduce them to lands which are outside their natural range. The Red-legged partridge has been introduced to Germany, Hungary, Belgium, Norway, Sweden, the Channel Islands, the Azores and (most successfully) the British Isles. The Rock partridge has been released in France, Germany and Britain.

These various introductions met with varying amounts of success. The first introductions to England in the seventeenth century and subsequently were among the most successful, and the species fared especially well in the eastern counties, where the soils are light and there is a low annual rainfall. The species' natural spread has also been helped by very extensive releasing of Red-legged partridges by sportsmen in many parts of Britain and Ireland. Some of these releases have resulted in the establishment of self-sustaining wild populations of Red-legged partridges, but others have failed to create a wild stock, especially in the northern and western parts of the British Isles.

### Characteristics

The Red-legged partridge and the Rock partridge are very closely related, and the most important distinguishing feature is the shape and marking of the black area on the bird's front which defines the distinctive white throat area. The Red-legged partridge also has a reddish-brown colour on its under-tail covert feathers. These partridges are small and plump birds, similar in general shape to the familiar Grey partridge, but rather larger. The Rock partridge is slightly larger than the Red-legged partridge. Typical weights for mature adults are in the range 500–700 grams, and the male is slightly larger than the female. The flight of the Red-legged partridge and the Rock partridge is very similar to that of the Grey partridge.

On the island of Crete and some other islands in the Aegean, the Asiatic species *Alectoris chukor* occurs. This resembles the Rock partridge, and some experts contend that they are races of the same species. The Rock partridge often occurs well above the timberline, and in the Alps it is found up to an altitude of 2,500 metres, whereas the Red-legged partridge lives on the lower slopes and in low-lying farmland.

The mature birds of both species are also exclusively plant and seed eaters, but the young chicks share the common fondness of most gamebirds and gallinaceous species for insects and other high protein invertebrate foods. Research has shown that this protein-rich diet is important if there is to be a good survival rate among recently hatched chicks. When they are well grown the birds' diet will comprise leaves, buds, seeds and berries, and where Red-legged partridges have been reared and released for shooting, the gamekeepers will hold the birds on their ground in winter by feeding extensively with wheat and barley.

In autumn and winter Red-legged partridges form small flocks, comprising up to 20–30 birds. In general this is a sedentary species, although there may be some local movements of the birds, influenced by weather and the availability of food.

### Reproduction

The coveys of Red-legged partridges break up in spring and pairing takes place. These birds are monogamous, and will pair for several years if they are not predated or shot. The Red-legged partridge often lays two clutches of eggs in two separate nests, the male incubating the first and the female the second. The average clutch numbers about 12 eggs, and the Red-legged partridge does not try to conceal its eggs by covering them with vegetation when they are left unattended. Predation can therefore be quite high, perhaps twice the daily rate suffered by the Grey partridge, but the presence of two clutches largely makes up for the higher rate of losses through predation.

The eggs weigh about 20 grams, and their dimensions are approximately 39–45 mm x 29–33 mm. The eggs are yellowish-white with a sparse speckling of reddish-brown flecks and spots. Incubation begins when all the eggs in each clutch have been laid, and lasts for 24–26 days. The chicks are lively and active as soon as they have dried, like the chicks of most gamebird and gallinaceous species. Within about two months the young will be fledged and fully grown, and the entire breeding season therefore lasts for about 3½–4 months.

# Quail

## *Coturnix coturnix*

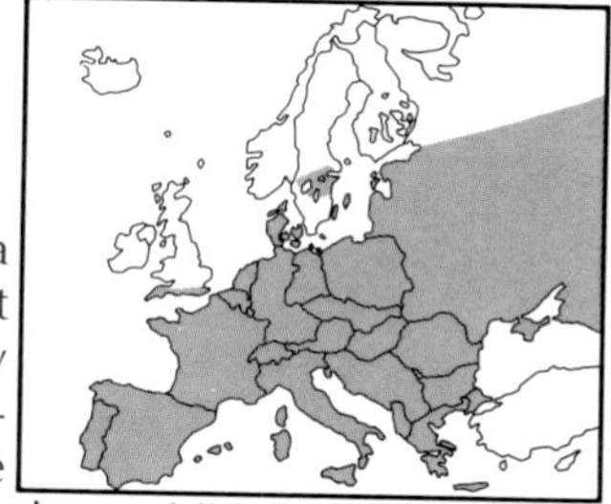

The quail is a resident of Europe, Asia and America, and it requires a climate that guarantees long, hot summers with plenty of sunshine. For this reason, the northernmost limits of the species' range in Europe is the southern shores of the Baltic Sea. In the middle of the nineteenth century it occurred much further north, in Sweden and Lapland. During this century quail have occurred periodically in the north of Europe but the general tendency has been for the species to retreat somewhat to the south.

In Britain and Ireland quail declined in numbers from the early nineteenth century onwards, but there have been a number of years since the 1940s in which unusually large numbers have been reported as calling and breeding in various parts of Britain. But the quail remains a scarce bird, with probably fewer than 50 breeding pairs each year.

The quail's body weight is in the range 70-134 grams, and its wing length is between 106 mm and 119 mm. The male and the female are alike in appearance. The quail is the smallest gallinaceous bird in Europe, with short legs, a small head and a very short tail. It has a rounded and plump body, rather like a miniature partridge. The male has two dark transverse bands on the throat, and these can vary in colour from dark chestnut-brown to yellowish-cream. The females and juveniles have a light-coloured throat.

The quail leads a furtive life, usually hidden in undergrowth, but the presence of quail is often betrayed by the male's repeated callings, often rendered as "*wet-my-lips, wet-my-lips*". The male will call by day and night.

The quail's diet comprises seeds and the green parts of plants. During the breeding season both the adults and the young chicks feed extensively on insects and larvae. They occur in open, cultivated land and in meadows with high vegetation. Their migration to their winter haunts in sub-Saharan Africa takes place by night.

### Reproduction

The quail is capable of breeding in its second summer, like the partridge. The female chooses the nesting site and makes a nesting scrape which she lines simply, using dried grasses and hay. She takes sole responsibility for incubating the clutch of 7–14 yellowish-brown eggs, which are flecked and speckled with brown, and which each weigh about 8–9 grams and measure 28–33 mm x 21–25 mm. The eggs are laid in May–June and incubation takes 16–18 days. The newly hatched chick weighs no more than 5 grams, and it will have fledged by the time it is three weeks old. The juveniles will be totally independent after a further 4–6 weeks.

# Grey partridge

## *Perdix perdix*

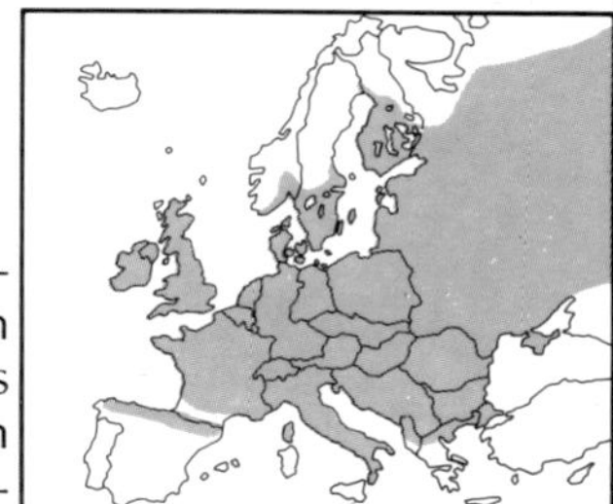

The Grey partridge is the most characteristic and familiar indigenous gamebird in Britain. Although its numbers and its sporting importance have latterly been eclipsed by the ubiquitous and non-native pheasant, the Grey partridge of British farmland still enjoys a special place in the affections of sportsmen and countrymen. In recent decades there has been a world-wide decline in the numbers of this bird, which is now rare or absent from many places where it was once numerous. The world population of partridges in the 1930s has been estimated at around 110 million birds, which by the 1980s had shrunk to less than 25 million.

The natural range of the Grey partridge and its closely-related subspecies extends across the British Isles, all of western Europe north of the Pyrenees and extending into southern Scandinavia, and eastwards across Asia to Mongolia and northern China. In North America the species was successfully introduced in the early part of this century and it became firmly established across much of the central parts of the United States and Canada. However, there has been a significant decline in partridge numbers in North America also, and it is probably true to say that the species has declined in every country where it is indigenous and to which it has been introduced.

In medieval times the partridge was a favourite quarry species for hunters with nets and using a setting dog to find the coveys. Trained hawks and falcons were also flown at partridges for sport. There were some conservation measures to protect partridges in the seventeenth and eighteenth centuries, and there are records of gamekeepers attempting to assess the spring breeding populations of partridges in central Europe as early as the 1690s.

In Britain the principal reasons for the decline of the partridges have been identified as loss of nesting habitat, excessive predation and a shortage of insect food for the young chicks, largely owing to the use of pesticide sprays.

## Characteristics

The partridge is a small, plump gamebird of the gallinaceous type, with short rounded wings and a short tail. Unlike the Red grouse it has unfeathered tarsi. At a distance the bird looks greyish in colour — hence the common name "Grey partridge" — and it has a distinctive and well defined marking, like an inverted horseshoe and coloured a rusty brown, on its breast. The young chicks are camouflaged in their down by colours of greyish-brown and yellow-beige with dark cross-bands.

The overall length of a partridge is 31–33 cm, and the males' average wing length is in the range 150–170 mm, while that of the female is 142–167 mm long. A fully grown partridge will weigh in the range 325–500 grams. The young birds and the adults differ noticeably in weight during the late summer and autumn, when the young birds weigh an average of 9%–13% less than the mature birds. When they hatch the partridge chicks weigh 7½–8½ grams, but they grow quickly and will triple their weight in the first week of life. By the age of three months they will weigh an average of 320–330 grams.

*In early spring one often sees birds running about uneasily in the flock. The mating game begins: a cock stretches his head and breast, expands his neck feathers, and runs toward a female. She meets him, angling her head forward and up. They stand breast to breast, and she extends her neck over his shoulder. Often several attempts are made before each bird finds the right partner. Once a pair is formed, it frequently leaves the flock.*

The weight is a good indicator of the partridge's physical well-being. The bird's weight drops if food is in short supply or if disease affects it. Starvation can occur in very severe winter weather when the ground becomes covered with frozen snow and food is difficult to find. Nevertheless, the partridge typically achieves its peak annual weight in December–January, which indicates that it probably lays up reserves of fat to meet its needs in severe winter weather. The female partridge also gains weight before she nests, but loses a good deal of her body weight when she lays her eggs and incubates them.

Young partridges become feathered early in life. They are capable of short, weak flights at the age of two weeks, by which time their flight feathers are sufficiently developed to do this. Young hen birds complete their moult into adult plumage by early October, and the older birds moult fully during the period July–September.

## Reproduction

Partridges break up from their winter coveys in early spring, and the birds form monogamous pairings, each pair with its own territory. The territorial nature of the partridge causes the birds to disperse, and densities of breeding partridges can vary greatly. In parts of Scandinavia a good local density has been given as 2 pairs per $km^2$, while the best European areas can achieve densities of up to 100 pairs per $km^2$.

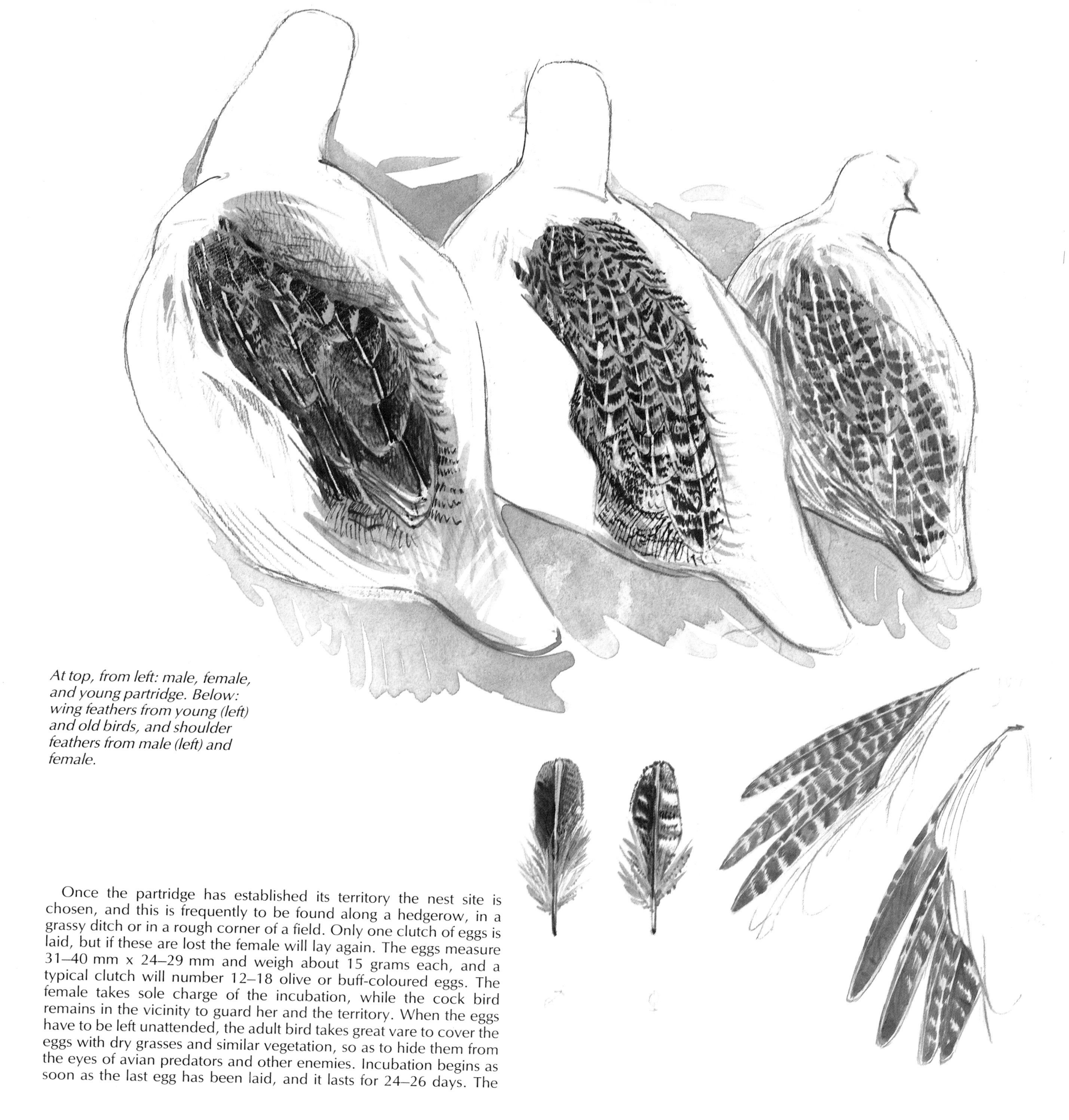

*At top, from left: male, female, and young partridge. Below: wing feathers from young (left) and old birds, and shoulder feathers from male (left) and female.*

Once the partridge has established its territory the nest site is chosen, and this is frequently to be found along a hedgerow, in a grassy ditch or in a rough corner of a field. Only one clutch of eggs is laid, but if these are lost the female will lay again. The eggs measure 31–40 mm x 24–29 mm and weigh about 15 grams each, and a typical clutch will number 12–18 olive or buff-coloured eggs. The female takes sole charge of the incubation, while the cock bird remains in the vicinity to guard her and the territory. When the eggs have to be left unattended, the adult bird takes great vare to cover the eggs with dry grasses and similar vegetation, so as to hide them from the eyes of avian predators and other enemies. Incubation begins as soon as the last egg has been laid, and it lasts for 24–26 days. The

eggs hatch simultaneously and once they are dry the newly hatched chicks are active and alert, ready to follow the adult bird and capable of feeding. At this early stage of their lives the young partridges' digestive system cannot deal with coarse cellulose material or plant matter, and it is important for the health and growth of the chicks that they should have access to plenty of insects and invertebrate larvae. Without this protein-rich diet their chances of survival are very poor.

The pair-bond between the male and female is very strong, and will last until one or other of them dies. Both adults play important roles in the guarding and rearing of their chicks, and the family group remains together until the larger winter coveys or packs form when the weather is colder and food less easy to find. In spring the coveys break up and the male birds move off in search of mates, and this helps to prevent a local population from excessive in-breeding.

The partridge breeds most successfully where there are suitable nesting sites, preferably situated in long grasses beneath the shelter of a hedgerow, and raised up on a bank so as to avoid the risk of flooding if there should be a sudden summer shower. There should be an abundance of insects in the vicinity of the nest, to provide a rich source of protein for the young birds, and the general area should afford the partridges some measure of protection from predators. Research has proved that strict control of foxes and crows can assist the survival of the nesting hen, her eggs and her growing brood.

## Senses and communication

The partridge relies for its safety on its good eyesight and its hearing. Sounds are important as a means of maintaining contact between the adult birds and also with the young chicks, and the young partridges begin to call and thereby establish contact with the hen bird before they hatch from the eggs. The adult partridge has a distinctive gravelly, harsh call, and the family group have a repertoire of calls which they use to maintain contact and to warn one another when danger threatens.

## Distribution and food

The partridge is widely distributed, as we have seen, and within Europe it is chiefly absent from Spain and Portugal, where its niche is filled by the red-legged partridge, and also from the far north of the continent, where conditions are unsuitable for it. Partridges are generally associated with low-lying farmland, and they fare well in areas which have a mixture of different crops interspersed with grassy ditches and well maintained hedgerows.

Partridges feed mainly on seeds, cereal grains and green plant material, and both adults and young will eat many forms of insects in the summer months. By the time the chicks are three months old they will have developed to the point where they can eat and digest plant food. The winter food requirements of the partridge have been estimated at about 65 grams per day.

There have been important recent changes in agriculture which have made it more difficult for partridges to survive, and which have undoubtedly contributed to its decline in many areas. Field sizes have been increased, hedgerows have been pulled out and good nesting habitat has been lost. There has also been a general swing towards autumn ploughing and winter-sown cereals, which means that partridges no longer have the benefits of stubble fields on which to feed during the winter.

Many attempts have been made to try and encourage an increase in the partridge population by releasing additional birds in the late summer, and by emphasising the need for "new blood". However, recent research in many countries has shown that the partridge will respond well to good habitat management, the careful control of predation, and the use of agrochemicals in such a way as not to deplete or destroy the insect species which are such a vital component in the diet of the young chicks.

## Population and subspecies

While all members of the *Perdix* family look quite similar, there are number of local races and subspecies which occur across the species' wide range in Europe and Asia. Of these the largest and the palest in colour are found in the eastern part of the range and are known scientifically as *Perdix p. robusta*. The darkest birds are found in Lower Saxony and in the Netherlands, and are known as *Perdix p. sphagnetorum*.

## Shooting: techniques and the law

Today in Europe most partridges are shot by sportsmen using shotguns and shooting over a pointing or setting dog. However, it is interesting to consider the various methods which have been used in the past to hunt and catch this highly prized species. Partridge netting, for example, is a very ancient art and one which was still practised in certain parts of Europe up to the middle of the last century. A bird dog was used to locate a covey of partridges and a net, often a bow-net, was set so as to ensnare the birds. Often a "stalking horse" was used to encourage the partridges to move into the net, for a horse was unlikely to cause the birds to take flight in panic. Traditionally the hen birds were taken from the nets and released unharmed, so as to contribute to the species' abundance in future years.

With the development of firearms, it became possible to use sporting guns with a flintlock action and firing a charge of small shot. The flintlock action was slow and often unreliable, and it was not uncommon for the birds to be shot on the ground. However, the art of "shooting flying" gradually developed, for it had become fashionable in Italy and gradually spread across Europe. Eventually the invention of the copper percussion cap made sporting guns more reliable and quicker to use, and birds could be shot on the wing. The usual technique was to use a pointer or setter to find and point the partridges, while the sportsman moved into position behind the dog. Then the birds were flushed and there was an opportunity of one or possibly two barrels being discharged at them.

Today the combination of a solitary Gun and a reliable bird dog remains common in many European countries, and various breeds of pointers and setters are used for this quarry, and also for grouse, pheasants and other game. It is unusual for any gundog to be kept solely for shooting partridges. In Britain the classic and most highly regarded way of shooting partridges was to have the birds driven forwards by teams of beaters, so that they fly over a line of standing Guns. This produced some very large bags of partridges as recently as the 1950s, but most driven shooting of partridges now relies heavily upon the rearing and releasing of birds, rather than on the wild population, and most gamekeepers find that the Red-legged partridge is more manageable than the indigenous Grey partridge.

# Rabbit

## *Oryctolagus cuniculus*

The rabbit is not strictly regarded as "game", but it has been netted, trapped, snared and shot for many hundreds of years by country people in many European lands. One of the favourite methods is to send one or more ferrets down into the warren of burrows, and thus to drive the rabbits out and into the meshes of bag-nets set across the holes. Many rabbits are caught annually by this method, but rabbits are also an exciting quarry for the sportsman with a shotgun or a rifle. Shooting bolting rabbits in close cover is a test of deft and speedy marksmanship, while stalking rabbits and shooting them with a small calibre rifle requires patience and fieldcraft.

It is difficult to estimate how many rabbits are now accounted for annually by ferreting and shooting, but the total must amount to many tens of thousands. This compares with annual totals of many millions which were netted, snared and shot in the years before the myxomatosis virus affected rabbit populations.

The popularity of the rabbit as a sporting species and for food is comparatively recent, when compared with other indigenous species of game and mammals. The rabbit is believed to have colonised most of Europe during the period prior to the last Ice Age. However, the renewed advance of the ice drove the species southwards, and after the last Ice Age the species is believed to have been restricted to north Africa, Spain and Portugal. It was gradually introduced to Europe north of the Pyrenees, and by medieval times rabbits were widely kept in hutches as an excellent meat resource. Wild rabbits were also prized as a source of food and as a sporting asset, and "rights of warren" were carefully guarded by those who were entitled to them. At one stage rabbits were so highly regarded that the price of a rabbit was recorded as equal to that of a pig!

Today introduced populations of rabbits are to be found in many parts of the world. Australia has had vast numbers of rabbits, amounting to plague proportions, and they are also to be found in New Zealand, in the USA, in South America, and in the West Indies, among other places.

### Characteristics

The rabbit is smaller than the hare, and looks a little plumper. It has relatively short ears, and if these are folded forward they do not reach to the tip of the nose. The ears are not tipped with black like those of hares, and the head is proportionally shorter than the hare's. The hind legs are also comparatively short.

Rabbits move by gentle hops or by darting and dashing quickly when danger threatens. Another initial reaction to danger is for rabbits to crouch down and freeze close to the ground. The rabbit's tail or "scut" is a bright white colour, especially when it is elevated in fear, and the upper side is brownish-black. The fur of the rabbit is greyish-brown, and the underparts are pale and white. Different local colour variations can appear, and some rabbits are yellowish-grey, some are darker than normal, and black (melanistic) individuals also occur.

The typical weight of a fully grown rabbit is between 1.5 and 2.0 kg, and it is most unlikely ever to exceed 3 kg. The males are about 10% larger than the females. The skull measures 68–75 mm. and the maximum skull length is at least a centimetre shorter than that of the smallest brown hare, and 5 mm shorter than the smallest arctic or alpine hare.

Rabbits are normally silent, although they will scream if they are injured or wounded. They can also be heard occasionally to grind their teeth. One of their reactions to danger is to drum on the ground with their hind legs, which is a danger signal which alerts other rabbits in the vicinity, above and below ground. They will also twitch their tails, and the white underside of the scut is used as a clear warning signal.

Rabbits appear to have a well developed sense of smell, and they use both urine and faeces as scent markings. In addition, the rabbit secretes scent in special glands in the chin, which are used to mark the ground and other objects.

Rabbits are gregarious and frequently live in colonies. A number of males and females form a group, and competition and aggression are found within the group and between rabbits living in close proximity to each other. Females will also compete with one another for burrows, especially in areas where only a limited number of burrows can be dug. In some warrens there is a considerable overlap in the territories of the various females, but in other colonies there may be very little such overlap.

Rabbits spend most of the daylight hours underground in their burrows and the interconnecting passages, and they are especially active in the early morning and towards dusk, and also during the night. They normally have fairly small territories, and rarely stray more than 400–500 metres from their burrows. Burrows and the rabbits' individual nests tend to be situated in banked or sloping ground, which reduces the likelihood of their being flooded. In particular rabbits are attracted to sloping areas of light, sandy soil, such as river banks and sand dunes, which are well drained and easily excavated. Tree roots and similar vegetation binds the soil together and prevents the burrows and underground chambers from caving in. The location of a rabbit warren is dictated both by the amount of nutritious grazing available nearby, and also the suitability of the soil for digging out the burrows. Rabbits are hardy and can withstand considerable extremes of cold, but their presence is dictated by the depth of the snow cover and the availability of food during a cold winter.

## Reproduction

The rabbit breeds quickly and frequently, and typically produce four or five litters of young at least every year. The young rabbits mature very quickly and come into breeding condition at a very early age. It is also possible for the female rabbit to come in season and be mated on the very same day as she gives birth to a litter of young. Although they are born blind, the young can see by the time they are ten days old, and they are weaned and capable of living independent lives by the time they are one month old. The youth and adolescence of a young rabbit therefore lasts approximately as long as the female's gestation period.

Since young females become sexually mature by the time they are six months old, and since each litter consists of 5–7 young and sometimes more, a local population of rabbits can grow very quickly indeed. In places where the rabbit population is very high, however, the annual production of young per female may drop to 10 or 11, while other females in less densely populated areas will produce 20–30 young in the course of the year. The newly born young rabbits weigh about 40–45 grams, and they are fully grown by the time they are one year old.

In continental Europe rabbits can reproduce all the year round, although the main breeding season begins in January–February and lasts until late August and September. The breeding season becomes shorter as rabbits become more numerous and the population density builds up. Cold weather with snow and stormy conditions can delay the start of the breeding season, and where rabbits are very numerous it can happen that the development of the foetuses stops and they are adsorbed into the mother's body, and in this way up to half of the possible young can be lost.

The young are born naked and blind, in underground nests which are prepared by the pregnant female somewhere in the underground passages and tunnels which comprise the warren. Occasionally a female will excavate a short, "blind" burrow away from the main warren, and give birth to her litter of young there, and this can also happen in gardens, nurseries and urban areas. The nest is lined with dead grasses and straw, and the adult female also adds white fur which she pulls from her abdomen and her breast. In small, single-family burrows the mother will cover up her young with nesting material and will also scratch up soil to cover the entrance to the nest hole. At night the female will stop feeding briefly and go to her litter of young and suckle them for about five minutes.

Female mature rabbits are in season about one day every week, and sexually receptive females will advertise the fact by releasing scent signals which the males will quickly detect. Rabbits are very promiscuous, and mating will often take place during the time the female is already heavily pregnant. A number of males may vie with one another for the attention of a female in season, and rival males may spend a good deal of time chasing one another.

## Population structure and mortality

The rabbit is a most important prey for many species of predators, and although individual rabbits are capable of living to the age of ten or more years, in the wild they are very unlikely to survive longer than a very few years. In large warrens and where the population of rabbits is high, a mortality rate of up to 60% has been recorded among young rabbits in their first year. Adult rabbits are estimated to suffer an average of 30% mortality per year, and it has been calculated that only 8% of one year rabbits will reach the age of six years. There is therefore a rapid turnover in the rabbit population.

The rabbit's natural enemies are stoats and polecats, which hunt them in their burrows and above ground, and buzzards, hawks and foxes which will take rabbits readily. However, the greatest depletion in rabbit numbers has been caused by the myxomatosis virus. This is carried from animal to animal by the rabbit flea, and the virus spreads rapidly through the population, especially where rabbits are numerous. In many places the disease has reduced the rabbit population by well over 90%, but it is believed that some rabbits can develop an immunity to the virus, or recover from it. But since the rabbits' rate of reproduction is limited by the density of the population, depleted populations can quickly recover their numbers again after a serious epidemic of myxomatosis. Rabbits can also suffer from coccidiosis, as do young hares, and this can cause heavy mortality where there is a dense population of rabbits.

## *Leporides*, or "rabbit-hares"

It is easy to tell at a glance that rabbits and hares must be quite closely related, and this can be demonstrated in many ways. Rabbits in captivity have been known to escape and to attract the attentions of wild hares, which have chased and courted them, and tried to mate with them. Nevertheless, in the wild rabbits and hares live side by side without any interbreeding. Research into the possibility of rabbits and hares breeding together was conducted by a French medical scientist, Dr. Broca, who has also been credited with discovering the speech centre in the human brain.

As a Frenchman Dr. Broca was possibly fond of rabbit meat, for it has been estimated that the French eat an annual average of 20 kg of rabbit meat each. Possibly Dr. Broca also sought a midway alternative to the pale, sweet flesh of the rabbit and the stronger and darker meat of the hare. Whatever his motivation, Dr. Broca is credited with the first person to succeed in interbreeding rabbits and hares in captivity. He called the result a *leporide*, and these hybrids were kept in cages and hutches, especially in France and Belgium.

# Badger

## *Meles meles*

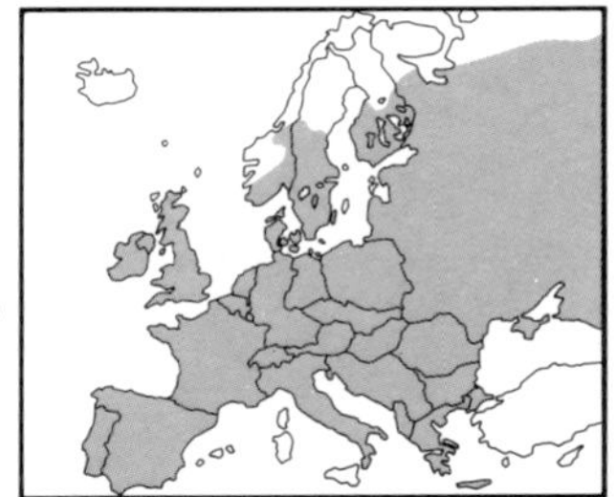

The badger has been designated by a host of nicknames and local terms of affection in every country in which it occurs. Many of these references draw parallels between the badger and the pig, or the bear, and up to the eighteenth century its Latin name was *Ursus meles* — "the bear badger". The pig-like qualities are retained in the fact that we refer to adult male badgers as "boars" and adult females as "sows". Early naturalists and scientists apparently had some difficulty in assigning the badger to the right family of animal relatives. It is now recognised, of course, that the badger is not closely related to either the bear or the pig, but that its closest relatives are the members of the marten family, like the pine marten, some species of weasels, and the wolverine.

In Britain a favourite colloquial name for the badger is "old Brock", which is a name derived from the Celtic language of old. According to an ancient belief, badgers are savage animals and likely to attack a man. It was therefore recommended that anyone going out to walk through the woods at night should fill their boots with pieces of charcoal. This was intended to fool the badger, for should he attack and bite your boot, he would hear the sound of the charcoal crunching and think that his bite had crushed a bone in your foot!

In the past badgers used to be killed for their meat, and its bristles have often been used in tufts to decorate hats and to make good-quality shaving brushes.

## Characteristics

The badger has an unmistakeable appearance, and its black and white head is especially easy to spot and identify. Its legs are short, which gives the animal a podgy appearance, and when on the move it appears to be half-walking and half-creeping, and it moves with a rolling gait. Its claws are especially long and powerful, and they are usually at their longest and sharpest in spring, when the badger emerges after his winter hibernation. Thereafter the claws are progressively worn down by the animal's digging and scratching for food. The badger has small ears and these are edged with white, and partially hidden by fur. The tip of the badger's nose is black and the eyes are small and set close together, and have a deep brown colour. The tail, which is 15–20 cm long, is greyish and long-haired.

The badger is related to a family of carnivorous predators, but its own eating habits are very varied and catholic. This adaptation has affected the structure of the badger's teeth. The incisor teeth have receded in the upper jaw and in the lower jaw they are rather blunt, while the rear molars are large and coarse. A badger has a maximum of 34 teeth, which conform to the formula

$$\frac{3\ 1\ 4\ (3)\ 1}{3\ 1\ 4\ (3)\ 1}$$

The male is usually larger than the female, although occasionally the sexes may look and weigh the same. Generally, however, the

male weighs 12%–13% more than the female, and in the autumn its weight may be 30%–50% higher than its weight in late spring and summer. This additional weight is attributable to layers of fat which the badgers lay down in preparation for the onset of winter. Individual weights of up to 20–25 kg have been recorded, and the recorded maximum seems to be 27 kg. By the time they are four or five months old, juvenile badgers may weigh as much as 9 kg, but they are not fully grown until they are two years old.

Badgers are active at twilight and during the night, but where they are not disturbed they may come above ground and move about by day. They usually live in small groups or colonies, comprising several adult males and females, and a number of young animals. Badgers will also live separately in pairs. The centre of their activities is their burrows or "setts", which they will use for year after year, often extending and enlarging them as time goes by. A large sett will comprise several floors, perhaps three or four all on different levels, and they are interconnected by various passageways, which in turn lead to several different exits. The entrance hole into a sett may extend underground for as much as 20 metres or more. Outside the entrances to the sett will be seen piles of excavated soil and miscellaneous waste materials, often mixed with straw, dry grasses and other bedding materials. These large setts are often located in areas with rocky soil and in places where the natural drainage of the soil is good. Favourite places are the edges of woods and among small plantations of trees.

The badger group lives in a territory or home range which varies in size according to the type of countryside, the nature of the soil, the availability of food and the number of badgers in the colony. A territory may be as small as 20–30 hectares or as large as 200–300 hectares, but it is generally reckoned that badgers will do most of their wandering and foraging within an area of one square kilometre. It is thought that adult males will travel furthest by night, perhaps up to 10–15 km.

Badgers are territorial, and will create a number of latrines or dunging places within their territories or on the boundaries. They will also mark these spots with scent secreted by their glands, and badger living together in one community will also mark one another with this scent.

Scents and scent marking play an important role in the way badgers communicate with one another, but their use of sounds is also important. The adult badger has quite a repertoire of cries, grunts and growls, while the young have a number of whining and hissing calls. The badger's sense of sight is much less well developed than its acute senses of smell and hearing.

In Europe the length of time for which badgers will hibernate varies according to the geographical location and the climatic characteristics. In the furthest north of their range badgers will hibernate longest, from October to April. In mild winters they remain active longer and emerge from hibernation sooner, and in temperate parts of the species' range badgers are often active by night throughout the winter.

## Reproduction

It is thought that badgers, which are monogamous, pair for life, and that the two sexes remain together throughout the year. Young badgers become sexually mature at the age of 1½ years, although the males may take longer to mature than the females. Mating takes place between April and August in continental Europe, and in the mild conditions of spring and autumn in the British Isles mating may take place as early as February and (exceptionally) as late as October. In the Nordic countries the badgers mate in May and June. It is supposed that this extended breeding period is to allow those females which have failed to become pregnant by their first mating to mate again. The actual length of the gestation period depends upon the phenomenon of delayed implantation, which may be for a period of from three to nine months. Once the blastocyte is implanted in the uterus, the development of the foetus only takes about two months.

Badgers produce just one litter of young each year, and sometimes they only breed in alternate years. The litter comprises two to four young, and these are born between January and March. Several adult males and females live together in the same sett, but the litters of young are placed in separate "nursery nests", lined with grass, moss and bracken. The young badgers' eyes are open after four weeks, but they remain underground for up to two months. They are suckled by the mother for up to three or four months, and she also feeds them by regurgitating half-digested food.

The badger cubs and their parents remain underground with the other members of the group until the following spring. Among communities of badgers the sexual ratio is even, but among the young badgers females predominate; and young males leave the community earlier than their female siblings.

## Population structure, occurrence and diet

Badgers live in relatively open and cultivated countryside, where woodlands of various sizes alternate with pastures and arable fields, and where the woods border open fields. Badgers are often found living close to man, and, like the fox, badgers can be found living in cities, in suburban gardens and in public parks. In Europe, and especially in areas such as the Alps, badgers will live at altitudes up to 1,800 metres above sea level. Occasionally exceptional individual badgers have been seen as high as 2,700 metres above sea level.

In parts of England where the habitat is ideal for badgers, densities of up to 56 badgers per 1,000 hectares have been recorded. The longest known lifespan of a badger is 15 years, but most die much younger. In particular, the mortality among young badgers in their first year of life may be as high as 50%.

Badgers are thoroughly omnivorous. Oats and other cereals, berries, fruit, clover and grass predominate in the diet, which will also comprise various small animals, including earthworms. Badgers will also eat insects, snails, molluscs and the young of other mammals such as voles and mice. Birds' eggs, young chicks and baby rabbits also figure in their diet. Badgers living in cities will forage and consume household garbage of many kinds.

## Distribution

Badgers can be found in most of Europe, but they are absent from the Mediterranean islands of Corsica, Sicily and Sardinia. They are also absent from northern Scandinavia, and their range extends eastwards across eastern Europe into Asia and as far as China, India and Burma.

# Weasels

The European members of the family *Mustelidae* include animals as large as the Wolverine (28 kg) and as small as the Weasel (80 grams). Three species — the [Pine] Marten, the Stoat and the Polecat — are familiar animals of the countryside. All three occur on the European mainland and, although none are found in Iceland, Stoats and Polecats are also found in Scandinavia and in the British Isles.

The distribution of these three animals is interesting. The [Pine] Marten is found in Denmark, and it belongs to the group of European species that did not have time to cross the ancient land bridge to Scandinavia from mainland Europe at the end of the Ice Age. It never found its way to those northern countries, nor to the British Isles. However, the Polecat and the Stoat did spread further north, and the Polecat has a presence in the southern part of Fennoscandia, while the Stoat can be found from southern Sweden right up to the Arctic Ocean.

These are all territorial animals, and they mark their territories with scented secretions from their anal glands. This scent is so powerful that it can sometimes also be detected by man. The size of individual territories is dictated by local conditions, and especially by the availability of food. The Stoat can have a territory of between 30 and 60 hectares. Stoats are promiscuous, and several females may live within the territory of one male. The females also have their territories, but like juvenile Stoats they can sometimes wander by night as far as 5–10 km.

In most members of the Weasel family there is a delayed implantation of the blastocyst which may last for 8–10 months. Once implanted, the actual gestation period is only 4–6 weeks.

## Stoat — *Mustela erminea*

The Stoat or Ermine is an animal with an elongated body and short legs. Its coat turns white in winter, while in summer it is reddish-brown with a creamy-white underside. In northern parts of the species range it can be flecked with white during the spring and autumn. The end of the tail remains black at all times of the year. The Stoat's ears are small and rounded, and the skull is flat, with comparatively small jaws. The tooth structure is the same as that of the Pine Marten.

A fully grown animal may weigh up to 450 grams, and the male is larger and heavier than the female.

The Stoat is both nocturnal and diurnal in its habits, but it tends to be most active at dawn and dusk. It likes areas of tumbled stones, ditches, wood piles and the tangled roots of trees. It moves rapidly and sinuously, and gives an impression of great alertness and curiosity. It often moves with a bounding movement, and when necessary it can both climb and swim.

The Stoat inhabits most of Europe apart from the Mediterranean areas. It also occurs in a wide range which extends across Asia to the Pacific Ocean, including the islands of the Japanese archipelago, and it is also present in Canada and in large areas of northern and western U.S.A. It lives in the Arctic zone, and it can also be found at altitudes up to 3,000 metres in the Alps.

The Weasel and the Pygmy or Snow Weasel are close relatives of the stoat. They occur throughout Europe with the exception of Ireland and Iceland, and they are to be found in a wide range extending across Asia to the Pacific and also across large parts of North America.

The Weasel can be distinguished from the Stoat by its smaller size and by its tail, which is much shorter and lacks the Stoat's distinctive black tip. The northern form moults to an all-white winter pelage, but

*Beech marten (left), polecat (bottom), and stoat in brown summer and white winter coats.*

the southern form keeps its reddish-brown coat all the year round. The cranium of the weasel (26–38 mm) is smaller than that of the Stoat (38–49 mm).

The stoat builds a small den lined with feathers, mammal hair and grass, and there the annual litter is born. The kittens are born both blind and helpless. They suckle for two months and their eyes are not open until they are 5–6 weeks old.

The stoat lives mostly on small rodents, but it will also take rabbits, small birds, eggs and young chicks. Normally it eats 30–60 grams of meat per day. When there is an abundant supply of small rodents the size of the litter can increase, and litters numbering up to 18 young have been recorded. Mating takes place in the spring (February–April), and the young are born between April and June. There is a second mating season in June and July. Apart from the period of delayed implantation, the actual gestation period is 28 days.

*Polecat*

## Polecat — *Mustela putorius*

The Polecat can be distinguished from the North American mink, which has been introduced to Europe, and also from the closely-related European River polecat (*Mustela lutreola*) by the colour of its head. The mink has a dark head with a white patch on the lower lip and on the chin. The same is true of the River polecat, although its upper lip is also white. The underfur of the mink is greyish-brown, while that of the River polecat is dark brown. The underfur of the polecat is yellow with dark blackish-brown guard-hairs, and the head and the edges of the ears have an obvious yellow-white pattern.

*Stoat*

The polecat can weigh up to 1.5 kg, and the males are the largest. The weight of a fully grown female will not exceed 1 kg. The length of the skull of a mature individual is 55.5–69.5 mm, and the indentation of the skull behind the eye is deeper than that of the mink.

The polecat is active at dusk and through the night. It occurs in a wide variety of different habitats, including continental mountain ranges as high as the snowline. It is fond of woodlands, but prefers open terrain with scattered plantations of trees, stone walls, ditches and streams. It often occurs in the vicinity of farms, villages and larger communities. It sneaks inside barns and poultry houses where it hunts for rats and chickens. In areas where there is a large wild population of rabbits, it lives in rabbit burrows and warrens where it hunts rabbits of all ages. Besides these types of prey, it also catches small rodents and young birds, steals eggs from nests, and it will also eat frogs, toads and occasionally fish. When nothing else is available this opportunist feeder will eat large insects and earthworms.

The polecat's area of distribution stretches from Wales and England in the west to the Ural mountains in the east. Polecats are ready to breed in their second year, and the females are in season in March and April, and occasionally slightly later. The males become sexually active a month or two earlier. The young are born between April and July after a gestation period of 40–43 days. The single annual litter comprises 3–8 young, and sometimes more if there is an exceptional abundance of food. The young are born blind and can see after about 4–5 weeks. They continue to suckle for a further month, and remain with their mother for several months more. They are considered fully grown and largely independent at the age of five months.

*Pine marten*

*Skull weight depends on body size, as is clear from these skull weights: Weasel (0.9 g), Ermine (2.6 g), Pine Marten (15.5 g), Badger (104 g), Wolverine (185 g).*

# Fox

## *Vulpes vulpes*

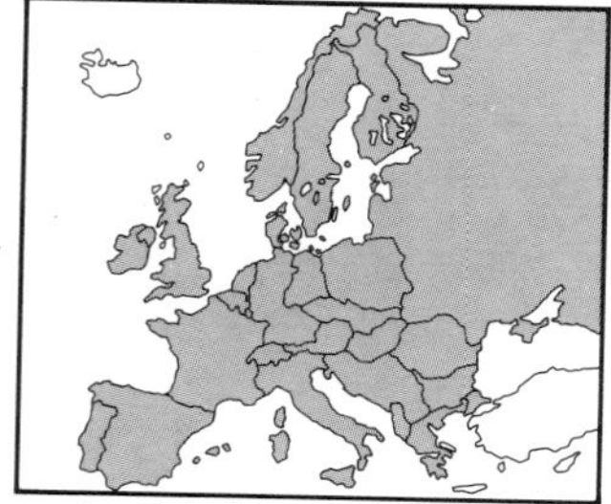

"Reynard the Fox" is a familiar character to everyone. This has long been the case, as we can tell from the many folk tales and old legends which exist. Nevertheless, the fox has never been a popular animal — apart, perhaps, from those who cherish him as a sporting quarry for hunting with hounds — because foxes are generally regarded as personifying craftiness, cunning and cruelty. Above all, in rural areas the fox has an age-old reputation as a chicken thief.

The fox is adaptable, and has learned to adapt to many different environments and living conditions. It can be found in open country and on arable land throughout Europe, and in commercial forestry plantations as well as in natural deciduous woodlands. During its nocturnal wanderings it will prowl around houses and farms looking for scraps of food or stealing chickens. The fox is always alert to man's ill-will, and most of their life in the wild is spent in hunting small rodents and mammals of various kinds.

In London, and in many other towns and cities in Europe, the fox has been able to adapt to a rather easier life than in the countryside, for it can scavenge for food. Foxes are frequently to be seen by night in towns and cities, and the London fox population has been estimated at close to 10,000 animals.

### Characteristics

The fox is a member of the large and varied dog family. It has triangular, pointed ears which stand erect, and it is notable for its long, bushy tail. Its muzzle is long and pointed, but its legs are comparatively short.

The colour of a fox's fur can vary a great deal. The normal markings, which have given the red fox its name, can be seen in the accompanying illustrations. The tip of the tail is usually white, at least in adult males, while the newborn cubs are greyish-black and lack any guard-hairs in their coats.

The male fox is larger than the vixen, weighing an average of 20% more than the female. Exceptional individual dog-foxes have been found to weigh up to 13–14 kg, but the normal weight is between 5 and 10 kg.

The fox makes a number of sounds. The males bark throughout the year, but chiefly during the mating season which lasts from December to March. They will also growl to indicate aggression. Cubs make a high pitched whining sound. In the dark a fox's eyes will reflect a beam of light from several hundred metres away, and the white tip of the tail tends also to catch the light.

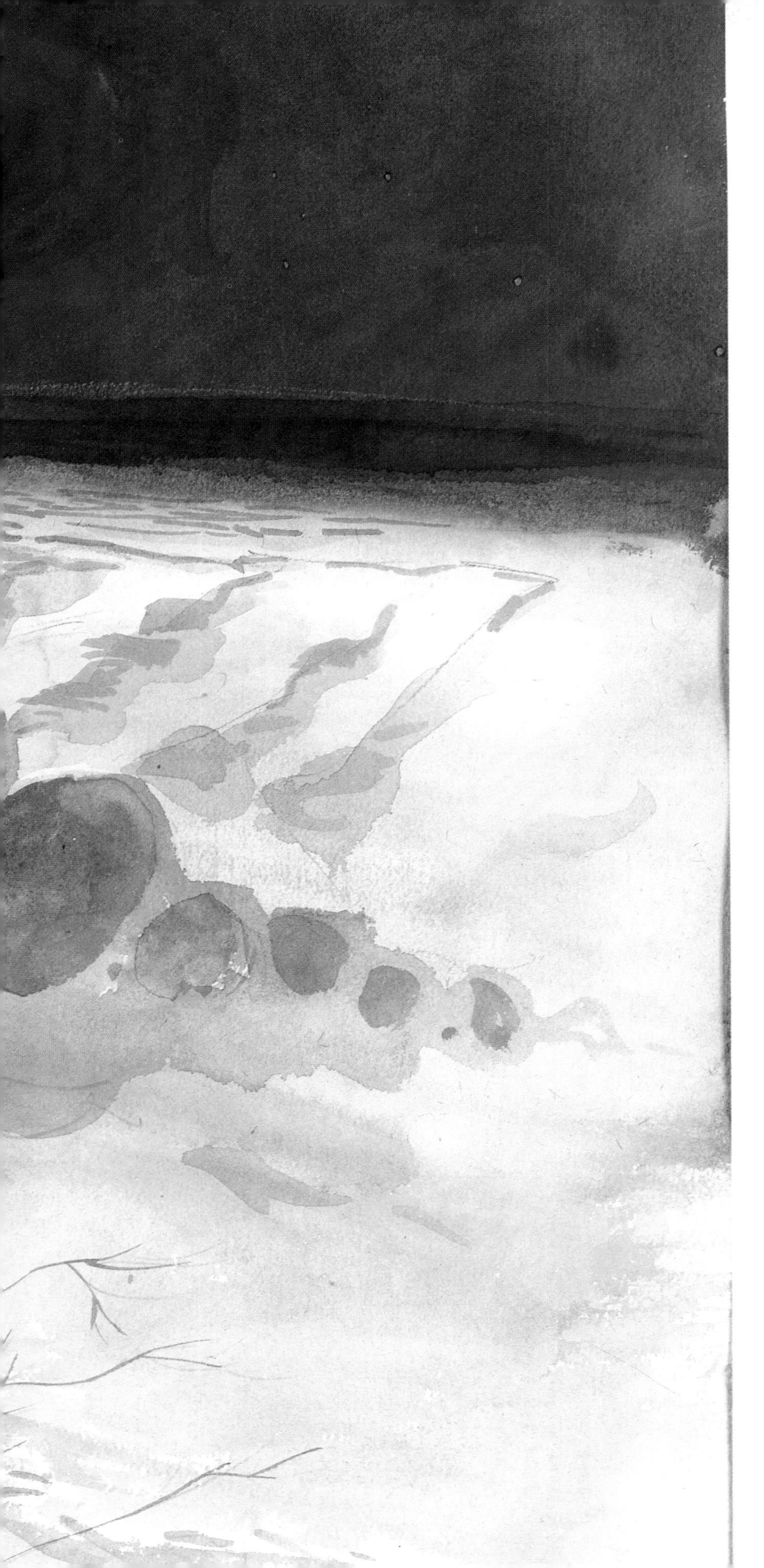

## Adaptions and way of life

The red fox is active mainly at night, although its activities are regulated by the different phases of the species' way of life. When there are hungry and growing cubs in the den the fox may be forced to hunt by daylight, and in winter foxes may also hunt by day. Heavy rain and snow seem to upset their activities, and foxes often tend to be rather inactive during bad weather.

The fox is an excellent swimmer, and he will also climb trees on occasions. He looks for resting places just above the ground, and he will often tend to rest by day on a spot which gives him a good view of the surrounding country. The fox will also sunbathe on flat rocks or in the middle of open fields. Foxes live most of their lives above ground, but where there is little cover the fox will resort to holes and hollows. When it is hunted and in imminent danger the fox will "go to ground" for safety.

Foxes are territorial animals, but the size of an individual fox's territory varies greatly. The food supply, the lie of the land and the density of the fox population all affect the size of the territory. Male foxes maintain larger territories than females, and individual territories can be as small as 5–10 $km^2$, or as large as 15–20 $km^2$. Foxes roam across their territories when searching for food or hunting, but also for the purposes of maintaining and defending them against the incursions of other foxes.

In summer foxes tend to live in family groups within the territory, and it is believed that each fox family has its own territory, and that these summer territories do not overlap. The boundaries are marked with urine and with secretions from the anal glands. The pads of the fox's feet also secrete a scent through special glands, and thereby show where the fox has been. (It is this scent which foxhounds follow when they are hunting a fox.) The fox's elongated droppings reflect the various foods on which the animal concerned has been feeding, and these are often deposited on prominent places, such as tree stumps, stones and other pieces of raised ground. Secretions from the anal glands give fox droppings their distinctive foxy smell.

The summertime family groups can sometimes comprise several females with one adult male, forming a kind of harem. It is thought that all such females are related, as females from the previous year's litter remain with their parents to help provide food for the new litter. Apart from this help, the food gathering responsibilities rest chiefly with the adult female than with the male.

The young males leave the family community and seek out their own territories. As with many other mammals, their wanderings are more extensive than those of the female. British studies have shown that young males covered an average distance of 14 km, while young females went only an average of 6.5 km. from their birthplaces. This dispersal takes place when the foxes are aged about eight months. (Some marked foxes have been recorded as travelling very much longer distances, some up to several hundred kilometres.)

## Reproduction

The fox is usually monogamous. Mating occurs in the period January–March, and somewhat earlier in central and southern parts of Europe. At mating time the male foxes are known to fly at one another and to fight each other while rearing up on their hind legs. This behaviour probably occurs when the foxes are forming new relationships with vixens. It is known that a pair of foxes can stay together for several successive years. The young are born in dens or earths, and the adults remain close by. As a result the territories are not fully utilised during the breeding season.

The annual litter is normally born between spring and early summer, after a gestation period of 52–53 days. For a variety of

*The male fox has a larger head than the lightly built female.*

reasons a number of females do not give birth, and absorption of the embryos is probably a more common cause than infertility. The proportion of females that do not give birth varies with the time and the place — in some areas 90% of females reproduce, in others only 20%. Like the domestic dog, females that do not breed can simulate pregnancy, and the mammary glands may become developed. A normal litter comprises 5–6 cubs, but the number can vary from one up to 10–12. The vixen has eight nipples, and the young cubs' chances of survival are directly related to their having ready access to their mother's milk.

The cubs weigh 100–130 grams at birth, and they are born blind. The eyes open after 10–14 days. The female stays with the litter both day and night for the first few weeks, and if the den is disturbed the vixen may move her cubs to another den not far away.

## Feeding habits

The red fox is omnivorous, and is highly opportunistic in his feeding habits. It will make use of whatever forms of nutrition are available to it within its territorial range. However, the fox is clearly and principally a carnivore and a predator. Its legs are comparatively short compared with other members of the dog family, but it can nevertheless run very fast and it is often able to outpace its prey. However, the fox's favourite hunting tactics are mainly to outwit their prey. The way foxes hunt voles is a case in point. The fox steals along ditches and across fields where the voles have their passages in the grass or under the snow. In a tense and motionless position the fox waits and listens until it can locate the vole. It then springs upwards in a sudden leap and pounces down upon the surprised and unsuspecting animal. A fox will sneak up on chicks and young mammals and complete the hunt with a sudden final dash. At other times it will dig out nests of voles and mice after first using its sense of hearing and smell to locate the animals.

The fox's food preferences have been studied in many parts of Europe. In Britain, in much of mainland Europe and in southern Scandinavia the fox lives mainly on small mammals and game, up to the size of large rabbits. A large part of the diet is composed of small rodents, especially among the northern populations and in highly cultivated agricultural areas. It seems to have a preference for the easily caught and relatively large field vole, but it also catches other voles and field mice. In Scandinavia the lemming is a very important prey species, especially in years when the lemming population is very high.

Along the sides of main roads and in urban areas the fox is mainly a scavenger which feeds on carrion and discarded rubbish. It will readily eat animals and birds killed by motor traffic, and it will also scavenge to find after-births when sheep and cattle are giving birth. The fox is also well known for the havoc it can wreak among domestic poultry such as chickens, ducks and geese, and in release pens full of young pheasants and partridges. Foxes will prowl around rubbish dumps, and it will occasionally kill more than it can eat or save. Like the badger, at some times it will feed heavily on earthworms and insects. Amphibians and the occasional lizard or snake will also figure in the fox's diet, and at times the fox will eat wild berries or soft fruits cultivated in commercial plots.

The fox's food requirements vary with the age of the animal, its size and the time of year. An adult fox needs a daily average of 0.5 kg of food, although it is capable of going for several days on very little food. Half a kilo equals about nine fully grown field voles, two large water voles or one very large brown rat. A medium-sized wild rabbit can sustain a fox for up to three days.

## Reproduction and food

Like all living organisms, grazing animals and the predators which prey on them live in a balanced relationship. Since animals usually seek their food where it is most easily found, the largest fox populations tend to be found on farmland where there is a scattering of woodland. However, their lives and those of urban foxes are very different from that of the woodland fox which inhabits the forests of northern Europe and Scandinavia.

There the foxes live mainly on small rodents found in natural forest habitats far removed from human habitations. In such areas shortages of small rodents is a much greater threat to the survival of the fox population than the activities of sportsmen or commercial hunters. In years such as 1966–67 when there was an abundance of rodents in these northern parts, 87% of the adult vixens had litters of cubs. The following year only 33% had cubs, owing to a drastic reduction in the numbers of lemmings, voles and mice. Furthermore, few of the cubs will survive in such bad years. One consequence is that the individual territories become very much larger or are left abandoned.

The settled areas of southern Sweden have large fox populations, and the foxes have a rich and varied diet available to them. Even so, many females, perhaps half of them, do not reproduce. The litters are larger here than in the far north, yet the cubs' mortality can also be high. The size of the litters varies between good and bad years, but the size of the territory remains largely unchanged. The fox populations of southern and central Sweden nevertheless produce a surplus of young foxes which, if they survive, will emigrate.

## Survival, numbers and lifespan

The largest fox populations of foxes are to be found in good agricultural areas and also in close proximity to cities and towns. Several studies of foxes provide the following figures: a Scottish survey showed one pair of foxes per 40 km$^2$, while in upland areas of Great Britain densities of one pair of foxes per 1.2–4.8 km$^2$ have been reported from Wales. In Germany the density has been found to vary between 0.3 and 2.0 foxes per km$^2$.

The winter population of foxes in Europe has been estimated at around 700,000 animals. It should be noted that there is a mortality rate of 50%–80% among young cubs in their first year. Many foxes are hunted or shot each year, and some estimated figures are: Britain 50–80,000; West Germany 100–130,000; and in Sweden 50–90,000.

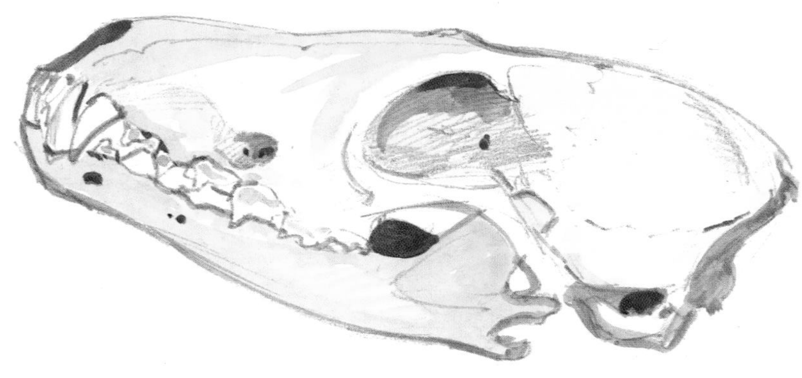

*Unlike the Arctic fox, the Red fox has pointed ears and a longer, more pointed nose.*

## Habitat and distribution

The fox is as adaptable in its choice of a living environment as in its choice of food. The European red fox therefore inhabits both unspoilt wilderness areas and highly populated urban and suburban areas. It is found on open grasslands, on the steppes of Asia and in desert and semi-desert areas as far south as the northern Sudan. In the border grasslands of the tundra and in northern Scandinavia it haunts upland moors and breeds in birch woods and among scrubland and osiers. In the Alps the fox occurs at altitudes up to 3,000 metres. It can survive in areas where the winter temperatures fall as low as −40°C, and in deserts where the temperature may soar as high as +40°C.

The fox's distribution includes all of Europe, Asia and North Africa. In Africa it is found only north of the Sahara, with the exception of the Sudan. In Asia it is found as far east as the Pacific coast, and on several offshore islands. It also exists in India as well as on the Arabian peninsula.

Within this huge distribution zone the fox has been influenced by many different environmental conditions, resulting in local races and subspecies with distinctive characteristics. Some researchers even regard the American red fox as a subspecies of the European red fox. If this is correct, the distribution area of the species also includes much of Canada and the United States. Other scientists, however, regard the American fox as a separate species, with the Latin name *Vulpes fulva*. The American red fox is found from the coasts of the Caribbean to the northern parts of Baffin Island, and it shows the same environmental preferences as the Eurasian red fox.

## Subspecies

World-wide there are 21 different species of fox. These are subdivided into four groups, mainly on the basis of the shape of the skull and the structure of the teeth. The genus *Vulpes*, to which the European red fox belongs, comprises 12 species which occur in Europe, Asia, Africa and North and South America. Of the red fox *Vulpes vulpes* there are almost 50 subspecies. The red fox of northern Europe is classified as the large reddish-brown nominate form, *Vulpes v. vulpes*, and the red fox of continental Europe is usually regarded as its own subspecies, *Vulpes. v. crucigera*. (No foxes occur in Australia or Antarctica.)

## Enemies, diseases and mortality

The foremost enemy of the red fox is man. It is estimated that one-third of all foxes fall victim to the sportsman's bullet and the hunter's snare. During recent times the killing of foxes has increased as changes in fashion have increased the demand for fox fur, and prices have risen accordingly. In spite of this, it is believed that in some areas, such as the British Isles, there has been a slight increase in the total fox population. The North American populations, however, have been reduced by hunting. In England and in Germany the killing of foxes is legal throughout the year, while in Sweden the fox is protected during the breeding season. In Scandinavia hunting probably has less influence on the fluctuations in the fox population than does the food supply, for periodic fluctuations in the numbers of small prey animals such as lemmings and voles cause short-term variation in the size of the fox population.

Disease also plays some part in regulating the total number of foxes. For example, the Swedish fox population has been severely affected by an epidemic of fox mange, caused by a skin parasite and resulting in dramatic reductions in some local populations of foxes.

Rabies is a much-feared virus which occurs throughout the world. Foxes and dogs are considered to be the prime carriers of the disease, together with some species of bats. The fox has an especially bad reputation as a rabies carrier in Europe, although the virus has also been found in association with many other species of wild animals. In mainland Europe, for example, it is quite common among roe deer. Loss of weight and aggressive behaviour lead to paralysis and the disruption of the entire central nervous system. The virus is spread by bites and in the saliva of affected animals. The incubation period can be long, often several months, but it usually lasts for only a few weeks.

In continental Europe rabies has caused dramatic declines in the fox populations in some areas, and in some regions up to 70% of the fox population has disappeared. Because the fox can transmit the rabies virus to domestic pets such as dogs and cats, and in some unfortunate cases directly to man, eradication campaigns have been undertaken. Scandinavia and Britain and Ireland are free from rabies, but in Denmark a few cases have been reported. For this reason foxes are intensively destroyed along the border with Germany, in an attempt to create a fox-free buffer zone.

Young foxes moving away from the family groups to find their own territories are at the greatest risk of being shot, or poisoned if a carcase is set out as bait. A cautious, territory-conscious fox can elude the hunter for a long time, and at the same time succeed in defending his territory against younger foxes. When this type of "clever" fox is shot or trapped, younger foxes may move into his territory and provide easy prey for shooters and trappers. The result of this is that excess foxes are destroyed, while the more hardy and permanent territorial populations of foxes remain largely unscathed.

Since younger foxes, especially males, roam widely during the winter months, it is difficult to get a clear picture of the sexual balance, the age structure or the size of the population by examining only the trapping and shooting statistics. The oldest wild foxes are believed to live to the age of 10–12 years, but captive individuals have lived rather longer than this.

## Hunting methods and hunting dogs

There were few regulations governing the hunting and shooting of foxes anywhere in Europe until the 1930s. It has generally been common practice in country areas to locate fox dens and to destroy the young cubs, and, if possible, the breeding adults. This is still an annual aspect of predation control in the Scottish highlands, where gamekeepers and stalkers do annual rounds of fox dens and breeding cairns with terriers, to bolt the adult foxes which are shot, and to kill the cubs underground.

Over much of Europe foxes are caught by trapping with steel traps and by snaring with running nooses of wire. Another form of fox control, which combines fieldcraft and marksmanship, is the calling of foxes by the use of a special fox-call or squeaker, to entice the fox to within range of the waiting shotgunner or rifleman. Foxes are also shot at night, especially in hill country, by sweeping the area with powerful spotlamps and shooting the foxes with a high-powered rifle.

A number of breeds of sporting dog have evolved because of the need to have good dogs for fox control. Among those which hunt the fox in the open by means of following its scent, the English foxhound is supreme. Dachshunds and terriers belong to a different category of dogs, those which are expected to be able to go to ground after a fox and to face it bravely underground. They need to have energy and aggression, and their task is either to bolt the fox from its den or to attack and kill it underground. For this reason these breeds need to be small, tough and full of stamina.

# Red deer

## *Cervus elaphus*

For as long as man has hunted in Europe, whether for meat or for the pleasure of the chase, the Red deer has been an extremely important species. Archaeological investigations have shown that this animal was an important quarry for early hunters living as far apart as the Mediterranean lands in the south and the Nordic lands to the far north. Its pursuit spans all civilisations from the primitive Stone Age cultures up to modern times. The overwhelming importance of the Red deer led to its being frequently depicted in early art and also in literature. Classical authors described the hunting of the Red deer in ancient Greece and in Roman times, while Nordic peoples from the Stone Age to the Bronze Age depicted the hunting of the Red deer in their rock carvings. Red deer hunting was frequently depicted in medieval tapestries and in paintings of hunting scenes, and early legislation in many countries reflected the red deer's supremacy as a sporting quarry, often reserved strictly for nobility and royalty.

### Characteristics

After the Elk, the Red deer and the Reindeer are Europe's largest deer species. The Red deer stag has branched antlers, which are more elevated and compact than those of the Reindeer. Only the male (stag) has antlers, and he is noticeably larger than the female (hind). His legs are long and powerful, and his body is large and heavy, with an almost mane-like growth of hair on his thick neck. By comparison the neck of a hind is longer and more elegantly slender. In both sexes, the rump patch is prominent and the hair has a yellowish-white colour, with a short tail.

The summer pelage of the Red deer is a uniform reddish-brown, but it becomes greyish-brown with the coming of winter, and this gives the animals an altogether darker appearance. The young calves are a light reddish-brown, heavily flecked with white on their neck, back and flanks. They can occasionally be a uniform reddish-brown, but this is very uncommon. At about two months old their coat changes as new fur grows, and the white flecking gradually fades and eventually disappears altogether.

The Red deer bears many similarities to various related species, such as the Sambar (*Cervus unicolor*) and the Sika deer (*Cervus nippon*) which occur in India and Asia. In areas where these species have been introduced by man, hybrids have sometimes occurred, such as the interbreeding of Red deer and Sika deer in Ireland, Scotland and New Zealand. Such hybridisation is also believed to happen in the wild as a natural occurrence where the ranges of the species overlap.

### Size and teeth

Red deer vary greatly in size, depending upon the subspecies and the distribution area, and also the local conditions. The North American stags are the largest, and these are said to achieve body weights up to

450 kg. This is in contrast to the stags on the open hill in the Scottish highlands, where a stag weighing 200 kg would be quite exceptionally heavy. An average for Red deer stags in Germany is 225 kg, and much the same for those in Denmark. The largest Red deer in Europe are found in Poland, Czechoslovakia, Hungary and Romania, while the very smallest Red deer live in Corsica, where they weigh 70–100 kg. In general hinds weigh 65%–70% as much as stags, while the newborn calf will weigh 6–7 kg. It takes 10–12 months for a calf to grow to the point where it achieves complete independence from its mother, and some Red deer may not be considered fully mature until they have reached the age of six or seven years.

Mature Red deer have 34 permanent teeth, with the tooth structure:

$$\frac{0\ 1\ 3\ 3}{3\ 1\ 3\ 3}$$

There are approximately 40 species of deer, and the Red deer shares their common characteristic of having no incisor teeth in the upper jaw. However, and in contrast to species such as Moose, Fallow deer and Roe deer, Red deer have canine teeth. (These only occur rarely among Roe deer.)

## Antlers

The male, or stag, has magnificent antlers. The points or tines grow from the main beam which extends backwards; while a pointed brow tine extends forwards from the base of the beam, just above the coronet. Above this are the "bey tine", which is often absent, and the third or "trey" tine. In many mature stags the beam extends upwards until it forks into a three-pointed cup shape, and twelve points, six on each antler, is generally regarded as typical of mature stags in the wild. A stag of this type is referred to as a "royal" stag in Britain and in various other countries, although there are various records of exceptional Red deer stags which have had antlers with many more points than this. This is especially true among the large Red deer of central and eastern Europe, and among park deer which are given large amounts of calcium-rich supplementary feeding.

The antlers of very large park stags and wild stags in central Europe can weigh up to 12–19 kg each, but 5–8 kg is a more typical average. The young male Red deer begins to develop his first set of antlers when he is 7–12 months old. The length of the antlers, the thickness of the beam and the number of tines gradually increases with age, so that by the time he is 4–5 years old a stag may have a head with 12 points. But for various reasons some stags' antlers do not develop in this way, usually owing to a deficiency of calcium in the diet, an injury, a hormonal disturbance or some genetic factor.

When stags begin to grow old — around the age of 10 years — their antlers gradually begin to become shorter and thinner, and with fewer tines. This process is known as "going back", and such stags are often selectively culled by sportsmen and deer managers.

Like other big game, the Red deer's antlers are prized as sportsmen's trophies, whose size and general conformation are assessed according to an objective formula, and which may be submitted for competition in hunting exhibitions. Apart from their personal importance to the individual sportsman, antlers also reflect the quality of deer populations and the richness of their habitat.

Red deer shed their antlers annually, and this usually occurs between February and May. Once the old antlers have been shed the new antler growth begins immediately, and stags may be seen with fully grown antlers as early as June and July. The times of shedding and regrowth vary, depending upon the geographical location, the weather, and the physical condition and age of the individual animal. Older stags shed their antlers earlier than young ones, and stags in good condition shed their antlers earlier than weak or sickly ones. By late summer the new set of antlers will be fully grown, and encased in a thick and furry skin, known as "velvet". This covering is richly supplied with blood vessels which promote the spectacular and speedy growth of new bone, and the velvet is gradually shed with the approach of autumn. Often the stags hasten the process by rubbing their antlers against trees and long vegetation. In general, the more mature stags clean their antlers and are in "hard horn" sooner than the young ones.

## Reproduction and the rutting season

Red deer stags reach sexual maturity by the time they are four or five years old, and hinds are mature by the age of three, and occasionally at two years old. The hinds tend to have a very high and prolonged rate of fertility, and consequently they can calve annually up to the age of 20 years and more. The age of sexual maturity and the period of the rutting season vary within different areas of the Red deer's very wide distribution.

In Britain and Europe the Red deer rut occurs during the period August–October. In Sweden, for example, it begins in late August and lasts for about one month, while in England and Scotland the deer rut mainly from late September until late October. During the rutting season the mature stags each gather a harem of hinds, and the size of the hind group will be directly related to the status of the stag and his physical ability to hold and defend his harem. A mature stag in his prime might hold anything from 10–30 hinds. A younger stag, or an older intruder, will occasionally try to intrude and approach a hind in season, but they are rarely successful and are driven away by the superior energy and size of the master stag. The master stag passes on his genetic characteristics to his offspring until his rutting days come to an end, either through old age, illness, accidental death or shooting. The sexually receptive hind is usually mated during her first period of oestrus, but if not there is a second period of oestrus about 18 days later. Gestation lasts for approximately eight months, which means that the calves (normally only one per hind) are born in late May or early June. At birth the calf is well developed, and it can stand on its feet when it is only a few hours old.

During calving the hind drives away her last year's yearling calf, but she often accepts it back again when the newborn calf is about a fortnight old. Hinds have been known to suckle their yearlings, but this usually occurs when she has not given birth to a new calf, or if she has lost it. Normally there are as many female calves as male calves, and thus the balance between the sexes is maintained. The male calf leaves its mother and leads an independent life during its second summer or autumn.

*Hungary is famous for its superb Red deer. Record antlers have often been noted there. In 1986 a new world record was set with antlers scoring 271 C.I.C. points, and weighing 15.195 kg. The previous record was also set in Hungary.*

## Senses and patterns of activity

All the Red deer's senses are highly developed. In particular, the sense of smell, hearing and sight, in that order, are the most highly developed, and they are important for relationships between the deer themselves, as well as being indicators of possible danger. Red deer are active at dawn and dusk, while the middle of the day is often spent lying up and ruminating. In upland areas the deer will often move up towards higher ground during the day, and they may also seek areas of heather or long bracken in which to lie and rest. Red deer in woodland are often active by night, especially if there is a bright moon.

Red deer give a short, sharp bark as a warning. Stags make only occasional barking noises, but hinds often bark repeatedly when they are alarmed. During the rutting season, the master stags utter a deep and resonant roaring bellow, which is their vocal proclamation that they are holding a harem of hinds, and a warning to rivals to keep away.

Old stags occasionally live solitary lives, but younger ones often form stag herds and the stags and hinds live separate lives except during the rutting season. Hinds with calves and half-grown followers normally form herds during the winter and spring.

Like many other species of deer, the Red deer secretes scents from glands at their hooves, below their eyes, and around the anus. It is also apparent that the hind emits an attractive scent when she is in season, and this obviously plays an important part in the rutting behaviour of both sexes. Stags mark their territories with glandular secretions and by rubbing their antlers against trees and stumps. The eye glands, however, may have a function for individual animals and for the herd which is not specifically related to tenure of a territory. Communication between the members of a herd is maintained by visual signals, including the light rump patch which is thought to have a special importance when the animals are running away from danger. Before taking flight, deer may show their unease by stretching their necks and staring towards the source of the danger, and by lifting their legs one at a time, stiffly and cautiously.

The population density for Red deer varies greatly depending upon the lie of the land, the climate, the vegetation and the extent to which shooting and hunting may put pressure upon the population. In the Scottish highlands densities have been recorded of between 13 and 77 animals per km$^2$. The home range for a mature stag is about 490 hectares, for the hind about 390.

Although the ratio of males to females is equal among newborn calves, mature hinds are usually more numerous than mature stags, owing to selective shooting or hunting of stags.

## Distribution

The Red deer is widely distributed over most of the palaearctic, and encompassing Europe and northern Africa. It is also found in Ireland and in the Hebrides, and it abounds in Scotland. To the east it occurs in Manchuria and China. In Norway it is found as far north as the Arctic Circle, while southern populations occur in Asia Minor, Iran, Afghanistan and Kashmir. However, there are large parts of central Asia from which the Red deer is absent (see distribution map).

The North American Wapiti (or Elk as it is commonly called in America) is also a subspecies of Red deer. It occurs mainly in the western U.S.A. and Canada (in the Rocky Mountains), where its northernmost range lies at a latitude of approximately 60°N. (In contrast, Scandinavian Red deer occur in Norway up to a latitude of 67°N.)

## Local populations and subspecies

As mentioned, the Red deer enjoys a very wide geographical distribution. Large populations are separated from one another by wide gaps, even among the European Red deer. In eastern Asia there are northern and southern distributions, as well as those in North America. Within these areas there are also local populations which are separate from each other.

In 1758 when the Swedish naturalist Linnaeus gave the Red deer its scientific name, he began with the wild Red deer of southern Sweden, which he called *Cervus elaphus*. When differences in other populations were noticed and identified, the deer of southern Sweden were given the subspecies name *Cervus e. elaphus*, and they thus became the nominate form of the species. The Scottish Red deer are isolated from the Red deer of mainland Europe, and they are noticeably smaller and lighter. But differences in the shape of the skull caused them to be designated a subspecies, *Cervus elaphus scoticus*. Also based on studies of the shape of the skull, other red deer were designated as the *atlanticus* subspecies, and the Red deer of central Europe became the subspecies *hippelaphus*, because of their size, length of antlers and number of antler tines. The Red deer of Corsica and Sardinia, the subspecies *corsicanus*, are less than half the size of central European Red deer. Another subspecies that has been described is that from the Iberian peninsula, the subspecies *maral*.

## Occurrence, habitat and diet

The Red deer is mainly a creature of woodlands, especially where these are combined with marshy areas and open glades, and where they are close to cultivated fields. Red deer thrive in transitional habitats between these types of area, and many Red deer in mountainous country live well above the treeline. In Scotland the main habitat is heather moorland and open, grassy hills; in Norway the coniferous and mixed woods; and in central Europe wooded and mountainous areas. Their diet varies from place to place, and also with the seasons. Grasses, sedges and young shoots are important all the year round, while heather and the shoots of coniferous trees are important locally, especially in winter. In woodland the diet is dominated by small twigs and shoots of bushes, as well as from deciduous and fir trees, and bark is also stripped and eaten. The red deer is both a grazer and a browser. In some places it will also raid cultivated fields and feed on crops, both cereals and roots. In woods where beech and oak occur Red deer will eat beechmast and acorns.

## Relations with man

In much of its range the Red deer has been held in very high regard as a sporting animal. Stone Age man hunted the Red deer and it was later to become a prized and specially protected quarry for kings and great landowners. It was accorded a much higher status than the Moose, which is Europe's largest big game animal. In medieval times the hunting of Red deer developed into a matter of considerable ritual and glamour. Small landowners and other countryfolk were forbidden to hunt or even to disturb the Red deer, which were reserved for those whose social position entitled them to hunt them. In some parts of Europe Red deer were over-hunted, and special deer parks were established to hold herds of Red deer. Deer in parks were often kept principally as a form of "meat on the hoof", and their pursuit was a matter of netting them and catching them up for slaughter. In some parks the deer were driven by lines of beaters towards hidden marksmen with bows and arrows, and later with sporting firearms.

Stalking the deer on foot was a development of early man's stealthy pursuit of wild game by sneaking up on it unawares. But hunting with hounds was a much more dramatic and lively affair, in which a selected deer was pursued with great noise and activity by specially trained hunting hounds, while the huntsmen followed on horseback. This ritualised form of deer hunting was widespread until the nineteenth century, and in its glamorous and ritualised form it is still carried on in France and Belgium. In south-west England Red deer are also hunted with hounds accompanied by mounted followers.

The Red deer has frequently come into conflict with forestry and agriculture, and this remains a problem in much of Britain and Europe. In some parts of medieval Europe the antipathy of foresters and farmers to the destructive deer led to a serious decline in the numbers of Red deer, and the future of the species was assisted by the maintenance of special deer parks. Today in Scotland and other parts of Britain there are organised attempts to breed Red deer in captivity as a means of meat production, and the animals are kept in enclosed parks and fed on a special diet to promote the rapid conversion of fodder into good, lean venison. In addition to deer farming, large moorland areas of the central, western and northern highlands are maintained with wild deer management as one of the principal forms of land use, and the deer generate an income for the various estates as a result of the fees which sportsmen will pay to go stalking wild Red deer on the hill.

Red deer have played a most significant part in the history of art, in paintings and many other forms of pictorial representation. Red deer antlers have always been used for many purposes, including knife handles and buttons, and deer hides are also much in demand for many uses.

## Enemies, mortality and lifespan

Man has been a major and very serious predator of Red deer in some parts of Europe, to the point where the very survival of local populations of Red deer has been in doubt. This is especially true of southern Sweden, of Corsica and of Algeria and Tunisia. Among wild animals the wolf is the most important predator, where they still exist in significant numbers; and both Golden eagles and foxes have been known to take young Red deer calves. However, the most significant cause of mortality in most of the Red deer's range is insufficient food owing to heavy snowfalls, and adverse weather conditions. In particular, a succession of very mild but wet winters in the western highlands of Scotland during the late 1980s caused heavy mortality among Red deer on the open hills.

There are reports of Red deer attaining the age of 30 years, although 20 years is generally thought to be the maximum natural lifespan for most wild Red deer, and a high proportion of deer die well before they reach that age. The average lifespan of a Scottish Red deer has been estimated at 5–7 years.

## A year with the Red deer

**Winter.** Red deer live in herds of differing sizes during the winter months, and this depends upon the density of the deer population. Where Red deer abound, it is usual to find the hinds with their calves and followers forming fairly large herds, perhaps of 100 animals or more. The stags usually wander in separate and rather smaller herds. Thus the sexes live separately in winter, and since the species has a matriarchal social organisation, the hinds and followers are led by an old and experienced hind.

The winter diet varies according to where the deer live, and snow cover can affect winter feeding, but the main source of food is a variety of low shrubs, shoots and twigs. Red deer will also strip the bark off growing trees, especially young fir trees.

The Red deer's winter pelage is greyish-brown. This winter coat begins to grow in September–October, and is usually complete by November. Animals in poor physical condition tend to keep their reddish summer coats for longer than strong and well-nourished individuals.

The stags' antlers are free of velvet during the autumn and winter, but they are shed from February onwards.

**Spring.** In areas where the deer moved down to the lower ground in winter, they will now gradually return to the grassy upland slopes and the heather moors, above the treeline. This is especially true in Scotland and in Norway.

Grasses, fresh young shoots, a variety of herbs and leaves are now prominent in their diet.

The hinds are still accompanied by their calves and some older followers. When the new crop of calves are born in May–June, the yearlings are temporarily rejected by their mothers, but are allowed to rejoin them after a short time.

The stags have shed their antlers and new ones are beginning to grow fast. The new antlers are covered in hairy skin, known as "velvet", and are of varied length and shape according to the age of the individual animals, their physical condition and the quality of food to which they have access while the antlers are growing.

**Summer.** Red deer are rather elusive in summer. Hinds and their calves and yearlings form distinct herds, although the yearlings gradually begin to fend for themselves while still remaining members of the herd. The deer graze actively at dawn and dusk, moving along regular paths between their daytime lying-up places and their feeding grounds. Deer in woodland are especially difficult to see, owing to the dense undergrowth of summer. In hill and moorland areas deer often move onto the highest ground, to escape the torments of midges and flies.

The summer diet consists mainly of fresh green growth. Deer in woodland will browse on fresh shoots and leaves of trees and bushes, but they will also graze on grasses and herbs growing in forest clearings. When close to rivers and lakes they will often feed among the bankside vegetation, especially reeds and rushes. On hill ground the deer will feed principally on young heather and hill grasses.

**Autumn.** The rut in Europe usually falls in the period August–October. Red deer are polygamous, and at this time powerful and dominant mature stags gather around them harems of hinds with which they will mate in succession as they come in season. Year after year these master stags will return to their favourite rutting sites where they proclaim their presence and dominance by frenzied physical activity and continual bellowing and roaring, particularly at night.

After mating, hinds leave the harems of the rutting stags and rejoin their yearlings and other females in hind herds.

After the rut the stags wander off individually, but they will gradually gather together in stag herds. However, some old individuals may continue to lead solitary lives.

*Smailer deer and yearling.*

# Fallow deer

## *Dama dama*

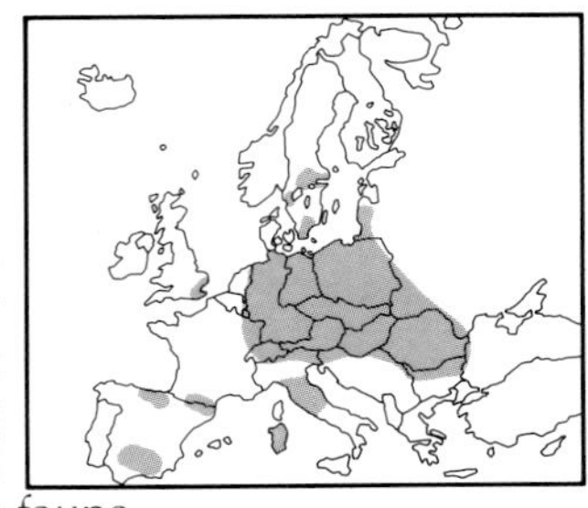

The Fallow deer comes originally from Asia Minor, but it has been successfully introduced into many parts of Europe. It has been a resident of Europe for so long that it is now accepted as a naturalised and permanent member of the European fauna.

Britain and Ireland have a particularly strong tradition of introducing non-native animals, birds and plants from other countries, and the Fallow deer is firmly established in the British Isles, in both feral populations and numerous enclosed park herds.

### Characteristics

Mature male Fallow deer (bucks) may be readily distinguished from Red deer by their palmated antlers, which are wide and flattened compared to the beam and tines of the mature Red stag. Younger bucks have shorter, more spiky antlers, somewhat similar to those of young Red deer stags. In contrast to the Roe deer, the Fallow lacks fully developed hard antlers in the spring and summer, but has antlers during the autumn and winter. The Fallow female (doe) lacks antlers.

The pelage of the Fallow deer can vary greatly from herd to herd, and from one individual to another. The common colour is reddish-brown with a flecking of large white spots, and there is a dark ridge of hair running along the spine. The pale rump patch is outlined with a dark border, and the tail is dark and relatively long.

The winter coat of the common Fallow is rather lighter and more greyish than the reddish-brown summer pelage, and both black and white Fallow occur quite regularly.

A mature Fallow buck will weigh up to 100 kg, and the average weight of a mature Fallow buck is about 40% more than that of a mature doe, which rarely exceeds 60–70 kg. Individual Fallow bucks weighing up to 125 kg have been reported from central Europe and from some park herds. At birth the buck fawns do not weigh much more than the doe fawns, and the average weight at birth is 4–5 kg, although weights as low as 2.5 kg are not uncommon. Fallow in captivity have been known to live for up to 20–25 years.

In January and February, when the male fawns are little more than six or seven months old, they begin to develop small spiked antlers which may measure 3–16 cm long. These are clean of velvet in late summer and are carried through the autumn and winter until they are eventually shed in late spring or early summer. New antlers begin to grow immediately after the old ones have been shed. As the deer gets older the antlers grow later in the season and are shed earlier in spring. Once they are clean of velvet the Fallow buck's antlers are carried from September until April–May. The antlers develop their flattened, palmate shape not earlier than the buck's third year, and usually when it is a little older. The antlers usually reach their peak of development when the deer is aged 8–10 years, provided the animal concerned has access to suitably nutritious and calcium-rich feeding. Mature antlers may measure 75–90 cm in length and weigh up to 5–7 kg.

Fallow deer are active mainly at dawn and dusk, but in areas where they feel safe they will move about by day. Older Fallow deer tend to be solitary, and to be almost totally nocturnal in their habits. Fallow deer often form large herds, and young bucks and does will also form separate herds. The Fallow deer rut takes place between September and November, and in the British Isles and in Mediterranean countries there may be signs of rutting activity as early as August. The dominant buck lays claim to a rutting area and marks his territory with rutting scrapes. He will attract and hold a harem of about ten hinds, and the characteristic noise of a rutting Fallow buck is a deep belching groan which is very different from the somewhat bovine bellow of a Red deer stag.

The Fallow deer's diet consists of various grasses, herbs, twigs and buds. In autumn they will also feed on beechmast and acorns. Fallow will also eat the growing shoots of young deciduous and coniferous trees, a habit which does not endear them to foresters and farmers, but they are considered less destructive in this respect than Red deer.

### Reproduction

As we have mentioned, rutting takes place in the autumn, and each Fallow doe is in season for about two weeks, during which time she may be mated on several occasions by the buck to whose harem she belongs. The gestation period lasts for approximately 230 days and the young, often twins and occasionally triplets, are born in June and July.

Although the newborn fawns can stand up and suckle soon after birth, they do not graze with their mother until they are two or three weeks old. The fawns suckle for up to four months, and occasionally for much longer. They remain with their mothers in separate doe herds which the female fawns will stay with. The buck fawns gradually leave the herd by the time they are about two years old. The Fallow are usually fully grown and sexually mature by the time they are three years old, even though the bucks may be sexually mature as early as one year old.

### Environment and distribution

Fallow deer like open countryside interspersed with deciduous woodlands. Large coverts or woodlands offer protection, but they should also be adjacent to open fields, meadows or glades which provide feeding areas.

The Fallow deer is found in Europe and adjacent parts of south-west Asia, including Iran and Israel. Iranian Fallow deer are larger and are often regarded as a separate species or subspecies of the Fallow deer, and are designated *Dama d. mesopotamica*. This form of the Fallow deer had been considered extinct until it was rediscovered about a century ago. The Fallow deer's post-glacial distribution area was south-east Asia and the eastern Mediterranean region. Thanks to many introductions since ancient times it is now found across most of Europe, although its distribution is very scattered. A map showing its distribution in Europe would reveal a number of evenly distributed populations quite isolated from one another.

In Europe the Fallow deer is believed to have been far more widespread prior to the last Ice Age. Archaeological discoveries from the last interglacial period reveal that Fallow deer were present in Britain at that time. During Roman times and in the Middle Ages the Fallow deer was reintroduced to the British Isles, and it is thought that the species was introduced to Germany between 300 A.D. and 700 A.D. Fallow deer do not occur in Norway or Finland, but populations of wild Fallow are to be found in many parts of southern and central Sweden, where it is the only introduced deer species.

*Fallow deer have flattened, palmated horns. The long tail, black on top, gives the white rump-patch a distinctive appearance. Two black crescents frame the patch at the top. The summer coat is red-brown with white spots. Only the buck has antlers, which are fully grown and clean in August–September. By then, the fawns are more than half-grown.*

## The fallow deer's relatives

Man has introduced a number of non-European deer species to different parts of Europe. One such example is the North American Whitetail deer, which is described later in a separate chapter. Others include the Axis or Spotted deer (*Cervus axis*), which originates in southern Asia; the Sika deer (*Cervus nippon*) which comes from eastern Asia; the Reeves Muntjac deer (*Muntiacus reevesi*) from south-east Asia; and the Chinese Water deer (*Hydropotes inermis*) which originates in eastern China and Korea. A close relative of the Reeves Muntjac is the Indian Muntjac or "jungle sheep" (*Muntiacus muntjac*), which was briefly introduced to England, and the Hog deer (*Axis porcinus*) which was introduced to Denmark, but without any lasting success. The Muntjac, the Chinese Water deer and the Sika all have flourishing wild populations in Great Britain, and all have been introduced to various parts of mainland Europe at various times. Wild populations of the Axis deer can be found in Yugoslavia.

Many of these deer species can breed very quickly, and there has been an important growth in deer farming in Europe in recent years. It is often suggested that suitable island habitats are the best places for introducing captive herds of non-native deer, since they are thereby isolated and their effects on the indigenous fauna can be minimised. All of the species mentioned above also exist in fenced parks and animal collections in most European countries.

The most successful introduced deer, in terms of the number of animals which are now to be found in the wild, must be the Whitetail deer, following its introduction to Finland. After this would come the Sika deer, of which thousands live wild in parts of Germany and Denmark, in Ireland, Scotland and various parts of England. The axis deer in Yugoslavia is effectively confined to the Istrian peninsula, where the population at one time numbered 1,000 animals, a density of 145 deer per $km^2$, and this inevitably led to over-grazing of the available habitat.

The introduction of non-native deer can have serious repercussions. Among the possible consequences are greatly increased damage to forestry, the possible spread of disease, and the possibility of the newly introduced deer interbreeding with or competing with the indigenous species.

# Roe deer

## *Capreolus capreolus*

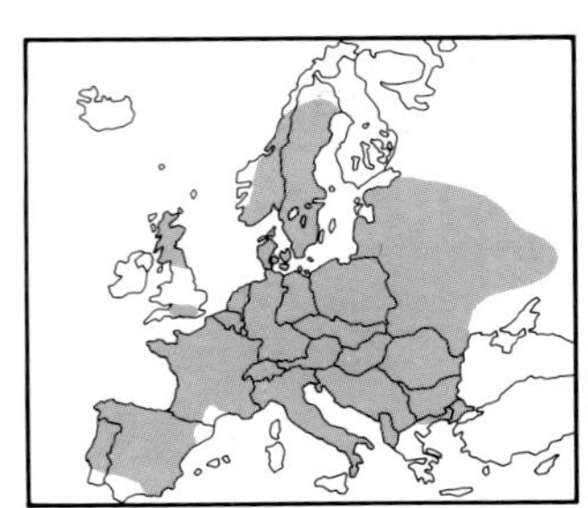

Throughout most of Europe the Roe deer is widely regarded as one of the most desirable sporting species in the "medium-sized game" category. In Sweden alone an estimated 100,000 Roe are shot each year. And yet it is interesting to consider that this has not always been the case, since the Roe deer has not always been so numerous or so widespread. Indeed, at some times in the past the Roe has been something of a rarity.

In continental Europe this little deer was usually fairly common, but it did not enjoy the exalted status which was accorded to the Red deer. From the Middle Ages onwards the royalty and nobility of Europe hunted the larger game animals, which included the Wild ox, the Elk, the Wild boar and, most important of all, the Red deer. "La grande chasse" was an elaborate and ritualised affair, but the Roe was not held in high regard, and the hunting of the Roe was often left to the common people. In England the Roe lost its former status under the terms of the Forest Laws from about the fourteenth century, but an exception to the general rule was Sweden, where royal edicts from the thirteenth century onwards reserved Roe as royal game only, which commoners were forbidden to hunt. It was not until the nineteenth century that the Roe became fair game for the Swedish sportsman.

### Characteristics

The Roe deer is the smallest species of indigenous deer in Europe. The buck is rather larger than the doe, and he bears antlers during the period from late spring until late autumn. The doe has no antlers. In general appearance the buck has a more solid looking and triangular shaped head than the doe. The Roe's summer pelage is a rich foxy-red, which changes in September and October to a muted greyish-brown. The young, called kids or fawns, are born with a patterning of pale spots of a light reddish-brown background, and this flecking disappears with age. The Roe's rump patch is yellowish-white in summer and a bright white in winter, and the Roe's muzzle is black and almost hairless.

The height at the shoulder is between 64 and 75 cm, but the "giant" Roe deer of Asia can be a tall as 89 cm. The length from nose to tail is 106–112 cm, and a mature Roebuck can weigh as much as 35 kg. However, a more usual average weight for a full grown animal is around 25 kg. By the late autumn the fawns will have achieved a weight of about 16 kg. (One Swedish Roe doe, with the exceptional number of four fully-formed unborn young, was found to weigh 29.5 kg when she was accidentally killed in the month of May.) The previous year's young are considered fully grown by the time they are 16 months old, by which time they are the same size and weight as the mature animals. The size and weight of individual Roe is in inverse proportion to the density of the population and the poverty of the diet.

The tooth structure for the milk teeth is

$$\frac{0\ 0\ 3\ 0}{3\ 1\ 3\ 1}$$

and for the permanent teeth it is

$$\frac{0\ 0\ 3\ 3}{3\ 1\ 3\ 3}$$

The Roe deer has no incisors in the upper jaw, although it can occasionally have canine teeth. There is a wide gap called a diastema between the canine teeth and the premolars of the lower jaw. It is possible to determine the age of Roe deer from some populations by examining cross-sections of the teeth, and by assessing the extent of the wear on the teeth, but this method of ageing remains somewhat controversial.

Roe deer can vary in appearance according to hereditary factors, age, physical condition and the animals' environment. It is incorrect to suppose that one can tell the age of a buck by the size of its antlers and the number of points they bear. Most mature Roe bucks have symmetrical antlers with three points each, but in some Asiatic populations there are often more, and individual antler size and conformation can vary greatly from animal to animal.

As early as their first autumn the male fawns can begin to develop small knobs on their heads measuring a centimetre or more, and the length, thickness and weight of the antlers will usually increase with age. A buck is usually a six-pointer by the time he is 2–3 years old. Older bucks shed their antlers between October and December and renew them in March–April, while the younger bucks shed theirs in November–December and grow a new set during April–May. When the antlers are fully grown and still covered with the hairy membrane of skin known as "velvet", the buck will begin to clean off the velvet. They will rub and thrash their antlers against the stems of trees, large bushes and tree stumps to scrub off the velvet. Antlered females have been recorded, but this is usually attributed to a hormonal imbalance.

The Roe fawn has a light reddish-brown coat, darker along the back and the withers and speckled with pale spots on the back, the sides and the rump. The spots fade about a month after birth and disappear completely in September–October when the first winter coat begins to emerge. Older animals have a foxy-red summer coat, and are greyish-brown in the winter. The bucks have a tuft of long hairs on the penis, while the doe, especially in her winter coat, has a pronounced tuft of white hair at the base of the rump patch. When a Roe is nervous or alarmed the white hairs on the rump patch are raised and appear very prominent. The winter coat is gradually moulted between March and May and is replaced by the summer coat, which then grows thicker from September through the middle of October, in preparation for the coming winter. The hairs of the winter coat are especially dense and coarse and provide the animal with good insulation.

### Reproduction

The Roe does come into season in late July and August, and at this time one may hear the bucks barking, a sharp and abrupt sound not unlike the bark of a terrier dog. The bucks will often be seen to chase individual does, although this is sometimes done in a sort of courtship play without mating following immediately. Sometimes the animals chase in circles around a bush or a stone in a rutting ring in preparation for mating. It is widely believed that there is a second, more subdued rutting period, or "false rut", in September–October. The development of the foetus is delayed and the blastocyst is not usually implanted in the wall of the uterus for about five months, which brings the total gestation period to approximately 290 days. Newborn fawns have been seen from late April to June, and the majority of young Roe are born in mid to late May.

The bucks are sexually mature by the time they are 14 months old, but the older bucks are more active at the beginning of the rutting season, while the younger bucks tend to mate later in the rut. The Roe doe invariably has twin fawns, but triplets are not unheard of, and does have been found dead with as many as five embryos in their uterus. Within an hour of their birth the young Roe fawns (or "kids") will have disappeared to lie down in long vegetation. The does visit them frequently and suckle them regularly.

## Adaptations and behaviour

The Roe deer has good eyesight, its sense of smell is exceptionally well developed, and the large and independently mobile ears provide excellent hearing. All these senses are important in the life of the Roe. The deer communicate among themselves by means of sound, scent markings and behaviour connected with movements and bodily posture. Both the buck and the doe can make sharp barking noises, and the doe can also utter a high-pitched *peeping* call. She will use this sound to call to the buck when she is mating, and it is also used later as a contact call with her young, which can also make a similar high-pitched call. Bucks will stamp their forefeet when they are excited or nervous, and both stamping and barking seem to play a part in the maintenance of the buck's territory.

Roe deer secrete scent from glands near the eyes and between the cleaves of the hoofs. The eye glands are used for territorial marking, when the head is twisted and rubbed against branches, tree trunks and stems. The glands in the hoofs make continuous markings as the deer moves about. Faeces and urine also convey information to other deer and to potential predators.

The white hairs of the Roe's rump patch stand erect when the animal is nervous or frightened. When it runs through high vegetation the Roe will move with leaps and bounds so that it can have a better view of its surroundings. When alarmed the Roe will also bark, but often when it has observed danger it tries to slip quietly away. Often Roe will stand completely still, particularly in thickets and bushy vegetation, cautious and alert, and inquisitive about the disturbance. It will turn its head towards the source of the disturbance and twitch its ears to try and catch any hint of sound.

## Habitat and diet

To the great pleasure of many naturalists and sportsmen, the Roe deer is now widespread and relatively common in many parts of Europe. Its wider distribution encompasses all of Europe and eastwards in a wide belt across Asia (between 45°N and 57°N latitude) as far as Manchuria and China. It also occurs in the south of its range in Korea and the Near East. Roe inhabit flat grasslands, upland moors and arable farmland, in addition to extensive tracts of forest. Often the finest populations of Roe are to be found in mixed habitats of woodland, clearings, farmland and small coverts.

The Roe deer's habitat does not extend to deserts, steppes or other inhospitable terrain. In the Alps they can be found at altitudes up to 2,400 metres. Roe have various main predators, including man.

Roe have a varied diet, which changes with the locality and the season of the year. They eat leaves, buds, young twigs and shoots from low-growing trees and shrubs, and the branches of trees and bushes such as rowan, sallow and juniper. They will graze on grass, clover and various herbaceous plants. They will eat acorns, beech-mast and even mushrooms, and they will also encroach on cereal crops and the plants in nurseries and gardens.

## Population and subspecies

Roe deer can differ in appearance and characteristics from one population to another. While the mature adult Roe deer of western Europe reach body weights of 25–35 kg, some Roe in east Asia attain twice that weight. Distinctive characteristics have separated the Roe deer of Europe (*Capreolus c. capreolus*) from the east Asian Roe (*Capreolus c. bedfordi*), which occurs in eastern China and Korea. In between there is the so-called Siberian Roe (*Capreolus c. sibericus*), which is found from the Moscow area eastwards to western China. What is called the "Giant Roe deer" (*Capreolus c. pygargus*) exists in central Asia, in the Tienshan Mountains.

The various Roe subspecies increase in size and weight the further east one goes. Even in eastern Europe Roe can regularly reach weights of up to 40 kg. But the subspecies are not merely indicated by larger sizes and greater body weights: pelage and antler conformation must also be considered. The giant Roe has antlers that can be as long as 45 cm. In addition, the Asian Roe deer often have four, five or even six points on each antler, like the giant Roe. The shape of the enlarged ring around the base of each antler, called the coronet, is also variable, as are the shape, size and position of the pearling above the coronets.

## Distribution and history in Britain

There are no Roe deer in Ireland, which became isolated as an island before the species moved westwards to colonise it. In the eighteenth century Roe were almost extinct in England, with the only large populations confined to Scotland. Introductions of Roe from France to Sussex and Wessex were highly successful, and Roe have now colonised large tracts of southern and eastern England. Large-scale commercial afforestation with conifers has helped the spread of Roe in many areas, and especially in the north of England and the Scottish borders. Roe have recently spread westwards beyond the Severn and into Wales, and it seems likely that the species will soon colonise suitable habitat throughout Great Britain.

Roe used to be hunted with packs of scenting hounds during the nineteenth century, especially in southern England, but no packs now do so. Prior to the 1940s few Roe were stalked and shot with a rifle for sport in Britain, except by a handful of enthusiasts. Many Roe were shot as vermin during the course of pheasant drives, and much cruelty and wounding was caused by the use of shotguns firing small shot. It was not until the enactment of the 1963 Deer Act that Roe were accorded a measure of close season protection, and rifles of a suitable calibre became mandatory in most cases.

Many British servicemen returned after the war having had experience of Roe stalking in Germany and central Europe, and from the late 1940s onwards the sport of woodland stalking gradually became more popular and widely practised. By the 1970s it had become generally acknowledged that the correct and humane way to control Roe deer populations was by selective shooting with a suitable rifle-cartridge combination.

## Roe stalking

In stalking Roe bucks the antlers are the principal trophy, both in Britain and Europe, and in many European countries there is a considerable amount of ritual attached to the stalking of Roe. Roe buck stalking is popular not only because of the skill required and the trophy antlers; it also enables the stalker to identify and follow individual bucks, to select those which should be culled, and then to pit his wits and his skills in fieldcraft and stalking against the highly developed senses of this wary and elusive quarry. This can be achieved either by sitting and waiting to intercept the buck as it moves around, or by actively stalking it on foot. Success calls for familiarity with the ways of Roe, with the land on which one is stalking, and with the individual habits and movements of the chosen buck. All these factors combine to make it an exciting sport, and the stalker has the additional pleasure of being out and about in the countryside in the peace and quiet of spring and summer, often at dawn and dusk when wildlife of many species is most active and interesting.

It is widely accepted, especially in Europe, that the dominant "master bucks" which have been identified and which hold known territories should not be shot until the time is right. Deer managers and foresters in southern Sweden claim that their carefully selective shooting has enabled them to maintain fine quality animals, with bucks which regularly produce antlers of gold medal quality. (A world record buck was shot in southern Sweden in 1982.) Antlers are assessed on an internationally accepted points system, which takes into account the size, shape and weight of the antlers, together with their colouration, pearling and symmetry, and a gold medal head calls for a minimum of 130 points on this system of assessment.

## Stalking and shooting methods

The methods and tactics used in Europe for shooting Roe deer vary considerably from country to country. However, there are three basic types of tacics: stalking on foot; "still-hunting" — waiting in ambush either on the ground or in a high stand; and moving the deer with dogs.

In continental Europe still-hunting is perhaps the most popular method during the Roe's rutting time in July–August. This gives the sportsman the chance to wait in hiding and to try to entice a suitable Roe buck within shot by imitating the calls of either a doe in season or a young fawn. In Britain most Roe stalkers proceed on foot, stalking in woodland and on adjacent farmland or moorland, and attempt to locate, identify and stalk to within rifle-shot of a suitable animal. Most shots are taken either in a standing or kneeling position, and most stalkers carry a long stalking pole or staff which acts as a monopod to steady the binoculars and the rifle.

Dogs are often used in Roe stalking. Many sportsmen and professional deer managers keep a dog which is capable of following the trail of a wounded deer, and of locating a carcase which may have fallen in dense undergrowth. Many breeds can be trained to do this, and some sportsmen stalk with a trained dog with them, to indicate the presence of deer upwind in cover.

In continental Europe dogs are widely used to locate Roe deer and to move them towards one or more riflemen waiting in ambush along woodland rides or in clearings. Dachshunds, beagles and setters are all popular for this work, and the object of the exercise is to sweep thoroughly through an area of Roe habitat, causing the deer to move steadily but gently ahead of the dogs and their handlers, and making them pass within easy shooting distance of the waiting riflemen. The dogs will detect not only the air scent of the deer drifting on the breeze, but also the scent left by the Roes' hoof glands. This can be an especially effective method of controlling the numbers of Roe does, which are shot in late autumn and winter, and which must be culled if the Roe population is to be kept under control.

## A year with the Roe deer

**Spring.** Roe give birth to their young during late May and early June. Occasionally triplets will be born, but this happens in less than 5% of cases. Twins are usual, in around 80% of cases, and occasionally only one fawn will be born.

By the time the fawns have been born, the mature bucks will have established and marked out their territories. By May most of the older bucks will have cleaned the velvet off their antlers, although some may not be fully free of velvet until late May. In spring the young bucks are driven out of the prime territories by the mature males, and there is usually a large scale movement of displaced young bucks at this time of year.

In good Roe habitat there will be obvious signs of their presence, especially on stems and branches where the bucks have rubbed their antlers, removing the velvet and also leaving scent from their facial glands. Within his territory a mature buck will try to hold a mature doe, and also any young does which he may be able to claim. Yearling deer drift away from their mothers, and it is not difficult for them to live an independent life at this time, when there is an abundance of fresh food available for them. In spring Roe will browse on the tender shoots and buds of various types of vegetation, and they are often to be found feeding in fields of young wheat and barley.

Although the doe normally gives birth to both her fawns at the one spot, it is not long before each youngster rests in its own hiding place, where it will be visited at regular intervals by the doe, so that it can suckle. This dispersal of the young is an effective means of defence from predators, and the young deer gives off little or no scent at this stage. When the fawns are bigger and stronger they rejoin their mother and the family group comes together again. At this time of year the adults undergo the hormonal changes which bring them into the breeding season.

**Summer.** The does come into season in late July and August, and at this time they do not wander far. The mature bucks tend to bark more than usual at this time of year, and the sexually receptive doe utters a high-pitched, plaintive call, which attracts the buck and also maintains contact with her fawns. In courtship the doe often takes the initiative by making close contact with the buck, and it is typical to see the buck chasing the doe, often at speed and in wide circles. These pursuits are often encouraged by the does, and a tired buck who lies down may be teased and even kicked by the doe into further chasing. The reedy, whistling *peep* calls of a doe in season are what many Roe stalkers try to imitate with specially made whistles and calling devices. Successful imitation may bring a buck bounding towards the concealed rifleman.

At this time of year the bucks are fiercely territorial, and if a strange buck enter another's territory a battle may result. It is not unknown

for a Roe buck to maim or kill an opponent in such encounters. Meanwhile, some of the displaced bucks may have paired up with young does and established territories of their own, but they carefully avoid the territories of the mature bucks.

Various studies have been made, to determine the average size of Roe territories, and this depends upon the quality of the habitat, the density of the population, the availability of food and other factors. A mature buck can have a territory of 7–8 hectares, while a number of does' home ranges may overlap parts of this area, and two or three does may range across the territory of one buck. Likewise, during her movements, a doe may wander through the territories of two or three different bucks. At one time it was thought that Roe were monogamous and that a buck mated with only one doe; but it is now understood that they are polygamous, and that a buck will mate with does whose movements bring them within his territory. Roe does are capable of breeding at 13–14 months old, and in ideal conditions mating will take place at this early age, which means that some does will produce their first fawns in their third summer, i.e. at two years old.

**Autumn.** During autumn the year's fawns remain with their mothers, who will sometimes continue to suckle them until well into the winter. The bucks can often be seen in the vicinity of the doe and her young, but their lives after the rut are rather more independent. There is no longer the same urgency about maintaining a territory, and territorial behaviour is therefore not so prominent; but the bucks tend usually to remain within the general area of the territories they held with such determination earlier. In autumn the Roe's diet consists of browsed material including twigs, shoots and leaves, but there may be ripe grain available in the harvest fields, and Roe will also eat beechmast and acorns.

Studies of Roe in Denmark have shown how Roe populations are made up at this time of year. It was estimated that young fawns accounted for 40% of the total numbers of Roe, with yearlings from the previous year accounting for a further 20%. A further 15% was composed of youngish deer between 2 and 3 years old. Of the 231 individual deer studied, only 0.9% of the population consisted of bucks aged 6 and 7 years, and it was concluded that bucks tend to have shorter lives than does, perhaps because their behaviour, combined with their desirability as sporting quarry, renders them more vulnerable, both during the rut and at other times of the year. This study indicated that mature bucks become fewer and fewer as autumn progresses.

**Winter.** By winter Roe deer numbers have been considerably reduced, owing to a combination of factors. The shooting of bucks during the period from April to October will have had an effect, as will deaths from natural causes among the older deer. Mortality among fawns in their first few months of life can be high, and winter mortality is often quite high among all age groups. In northern Europe it has been estimated that 60% of all Roe deer found dead in winter had died of starvation, and yet many individuals had their stomachs full of undigested fibrous plant material. It is thought that severe winter conditions can force Roe to eat woody twigs and other items which are too coarse for them to digest, as their digestive systems are not suited to breaking down very rough forms of cellulose.

In countries which suffer severe winter weather and have heavy snow cover, supplementary feeding is often used as a means of minimising the rate of winter deaths. In northern Europe and Scandinavia success has been achieved by putting out quantities of cut tops of birch, rowan and aspen, together with clover and hay. Roe will also eat animal feed pellets, cattle cake, silage, surplus root vegetables, apples and similar material.

Roe do not normally live in herds, and are usually found in small family groups, but severe winter weather may cause a number of such groups to band together and temporary winter herds of 10–30 animals are not uncommon. Where artificial feeding points have been established and are used regularly by the deer, it is not unusual to see these numbers of Roe gathered together to feed. Studies of this behaviour have revealed that there is a hierarchy or pecking order among Roe under such conditions, and the least dominant individuals seem to be those which are furthest away from their summer and autumn territories. Furthermore, the larger animals will tend to take precedence, and an antlered buck will take temporary priority over one which has already shed its antlers.

## Roe deer glossary

Buck: a male Roe deer.

Coronet: the bumpy and irregular ridge or ring at the base of each antler.

Doe: female Roe deer.

Fawn: a young Roe deer.

Going back: used to describe an old animal, especially a buck whose antler conformation is deteriorating with old age.

Kid: an alternative for Fawn.

Pedicle: the bony excrescence of the buck's skull on which the antler develops.

Rump patch: the large white area on the Roe's rump in winter, sometimes referred to as the "target".

Rutting ring: a circle of up to 3–4 metres in diameter, often around a bush or a rock, in which the buck chases the doe during the rut.

Six-pointer: a buck with three points on each antler, the usual maximum for European Roe.

Velvet: the membrane of skin and short hair which covers the growing antlers.

# Wild boar

## *Sus scrofa*

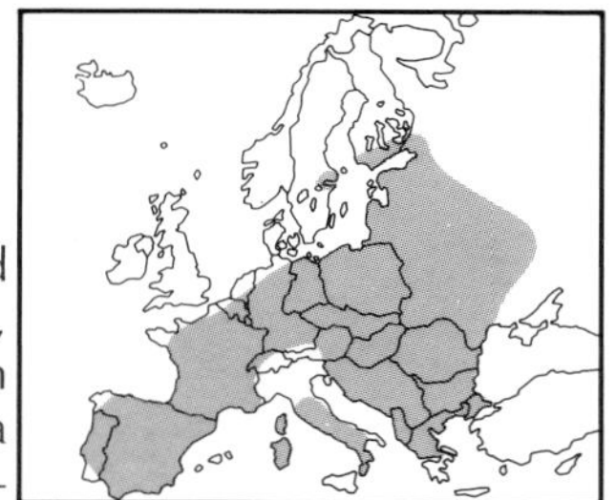

The European wild pig is usually referred to, regardless of its sex, as the "wild boar", and like several other species of animals in this section it is primarily regarded as a forest animal, principally mature deciduous woodlands. However, many of these species are also found on the fringes of cultivated agricultural land, to which they are attracted by the food available there. This is especially true in the case of the wild boar.

In much of Europe wild boar have been persecuted because of the damage they can do to growing crops, and also because at rutting time they are sometimes attracted to domestic pigs. In Scandinavia it is believed that the wild boar had been exterminated before the Middle Ages, and in Britain and Ireland the last wild boar were killed in the 17th century. In Denmark they were wiped out about a century later.

However, the essential nucleus of the continental European population survived, mainly in the large tracts of forest in central and northern Europe. In recent times attempts have been made to reintroduce wild boar to some of their former haunts, and there is now a small population of wild boar in southern Sweden. Boar have also been reintroduced to the Netherlands, and in the 18th century there was a proposal to reintroduce wild boar to Britain, but this was successfully opposed by the farming community. There are now no wild boar in the wild anywhere in the British Isles.

Since the Second World War the wild populations in much of Europe have increased in numbers, and the wild boar is now a highly esteemed sporting quarry. In West Germany alone an estimated 30,000 wild boar are shot annually, and there are large populations in the forests of France, where they are hunted with hounds and mounted followers in certain areas, and large numbers are also shot, both by sportsmen and to minimise damage to crops and vineyards. Around the Baltic the numbers of wild boar have increased steadily since the 1960s, and there has been a natural expansion of the species' range, which has included the recolonisation of parts of Finland.

### Characteristics

The wild boar is similar to the domestic forms of pig, but it has longer legs and a proportionately shorter body. The head is also disproportionately large. The skin is dark and covered with greyish and brownish-black shaggy hair. There is a long ridge of dense bristles which runs along the ridge of the animal's spine, and these bristles can measure up to 15–20 cm long. The underfur is greyish-brown. The tail, which is up to 28 cm long and tufted at the end, is held stiffly upright when the animal is frightened or taking flight. Unlike the domestic pig, the wild boar's ears are pointed and carried upright.

The mature males or tuskers are the largest, and can weigh up to 230 kg, and there are some records of individual animals weighing as much as 350 kg. By contrast, the female weighs less than 100 kg. A large boar will stand one metre high at the shoulder and the body of a large male will measure 1.85 metres long. The largest wild boar in Europe are thought to be in the Carpathian mountains.

The canine teeth form long tusks, which twist outwards, and this is especially pronounced in the lower jaws of old males, whose canine

teeth grow continuously during their lives. The molar teeth are irregular and jagged, and the pattern of the 44 teeth is:

$$\frac{3 \quad 1 \quad 4 \quad 3}{3 \quad 1 \quad 4 \quad 3}$$

The young piglets are lighter than the adults, and they have a dark longitudinal stripe along the middle of the back, and there are a number of other longitudinal stripes along the animal's flanks, three of which are yellowish-white in colour and four are brown. These bands often break up into rows of pale flecks on the animal's hind quarters. The young retain this patterning of pelage until they are about 10–11 months old.

The wild boar is a hoofed mammal, and each foot has two true hooves upon which its weight is carried, and there are two smaller vestigial hooves positioned higher up the leg. The wild boar therefore belongs to the group of even-toed ungulate mammals. The distance between the footprints of a walking animal is about 45 cm.

## Reproduction

Males and females live relatively separate lives during most of the year. However, during the mating season, which is normally between October and April, the males join up with groups of females and their followers, and there are violent clashes between sexually mature rival males. When fighting, wild boars usually direct their blows at the front flanks of the opposing animal, and during the mating season rivals will inflict substantial injuries on one another with their tusks. A successful dominant male can find he is able to mate with a succession of females, perhaps as many as ten in one season. The young females become sexually mature about six months earlier than the males, and some will mate even before their youthful stripes and bands have disappeared, to be replaced with the adult coat. Other females will reach maturity at the age of 1½ years, which is when the young males usually become sexually mature, but the young males will not be capable of competing with older males for females until they are at least one year older.

The gestation period is 115–120 days, and the young are born between February and July, and sometimes later because of a delayed rut. The sow builds a simple nest for her young, and it is unusual for a litter to exceed 12 in number, which seems well suited to the female's six pairs of nipples. However, most litters are much smaller than this. At birth the piglets weigh about 1 kg and by December–January they will weigh 15–20 kg. The eyes of the young are open within 24 hours of birth, and the piglets remain in or close to their nest for the first 2–3 weeks of their life. Young females have a longer gestation period and give birth to smaller litters of young than the mature sows.

The young are small by comparison with their mother, and they are left alone in their nests, which are usually made in a hollow and lined with twigs, leaves and grasses. There they often fall victim to predators. Cannibalism is not uncommon, and a fox can take a very young piglet. Where they exist, wolves are significant predators, and studies of wild boar in Poland have shown that 20%–34% of young wild boar die within three months of birth.

Weather conditions in winter, and access to food, are thought to be the principal factors which regulate wild boar populations. During the winter the amount of snow cover can make it difficult for the animals to reach their food. In addition, fluctuations in the abundance of acorns are believed to be able to affect breeding success by up to 100%. The timing of the rut is also thought to be a significant factor, and a succession of good acorn years in Germany in the 1960s is believed to have contributed to a population explosion among the wild boar.

The young stay with their mother until the next mating period, when they are driven away by the male; but they are reunited with their mother after mating. However, it is not long before the young males leave the family group.

## Adaptations and way of life

Like many other hoofed animals, the wild boar is a herd animal. Among wild boars, the herds are chiefly composed of mother sows with their litters of young and some young females. The old boars are normally solitary, and only join the herds at mating time. At this time, between October and April, the closely knit herds are dissolved. The yearling males leave their mothers, although the females often remain. The herds are soon reconstituted, and the young males form their own groupings.

As is the case with most mammals, it is the males who tend to undertake any long distance movements. However, wild boar herds and groups tend to wander within an established home range which may be as large as 10–20 km$^2$. When a herd of females becomes too large it splits up into smaller groups which seldom wander far from the original home range. Studies of wild boar populations have revealed that two-thirds of the animals never wander further than 5 km away from the home range in which they were born and reared, and only 10% of wild boar travelled as far as 10 km away.

Wild boar are not very particular about their food; they are omnivorous. Perhaps their favourite foods are acorns and beechmast, and a variety of mushrooms, fungi and roots, but they will eat all kinds of vegetable matter. They often raid fields of root crops and cereals, and also vineyards, where they can do devastating damage. Even though they are predominantly vegetarian, they do not ignore carrion and carcases. They will root out nests of mice and young rabbits, and will root for insects, worms and small animals with their pointed snouts and powerful tusks, which they use to poke and grub in the earth. Defenders of the species have suggested that they can be helpful to forestry by "cultivating" and aerating the soil, while at the same time eating the larvae of insects which can be harmful to growing trees.

The wild boar is to be found in various kinds of woodlands, among conifers, deciduous trees and in mixed forests, but they prefer woods which are adjacent to arable farmland. They favour dense young forests, and woods of oak and beech interspersed with dense stands of conifers. They like damp and marshy places, and in the daytime will sometimes lie up in beds of reeds and osiers. At night it will visit cultivated fields in search of food. In some areas the wild boar lives in steppe country where there is occasional cover in the form of dense shrubby vegetation, and in other areas such as the Tienshan mountains it occurs at altitudes up to 3,000 metres. It is found in the Alps at heights up to 1,700 metres.

The wild boar is widely distributed across mainland Europe, and it also occurs in north Africa and across the greater part of Asia as far as the Pacific coast and the islands of Japan and Taiwan. It is found in southern Asia, Sri Lanka, Java and Sumatra. It has also been imported into Australia, New Zealand, the United States and Argentina.

Their elusive habits and secluded way of life make it difficult to estimate the present population of wild boar. In 1968 it was estimated that the spring population of wild boar in West Germany was 45,000 animals. However, during the ensuing hunting season (1968–69) approximately 24,000 wild boar were shot. Between the 1930s and the present the wild boar population of West Germany has greatly increased, and the numbers shot have almost tripled. In many other European countries the wild boar populations have similarly increased and spread. As a consequence, previously isolated populations of wild boar have merged and their general distribution has become more extensive. This is most obvious in the eastern Baltic area, from which the wild boar have spread north-eastwards into Finland, which now has a flourishing population. Individuals also occur in central Finland, and the distribution is believed to be expanding by about 50 km per year. Wild boar have also spread southwards into Austria, Hungary and the Balkans, where they had earlier been exterminated.

*Sucking-pigs.*

# Woodpigeon

## *Columba palumbus*

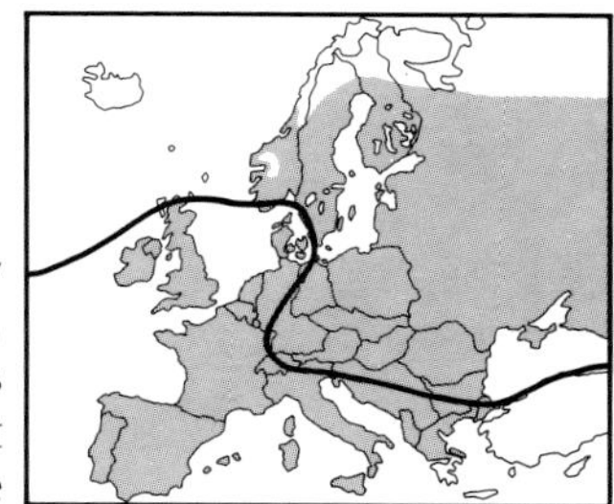

In some of the nineteenth century books about European birds and wildlife, we are told that the woodpigeon lives "far from human habitation" and that one of the birds' favourite foods is "the seeds of fir and pine trees". This paints a very different picture compared with what we know of the woodpigeon of the 1990s, when the species is a familiar inhabitant of city parks and suburban gardens, as well as being one of the most abundant birds on farmland. Nor do we tend to think of woodpigeons eating the seeds of coniferous trees.

It may be that the woodpigeon has changed its behaviour and feeding patterns during the last century and more. Other common birds have changed their ways, including the blackbird and the fieldfare; and several gull species have become much commoner and are now frequently to be seen far inland and feeding on such places as rubbish dumps.

There are many possible reasons why various animal and bird species choose to live around towns and cities, and in other environments where there is plenty of human activity. Some simply remain in their familiar environments and adjust to the new conditions as towns and cities are built around them. Others immigrate, and species which do so have apparently discovered some advantages in the urban and suburban environments which encourage them to stay and to increase in numbers. Blackbirds, for example, find winter conditions very much more congenial in suburban gardens than in the open countryside.

Woodpigeons in and around towns and cities will often breed prolifically and form more or less sedentary populations. Similarly the huge numbers of woodpigeons which live and breed in the British countryside tend not to migrate far, moving locally in flocks depending on the available feeding and the prevailing weather conditions. Woodpigeons from Scandinavia and northern Europe, however, are very migratory, and many of them migrate southwards in autumn to over-winter in southern France or across the Pyrenees in Spain and even as far south as northern Africa.

### Characteristics

The woodpigeon is the largest member of the family of doves and pigeons to be found in Europe. A fully grown woodpigeon will weigh between 420 and 615 grams. Adults are estimated to weigh about 14%–17% more than recently fledged and juvenile birds. An unmistakeable identifying characteristic for the woodpigeon when it is in flight is the prominent white flash on each wing. The adult birds have a speckling of white spots on the throat, but this is absent in birds younger than 3–4 months, and they also lack the iridescent metallic lustre on the sides of the throat. The young birds resemble the adults in their first winter, although the colour of the legs is paler, and there may be some tell-tale yellowish-brown edging to the feathers. The young woodpigeon has a yellowish-white iris to its eye, while the adult bird's eye is a bright yellow and its legs are a deep purplish-red.

*Old Stock dove (above), old (left) and young woodpigeons.*

The sounds made by the woodpigeon include a clattering of wings as the bird takes flight, and this also occurs in the territorial and courtship flights of the males. The dull, gentle coo-ing of the woodpigeon is a familiar sound, and chicks in the nest call with a high-pitched *peep*-ing sound.

In flight the silhouette of the woodpigeon is similar to that of the domesticated or loft pigeon, but larger and with a considerably longer tail. When perching the woodpigeon holds itself erect with its head held high.

## Adaptations and behaviour

Like other pigeons and doves, the woodpigeon has a tightly feathered and smooth plumage, which has an abundant secretion of talc-like powdering. The first moult takes place when the young birds are 5–7 weeks old, and this begins with the innermost primaries. By the time it is eight months old the young pigeon will have its full adult plumage, although some young birds stop moulting in the winter and begin again in the spring.

The adult birds moult their feathers in a systematic order. The moulting season begins in early summer and is complete by October. The first feathers to be moulted are the innermost of the bird's ten primary feathers, followed consecutively by the other primaries. The first wing covert feathers are shed after the first four primaries have gone, and the coverts are shed from two starting points, the first and the fifth covert. The twelve tail feathers are shed in a specific order, and symmetrically, beginning with the two central and longest tail feathers. The systematic and symmetrical pattern of moulting gives the bird generally good flying ability throughout the moult, and it is common in many other species. By contrast, most ducks tend to shed all their feathers at one time and then spend a period of flightlessness in eclipse plumage before they resume their new plumage.

It is thought that woodpigeons have the same colour perceptions as man. Their eyesight is very keen, as also is their sense of hearing. They are especially quick to detect movement, although they will often fail to notice a bird of prey which perches motionless or a human figure which stands still. Their white wing flashes are an important means of signalling to one another, and play an essential role in maintaining the coherence of the flock. Non-breeding birds will often form large flocks in summer.

The male woodpigeon performs a display flight, which often takes place at some distance from the nesting site, and this consists of a curving, upward flight with sweeping wingbeats, followed by a gentle downward gliding flight on set wings.

British woodpigeons, as we have noted, are mainly sedentary, but those from northern parts of Europe are migratory. Woodpigeons which have spent the summer in Scandinavia and northern Europe migrate due southwards across the Pyrenees, where they winter in south-west Europe. The Pyrenean migration involves millions of birds and can provide quite a spectacle, as flocks of migrating pigeon converge on high passes through the mountains, and also along the coast. This passage reaches its height in mid-October and large numbers of woodpigeons are shot by French sportsmen as they fly across south-west France.

## Reproduction

The woodpigeon is monogamous, and the birds remain paired for the duration of the breeding season, and occasionally for two years or more. The woodpigeon is able to breed in its second year, and in some parts of western Europe they will breed almost all the year round. Single pairs have been observed to rear three or four broods of young in succession, but this is usually among the sedentary populations of the British Isles and western Europe. The woodpigeons of northern Europe tend to have fewer broods of young.

The male woodpigeon selects a breeding territory and attracts a female by his display flights. The nest is a simple platform of loosely woven twigs placed on horizontal branches or in the fork of a tree. In areas of mixed woodland they will choose to nest in conifers, but will also nest in deciduous woods. Woodpigeons will also nest occasionally on cliffs, in bushes on treeless islands, and occasionally on the ground. However, most nests are placed about 3–4 metres up in a tree, and they are built by both partners.

Nest-building may take a few days, and the eggs will be laid over a period of 2–3 days. Thereafter, incubation takes 16–17 days, and both partners share the incubation, with the female incubating mainly at night and in the morning, while the male tends to incubate during the main part of the day. The young pigeons are ready to leave the nest by the time they are about a month old. It therefore takes 7–8 weeks for one brood to be produced. The young birds may be fed by both their parents for some weeks after they have flown the nest.

Woodpigeon eggs weigh about 18 grams and the average weight of a newly hatched chick will be about 14 grams. The newly hatched youngster is almost naked and has only a few sparse, yellowish wisps of down. The bill is unusually long and prominent, and the young bird's eyes are closed for 4–6 days after they hatch. When the chicks are fully grown they will weigh at least ten times their weight at hatching.

## Occurrence, environment and distribution

Woodpigeons live in practically all types of open woodland, in cultivated countryside and in urban and suburban environments. However, they do show a predeliction for nesting in coniferous trees, and for choosing nesting trees along the edge of a wood. It tends to avoid nesting in the middle of dense woods, but will frequent open areas, clearings and glades. It will fly long distances from the nest site to find food, and for this reason it is not uncommon to find breeding and non-breeding pigeons feeding together where there is suitable food. In winter pigeons like to live and feed close to oak and beech woods, where they will feed eagerly on acorns and beechmast, and they will also feed in flocks on open fields of clover, rape and other crops.

The woodpigeon is found chiefly in mainland Europe and eastwards as far as the Ural mountains and northern India. It nests as far north as 60°N latitude in Fennoscandia and Russia, and it breeds in the Orkney Islands and in the islands off the west coast of Scotland and Ireland. It will also breed as far south as the northern coasts of the Mediterranean, and it is to be found in Asia Minor, Morocco, Algeria, Tunisia and in the islands of Madeira and the Azores.

During the last hundred years the woodpigeon has expanded its range northwards, in addition to moving into environments close to man. It is interesting to note that while woodpigeons in the open countryside are usually elusive and wary, those which breed in suburbia and in city parks appear to lose much of their fear of human disturbance.

A number of different subspecies of the woodpigeon have been described, and five of these are recognised. The woodpigeon which occurs in Europe is considered the nominate form — *Columba palumbus palumbus*. The woodpigeons of the Atlantic islands are known as *C. p. azorica* and *C. p. maderensis*, and these are darker

birds with smaller spots on the throat. However, in the south-eastern part of the species range the woodpigeons are lighter in colour, and are considered to form two separate subspecies — *C. p. iranica*, found principally in Iran, and *C. p. casiotis*, found in the south-western USSR, Afghanistan and north-western India.

## Numbers and population structure

The European population of woodpigeons is numbered in tens of millions of breeding pairs, and the pair density varies with the type of habitat available. In Scandinavian forests 3–6 pairs per $km^2$ have been estimated, while in central Europe an average pair density of 0.5–1.5 breeding pairs per $km^2$ is reckoned. Local breeding densities can be very high, and this includes the parkland areas of towns and cities. An example of this is that up to 190 breeding pairs of woodpigeons per $km^2$ have been counted in Berlin.

The mortality rate is high, and among birds in their first year is estimated at 55%–70% Mortality among adults has been put at around 40%, and shooting is a significant cause of woodpigeon mortality. In exceptional cases wild woodpigeons can live for up to 15–16 years.

Apart from the large numbers of woodpigeons shot on migration in the Pyrenees, many are also shot in Denmark and on islands off the Danish coast, where an annual bag of 300–350,000 birds are shot. This represents 7–8 pigeons per $km^2$ of the entire country. About 90,000 woodpigeons are shot annually in Norway and slightly more in Sweden, but the British Isles accounts for by far the largest numbers shot, which have been estimated at one million birds annually.

## Feeding and food requirements

The pigeon takes advantage of foods which are common and easy to find, and there is a definite connection between the birds' nutritional requirements, the feeding time available to them, and the type of food. Pigeons will feed on grain, seeds, acorns, green vegetable matter, berries and insects. What they eat is usually dictated by what is available and accessible.

Pigeons will feed in sizeable flocks where there is rich feeding to be had, and this includes cereal fields, vineyards and fruit nurseries, and oak and beech woods in autumn and early winter. The flock behaviour of the birds helps them to know where food is to be had, and small flocks of pigeons will also reconnoitre areas in search of new sources of food.

It is unusual to find more than one type of food in an individual pigeon's crop, for the birds tend to feed exclusively on one type of food at a time, for this is more economical and energy-efficient. Only when food becomes scarce, and the amount of time required to search for it becomes too long in relation to the nutritional value, does the woodpigeon begin to feed on a variety of foods.

During early spring the birds' diet is dominated by tender, green shoots and other vegetable matter. In April and May the birds will feed on spring-sown grain and other seeds, and on green vegetation until the cereal harvest in July–August. Acorns, beechmast and left-over grain predominate in the autumn.

The woodpigeon's bill is a tool for selecting and collecting. Pigeons swallow their food directly and it is then processed in the bird's long digestive tract. Through the gullet the food passes into the digestive system, and the bird's crop is the first stage in the process of breaking down what has been eaten. It is in the crop that the cheese-like "pigeon's milk" is produced, a partially pre-digested source of food with which the adult birds feed their young by regurgitating it.

The pigeon's gizzard is where tough and fibrous material is ground up and macerated into a more easily digested pulp, often with the help of small stones and gravel in addition to the rough and muscular surface of the gizzard membrane. It has been discovered that pigeons which are feeding on grain have relatively little need of stones and gravel compared with those which eat acorns and berries, while pigeons feeding on green plant leaves, such as oilseed rape, have most stones of all.

Non-breeding pigeons search for food in the mornings and the early afternoon, retiring to their roosts with crops full of food. The crop then acts as a storage place for food which is gradually digested while the birds rest at night, which is a practical and efficient way of maintaining the supply of nutrients to the bird's system. The crop can contain a maximum of 80 grams of food, and individual woodpigeons have been found with between 30 and 40 whole acorns in their crops.

The woodpigeon has a rapid metabolic rate, like many other species of birds, and it therefore requires large daily quantities of food. A woodpigeon consumes an average of 20% of its own body weight in food each day, and a bird with a bursting crop is not as strong or agile a flier as one with an empty crop. It has been noted that migrating woodpigeons have empty stomachs while they are on the wing, feeding only when they stop which is usually in the mid-afternoon.

## Closely-related species

The woodpigeon's closest European relatives are the Stock dove (*Columba oenas*) and the Rock dove (*Columba livia*). Both are greyish, but they lack the white throat spots and the white wing flashes of the woodpigeon. Instead they have double, black wing-bands, which are especially prominent in the case of the Rock dove, which also has a most prominent white upper rump, while that of the woodpigeon and the Stock dove are grey.

Beside these two species of pigeons, there are three species of doves which occur in Europe — the Turtle dove (*Streptopelia turtur*), the Oriental dove (*Streptopelia orientalis*), and the Collared dove (*Streptopelia decaocto*). The Rock pigeon and the Turtle dove are medium-sized birds, and both smaller and sleeker than the woodpigeon. Their body weights are in the range 250–300 grams. The Rock dove exists mainly around the Mediterranean, and is found in good numbers along rocky coastlines in Britain and Ireland. Feral pigeons, the descendants of domesticated Rock doves, are commonly seen in the squares and parks of many of Europe's largest cities.

The Stock dove occurs within much the same range as the woodpigeon, but its distribution does not extend as far north as its larger cousin, nor does it occur in Portugal, Greece, much of Italy and north-west Africa, where the woodpigeon is present. Otherwise the Turtle dove exists all over Europe south of the Baltic Sea.

Since the 1930s, when its European range extended only to a few parts of the Balkans and Turkey, the Collared dove has spread with amazing rapidity over large parts of Europe. The first breeding in Britain was confirmed in 1955, and by 1972 it was estimated that the British breeding population was 30–40,000 breeding pairs.

*Collared dove and Turtle dove (below).*

# Passerines

It is not uncommon to find references in wildlife and sporting books of the nineteenth century to the large numbers of northern breeding birds which were shot, netted and trapped in their southern winter quarters and along their migration routes, especially in France and Italy. Often the birds were handled in commercial quantities, to be plucked, dressed and cooked for export to their countries as delicacies. The Scandinavian writer Charles Emil Haghdal, in his cookery book published in 1879, remarked that "in this way we have the pleasure of seeing these birds return again rather sooner than we might have expected; and they even have the courtesy to be cooked when they come".

This sounds cynical, but at that date the shooting and trapping of countless small birds was commonplace and legal throughout most of Europe. Even today, over a century later, many hundreds of thousands of small passerines and songbirds are shot and netted in certain parts of Europe, especially in France, Italy and Cyprus.

The idea of hunting small birds as delicacies for the table is nothing new, and there are records of this from ancient times. At various times in history there has been considerable culinary interest in starlings, blackbirds, fieldfares, thrushes, waxwings, cross-bills, bullfinches, wheatears and ortolans. The waxwing was included in this list because it sometimes occurred in very large flocks, was easy to net in very big numbers, and in autumn the birds were very fat and in prime condition for the table. Waxwings were netted and then sold in bunches, for one bird weighing barely 50 grams could scarcely be sold on its own.

Many small birds were caught in snares made of running nooses of horsehair and placed in bushes, and the tenaciously sticky substance known as bird-lime was also widely used. September was a favourite time to trap and net migrants, and this was often done in areas which had plenty of berry-laden bushes and shrubs to which the birds were attracted. To take as an example only one species, at the end of the last century it was estimated that about 1.2 million fieldfares were caught by commerical netters and trappers each autumn. Despite this huge annual toll, it is interesting and encouraging to note that the fieldfare has increased considerably in numbers and has extended its range over much of Europe.

Since the 1970s there has been widespread and growing criticism of the practice of shooting and trapping migrant passerines and songbirds, and this trend has been given added force by important documents such as the Directive on Wild Birds issued by the E.E.C. in 1979.

*Blackbird*

# Crows

Of all the various members of the crow family, one species is especially controversial — the "true crow", otherwise known as the European Carrion crow or the Hooded crow. Some commentators have actually deemed this bird fit for human consumption! But crows around towns and cities frequent rubbish dumps and keep company with rats and gulls, and it is widely thought that they can spread salmonella. Are crows merely an unhygienic nuisance or do they perform a useful task by cleaning up waste and carrion? Are they a harmless species, or are they destructive robbers of other birds' nests and predators on sheep and lambs? The questions are numerous, and opinions are often sharply divided.

Anyone who visits the haunts of nesting ducks, gulls or wading birds in spring has probably seen crows being mobbed by various territorial breeding birds. This often occurs where crows are guilty of taking other birds' eggs, and there are many instances where the numbers of broken and sucked-out egg shells can be very large. This naturally gives the impression that the crows have caused catastrophic losses among the other species, but in a good many cases the birds whose eggs and young fall victim to crows can compensate for the losses and sustain generally good breeding success.

Historically, crows have often been regarded as birds which survive and thrive on mankind's debris and refuse — even to the extent of feeding on the unburied corpses on battlefields. Periodically, naturalists and sportsmen have tried to initiate campaigns to exterminate crow populations, but have such measures really been beneficial in increas-

*Crow,* Corvus corone cornix.

ing the numbers of small birds and gamebirds? The evidence from scientists and ecologists is conflicting. In one well-known study, the predation of willow grouse and black grouse by crows was investigated on a Norwegian island. In one half of the island the crows were exterminated, while in the other sector they were left undisturbed. The unexpected result was that the grouse in both areas had the same breeding success, for it was found that other predatory birds, especially large gulls, moved into the area which had been cleared of crows, and took a toll of eggs and young chicks. On the other hand, recent studies in the south of England have shown that wild game and most species of small birds tend to fare much better in areas where crows and other corvids are strictly controlled, especially in the spring.

But local conditions can clearly affect the impact of crows on other species. In southern Sweden it was observed that crows took a high proportion of eggs from the first clutches laid by pheasants. Nevertheless (and most surprisingly) this seemed not only to have no damaging effect on pheasant numbers but to be positively beneficial to the pheasants! It was eventually discovered that the non-native, introduced pheasants had not adjusted properly to the breeding conditions in this area of Sweden, and had a tendency to lay their eggs too early, before the vegetation had grown long enough to provide adequate cover for the nests. Second clutches fared much better, for there was ample cover to conceal the eggs and young chicks from the sight of crows and other predators, and also the later hatched chicks had a greater abundance of insects and other protein-rich foods to promote their growth and survival. Second clutches were found to have a much more successful rate of hatching and fledging than first clutches.

Another Scandinavian study showed that the rigorous control of corvids in an area of woodland resulted in greatly increased numbers of pigeons, thrushes and blackbirds. However, it was also noted that the numbers of robins declined markedly, and it appeared that the robins, which had been unaffected by the crows' depredations, were being pushed out by unfair competition from the increased numbers of thrushes. Meanwhile, experience in Britain with game management on farmland and on upland moors tends to indicate that year-round control of crows and magpies is an important element in promoting the breeding success of gamebirds. All of this tends to emphasise the need for full and careful study of crows and their effects on other species which share the same environment, especially with regard to local conditions.

*Jay*

## Jay — *Garrulus glandarius*

The jay is perhaps the most attractive looking of the crow family, with a handsome cinnamon colour and a vivid flash of blue in each wing. This tends to be a woodland bird, and it used to be heavily persecuted by gamekeepers as a suspected egg thief and predator of young gamebirds. Today it is not regarded as a significant predator of game or other birds.

It is interesting to note that northern European sportsmen of the nineteenth century regarded the jay as good to eat, and recipes for cooking it can still be found in old sporting books. Some also give detailed descriptions of the best ways of attracting jays and catching them with snares or limed twigs. Like most corvids, jays can be decoyed to a dummy bird or a stuffed specimen, and they can also be called by a successful imitation of their calls. Another old method of decoying jays within netting or shooting distance was to place a tame owl or a carved imitation one in a locality which jays were known to frequent, for their natural instincts are to mob the owl. Some senior sportsmen in Scandinavia and northern Europe still occasionally shoot jays for the table.

*Magpie*

## Magpie — *Pica pica*

The magpie is a handsome and attractively marked member of the crow family, but it is an inveterate thief of other birds' eggs and chicks, and careful studies of this species have confirmed its reputation as a destructive predator of gamebirds and songbirds alike. It can be trapped in a cage-trap, using another captive bird to attract it, and where magpies numbers are effectively controlled there is invariably a marked increase in the abundance of other birds, especially songbirds and gamebirds.

The magpie is a focus for a great deal of country lore, superstition and legend. In Britain we are familiar with the rhyme, "One for sorrow, two for joy . . .", and in the Nordic countries they are associated with witches. Magpies tend to be sedentary birds, although they will make short local migrations in winter, and may flock together in quite large numbers at communal roosts. In Britain magpies have become increasingly common on farmland and in suburbia, where they have caused significant local declines in the numbers of passerines and songbirds.

## Jackdaw — *Corvus monedula*

The jackdaw is one of the smallest of the all-black group of crows. It weighs around 200 grams, and its diet consists primarily of worms, snails, insects, cereal grains and weed seeds. Folklore throughout Europe is full of references to the jackdaw, and it is a bird which has always had a close association with human dwellings and settlements. They are fond of nesting on and around prominent buildings, and many a chimney has been blocked by a pair of jackdaws building a nest in it. Jackdaws are generally regarded as friendly, innocuous birds, and in folklore they are often used as examples of marital fidelity because of their tendency to form close pair bonds.

## Rook — *Corvus frugilegus*

The rook is more inclined to form large flocks than other European corvids. It nests high in the treetops, and prefers to nest in colonies where a number of suitable nesting trees grow close together, preferably not far from open farmland. When searching for food, as well as during their migrations, the rooks gather together in large flocks on fields and pastures. They are generally sedentary in those parts of Europe which have a mild climate in winter, such as the British Isles, but the majority of rooks from the Scandinavian countries move southwards at least as far as southern Sweden and Denmark.

In the British Isles rooks are common and widespread, especially where there is a mixture of small woodlands and open farmland. It tends to avoid high moorland and also densely afforested areas. The rooks' favourite nesting trees have always been tall elms, but the ravages of Dutch elm disease have eliminated these, although the effect of this on the rook population has not been fully assessed. In the 1980s the British and Irish population of rooks was estimated at around 1½ million breeding pairs.

## Raven — *Corvus corax*

The raven is the largest European representative of the crow family, a large and impressive looking bird with a deep and resonant croak and a surprisingly wide repertoire of vocal calls. Although it is widespread in most of Europe, the raven is probably more familiar to most people by its name and reputation, rather than its appearance.

From ancient times the raven has been regarded as a bird of mystery and omens, a familiar sight on battlefields, at places of sacrifice and anywhere that carrion flesh can be found. Deer stalkers, especially on open moorland, regard it as a good omen to see a raven when they set out after deer, for ravens are fond of eating the gralloch which is left behind when a deer is gutted, and to see a raven may therefore betoken a successful day's stalking.

# Woodland and forest animals

The land area of the European continent amounts to about 11 million km$^2$, and of this approximately 32% — almost one third — is woodland and forest. Forests and woodlands can be of many types. In lowland areas we can find carefully managed woodlands, where man's activities have divided the landscape into a mosaic of woods and fields of arable crops and pastures, with the woodland areas laid out with distinct boundaries and internal rides, clearings and roadways. There are also mountain forests, and forests which have retained their wild and primitive character, virtually untouched by man's influence. Some of the original wild forest of Europe, such as the Bielawieza Forest on the borders of Poland and Russia, have been set aside as protected areas of great environmental and ecological importance. This particular forest was established in the Middle Ages, and is in its turn the product of an even more ancient forest. The same can also be said of the Abruzzi Forest Reserve in the Dinarian mountains of Italy.

The Bielawieza forest has maintained the last surviving European bison, while the Abruzzi reserve has likewise been a haven for the European brown bear.

In northern Europe there is a long tradition of planned forestry management, and at the end of the nineteenth century an important Swedish government report emphasised the need to conduct forestry operations in accordance with the natural laws of forest ecology. This was an enlightened and far-sighted attitude. About half the land area of Sweden is afforested, and about one tenth of this consists of wild indigenous forest in mountainous and remote areas. Finland is the most heavily afforested country in Europe, and almost three-quarters of its land surface consists of forests. Recent experience has shown that modern mechanised forestry has a crucial impact on the forests and on the populations of wildlife which live there. In addition, the recent phenomenon of acid rain has made it clear that forestry can be dramatically affected by man's industrial activities which take place far away from the forests.

The clear-felling of forest areas has been of direct benefit to some species, such as the Scandinavian moose, which has thereby been given access to a new food supply in the form of young leafy trees which emerge as secondary growth. Many species of woodland birds have benefited too, for they have access to a newly abundant variety of insect foods, which are important for the survival of their young. However, tree felling can obviously have adverse effects upon certain other species, including woodpeckers, some species of owl, and certain birds of prey.

The forests which are most attractive to wildlife tend to be those which are composed of a mixture of tree and shrub species, with a good understorey of shrubby growth. Many species favour the edges of woodland areas, where they are broken up by glades and clearings, and many make use of damp and marshy areas. Unfortunately, this style of forest tends not to be profitable for commercial timber production. The creation of ditches and effective forest drains remove the damp and marshy oases which are so attractive to many species in what are otherwise large areas of unbroken and monotonous forest. During the fifty years up to 1985, Sweden alone lost almost 10% of its old mixed forests.

A well-balanced forest environment generally comprises a variety of topography, trees and other vegetation, and the type and wetness of the soils will also vary from one area to another. However, northern Europe has large areas of homogenous forest, of which a good example is the taiga. This is the large zone of conifers which lies in a long band across Europe and northern Asia, stretching from the Atlantic in the west to the Pacific Ocean in the east. In the taiga forest the ground is often boggy and the trees border streams, lakes and other open watercourses. The difference in forest types can be seen in northern Europe, where the southern coniferous forests give way to the more remote northern forests.

# Red squirrel

## *Sciurus vulgaris*

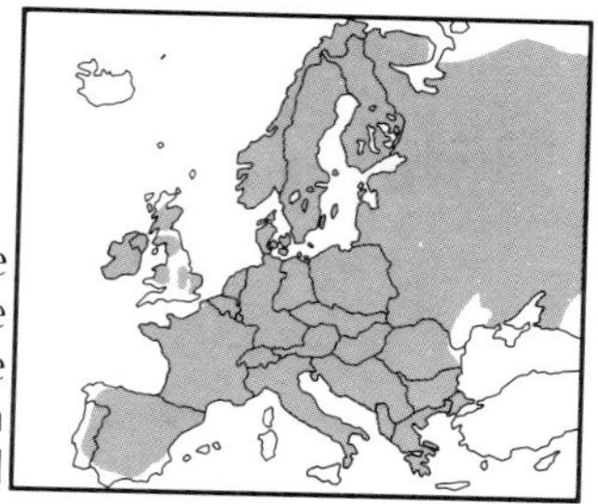

Red squirrel populations in Europe have always been liable to extreme fluctuations. From historical times these ups and downs can be seen reflected in the records of hunters and trappers, and also in the records of those who traded in squirrel pelts.

Where natural foods are scarce, especially after years in which a large number of young have been born, there can be large-scale migrations of squirrels. Othere die or fall victim to predators when starvation forces them to be less cautious than normal. The celebrated naturalist Linnaeus commented on squirrel population variation after he visited Gotland in 1741, when he remarked that the local population of squirrels would steadily increase over a period of seven years, which would then be followed by mass emigration.

Many zoologists have commented that some European populations of the squirrel are rather like those of lemmings, which are prone to marked increases and decreases in numbers, combined with mass migration. Folklore went even further, and described how migrating squirrels would take to the sea and would swim using their large bushy tails as sails, while having nothing to use as a rudder!

### Characteristics

The squirrel is so well known that it will be instantly recognised from a picture. However, it should be noted that there are wide variations in the colour of the Red squirrel's fur. It can be almost black, or "melanistic", and some individuals are flecked with white. Dark-coloured forms are considered common and typical in northern and central Europe, perhaps because a dark pelage suits best in the predominantly dark conifer forests of these areas. It is interesting to note the difference between, for example, the foxy-red and red-tailed squirrels of Scotland and northern England and the much darker squirrels of Scandinavian and alpine forests. In general, three types of colouration are recognised — black-tailed and dark; red-tailed and light; and brown-tailed. The Red squirrel's eyes are comparatively large and dark.

Northern squirrels change their red-brown summer pelage for a more sombre grey in winter, while more southerly populations keep their reddish-brown colour throughout the year. In winter all squirrels have denser fur with a thicker layer of underfur, and even the toes are protected by additional growth of hair which effectively protects the animal's feet from the frozen ground. Squirrels moult in spring and autumn, and in spring the fur is moulted from the front backwards. In autumn, however, the hairs at the back are shed first. The distinctive ear-tufts are shed in the spring, and therefore the squirrel in summer appears to have smaller, more rounded ears. Young individuals have shorter ear-tufts and less bushy tails than mature squirrels, and the hairs of the tail are thickest along the sides of the tail. The Red squirrel's tail represents about 75–80% of the animal's total body length.

The male and female are similar in size and in appearance, and a full grown squirrel will weigh 200–480 grams. In the hand, squirrels can easily be sexed by examination. The front teeth of the squirrel are reddish-brown in colour and are quite long; and they grow continually throughout the animal's life. There are 22 teeth which conform to the following pattern:

$$\frac{1\quad 0\quad 2\quad 3}{1\quad 0\quad 1\quad 3}$$

Both fore and hind feet are equipped with long, strong and sharply pointed claws. This, combined with their ability to grip with their toes, is an obvious adaptation to a life among the branches of trees. Even the squirrel's prominent tail is used as a prehensile limb. When squirrels peel pine cones or gnaw at hazel nuts, they tend to sit up and balance on their hind legs and their tail while grasping the food in their forepaws. The forepaw has four principal toes and a small "thumb". The hind feet lack these "thumbs".

### Behaviour

Like the nuthatch, the Red squirrel can move up and down tree trunks and branches. They can also swing between quite small branches and large twigs, and often take long aerial leaps from tree to tree. In doing this the squirrel spreads out its legs and uses its long tail to balance it in flight. Squirrels are also good swimmers, but only rarely do they use this ability.

Squirrels usually lead solitary lives except during the mating season. The males build and occupy their own dreys, while the females and the young have their own dreys and remain together as a group for almost two months. The Red squirrel is active by day and rests by night, and this makes it one of the few rodents which is not nocturnal. In stormy weather, intense cold and heavy snow squirrels are inactive, and they can sometimes sleep for several days in their well-insulated winter dreys.

Squirrels make a number of different sounds, the most common of which is a smacking, chattering sound which can be heard when they are excited. They often peep out from behind tree trunks, make a smacking noise and at the same time take small hops, rather as though they were jumping on the spot. They will also sometimes thump on the ground with their tails. They can sit absolutely still for long periods if danger threatens, and this is a common defensive ploy when a squirrel finds itself exposed in the open.

The Red squirrel has long, sensitive whiskers, and its senses of sight, smell and hearing are very acute. The eyes are large and set high on the head, and the nose is prominent and pointed.

Squirrels live most of their lives in trees, but they also spend a good deal of time on the ground. This is done partly to gather and store food in the form of acorns and nuts, which they hide in underground caches, and partly to move from tree to tree if they are widely spaced. When possible squirrels prefer to follow the lines of fences and hedges, and move along in long, quick leaps.

### Reproduction

Squirrels give birth between March and September, although the majority of litters are born in the period April–August. Under ideal conditions there can be as many as three litters in one year, but two is more usual. Young females breeding for the first time usually have only one litter of young, and in years when the food supply is very poor there may be no breeding at all. A mature male will probably mate with a succession of females, and the sexes only come together for mating. Mating usually takes place between January and April, but mating in December, May and June is not unknown.

The period of gestation is 36–40 days, and the number of young in the litter will depend upon the age of the female, the geographical location, the time of year and the availability of food; but on average there are 3–4 young per litter. The female has eight nipples, and litters of up to 6–7 young are not uncommon. Ten young is probably the largest litter recorded.

At birth the young squirrels are naked and blind, and weigh 8–15 grams; and their eyes do not open until they are a month old. The female gives birth in a well-lined and insulated drey which she constructs herself, although it is not uncommon for squirrels to make use of an abandoned crow's nest, a hole in a tree or even a bird nesting box. The female's drey is often a spherical nest of twigs high in a suitable tree. If a drey should be unduly disturbed, the mother will move her young to a safer and quieter place.

The young suckle until they are about two months old, but at about six weeks old they begin to venture out from the drey to explore their surroundings and to begin collecting food for themselves. They remain with the mother in the drey at night and become fully independent in the spring following their birth. They are sexually mature at the age of one year.

The Red squirrel is a rodent, and therefore belongs to a group of mammals which has been the most widespread and successful. Of the 4,500 known mammal species, no fewer than 1,650 (i.e. 37%) belong to the rodent group. The squirrel holds an interesting and important central position in this group of mammals, for taxonomists believe it is a more or less direct descendant of the most ancient rodents. For example, the hare belongs to this same large family of animals, but it is not a true rodent and its origins can be traced back through fossil remains for some three million years. By comparison, the squirrel can be traced back for no less than 38 million years, through various fossil forms which show a clear family resemblance to today's squirrels.

## Habitat and diet

The squirrel is a woodland animal found in areas close to the zone of large Eurasian coniferous and mixed forests. It is well known in many parts of its range as a creature of parks and even private gardens.

The squirrel subsists on a varied diet of fruits, seeds, nuts, acorns, beechmast, berries and various buds. Mushrooms, insect larvae, snails and occasionally baby birds and birds' eggs may also occasionally form part of its diet. Over much of its Eurasian range the squirrel is heavily dependent upon the availability of the cones of pine and fir trees, and the abundance or scarcity of these can cause the squirrel population to increase and decrease.

The squirrel's habit of hiding stores of food is well known. This habit develops while the squirrel is still young, when it learns to hide whatever surplus food there may be. These caches of food are usually buried in the ground and are generally relocated by a combination of memory and the animal's sense of smell. The importance of smell is especially evident when squirrels are able to locate their caches despite the ground being covered with several inches of snow. As is the case with some titmice in conifer forests, squirrels will take advantage of each others' food stores. Because all squirrels hide their food caches in similar kinds of places, all the members of the population benefit from this shared feeding. The food supplies are easy to find, nothing is wasted, and each animal contributes what it can and benefits accordingly. A squirrel can peel more than 150 pine cones in a single day. Squirrels have occasionally been accused of predating heavily upon nests of baby birds. However, a recent study examined the stomach contents of 1,600 squirrels, and only four individuals were found to have eaten baby birds or eggs.

## Distribution

The squirrel is found in most parts of Europe, although it is absent from the large islands of the Mediterranean. It also occurs widely in Asia, especially in the area of the taiga, the northern conifer belt that stretches across Asia to the shores of the Pacific.

Squirrels occur in Japan and on some other islands off the Pacific coast, and they are also found in the mountain areas of central Asia. Within this vast range the Red squirrel has evolved into about 40 different subspecies and local races. These tend to occur in areas where the squirrel populations are more or less genetically isolated from each other. The squirrel of Scandinavia, first scientifically described by Linnaeus in 1758, is known as *Sciurus vulgaris vulgaris*, while that of mainland continental Europe is designated as a separate subspecies, *Sciurus v. fuscoater*.

In many parts of the world the squirrel inhabits afforested mountain areas, and in the Alps it is found at altitudes up to 2,200 metres. In Fennoscandia it is found sporadically in the birch forests and in some mountain forests.

## Related species

Squirrels of various species are found all over the world, although they are absent from Australia. The European Red squirrel is closely related to the squirrels of North America, of which there are now fewer than six species. All belong to the same genus, *Sciurus*, but the North American Red squirrel belongs to the *Tamiasciurus* genus.

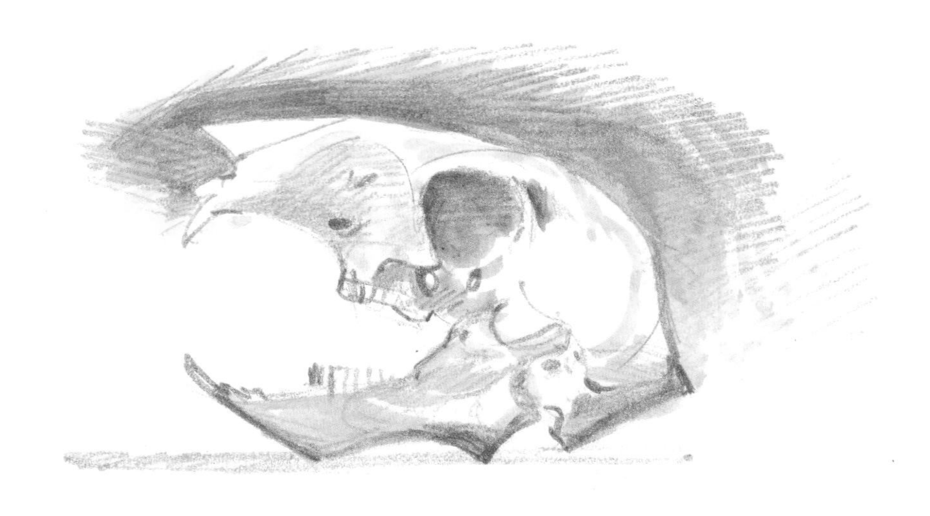

*The squirrel's skull reveals its specialized way of eating. The attachment for the large jaw muscles is clearly visible. Between the front and molar teeth is a conspicuous gap. Like other rodents, the squirrel has only two chisel-shaped incisors in each jaw. (Hares have four.) The front teeth continue to grow throughout life, compensating for the hard wear on them.*

## Predators and lifespan

The squirrel's principal enemy is man. In much of northern Russia, for example, the squirrel has considerable economic importance as a fur-bearing animal. Despite the general popularity of the squirrel, it has been exterminated in some parts of Europe. It is still found in Ireland and in parts of northern England and Scotland, but much of its former range in the British Isles is now occupied by the non-native Grey squirrel (*Sciurus carolinensis*), which was imported from North America in 1889. The first Grey squirrels were released at Woburn in Bedfordshire, and from there they have spread in all directions, and now populations exist in England, Wales, Scotland and Ireland. It has also been introduced to various locations in Europe and in the U.S.S.R.

Among the natural enemies of the squirrel are the Pine Marten and the Goshawk. When conditions are ideal for the squirrel, and squirrel numbers are therefore high, these two predators will feed mainly on squirrels. This is especially true of the squirrel's northern range in Europe and Asia.

Squirrels have been known to live for as long as 18–20 years, but the average lifespan is very much shorter.

## The squirrel in Scandinavia

The squirrel is an important and widespread animal in Scandinavia, and it has been the subject of a good deal of study by naturalists, sportsmen and also those involved in the fur trade.

Squirrels are found throughout Scandinavia, from the southern coasts to the northernmost parts of Lapland. It has always been an important and highly esteemed sporting animal, and up to recent times squirrel meat was generally popular. The great Swedish naturalist Linnaeus was not especially enthusiastic about the flavour of cooked squirrel, but he described it as being rather white fleshed and not dissimilar in taste to rabbit. Other writers and commentators have remarked that it is rather similar to chicken, but some old accounts suggested that a surfeit of squirrel meat could cause epilepsy. But the rather doubtful culinary importance of the squirrel was always overshadowed by the value of its pelt. Peter Thunberg, a pupil of Linnaeus, wrote that "the pelts are small and not very strong, but they are commonly used for lining clothes, for muffs, and for trimming clothes."

Inevitably the squirrel's thick winter coat was more highly prized than its summer pelage, and the squirrels from the most northerly areas of Scandinavia were especially coveted. In these northern regions squirrel hunting was an important local activity and source of income, and an early nineteenth century writer commented that "a good bird and squirrel dog is to be found in every household. In the legal settling of a country estate a good dog of this type is considered of equal value to a good breeding cow."

When the fur trade in Scandinavia was at its height in the last century, sportsmen and commercial trappers accounted for up to 200,000 squirrel pelts in a single year. Squirrel pelts and meat are no longer especially sought after, but some Scandinavian sportsmen still pursue squirrels in the traditional way, with a specially trained dog. An estimated 1,500 squirrels are shot in this way annually in Sweden.

# Brown bear

## *Ursus arctos*

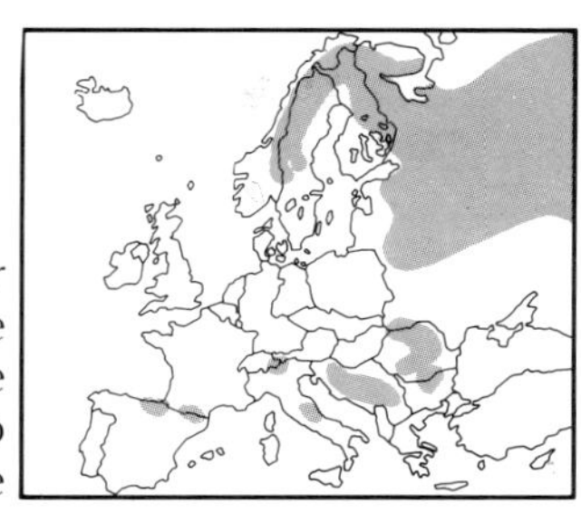

The present day status of the Brown bear in Europe is a direct consequence of the pressures which man has put upon the species and its habitats, and it also reflects the extent to which bears have been hunted throughout history.

An old northern European hunting manual gives details of a number of ways in which one can go about hunting a bear, the most important of which were: in a battue, where the bear is driven towards a waiting rifleman; hunting it with beaters, dogs and a number of shooters; hunting with hounds; "still hunting" or waiting in ambush; tracking a bear along its accustomed paths and tracks; and setting out a carrion bait to attract a bear.

These and other hunting methods, combined with human disturbance and loss of suitable habitat, caused the Brown bear to become extinct in much of its former haunts. By the end of the nineteenth century the last bears in Germany were extinct, and the last bears in the Swiss and French Alps died out in the 1930s. By the 1960s it was estimated that the entire Alpine population might number not more than 10–20 individuals. Despite continued hunting in parts of the Alps, the 1970s population was estimated at about the same number, but the threat of extinction is always present when any species becomes so depleted.

Scandinavia has a long tradition of bear hunting, and this activity was traditionally regarded as a test of the sportsman's courage. In late autumn, before the bears retired into winter hibernation, hunters would attempt to track them through the snow to their winter dens. The right to hunt bears by any means and at any time was every person's inalienable right. Once a bear had been tracked to its den, the hunter could either kill it himself or sell the right to kill it to another bear hunter. A number of individuals made names for themselves as bear hunters, including the English sportsman Llewellyn Lloyd, who wrote in detail about wildlife and sport in nineteenth century Scandinavia. Another notable bear hunter was the royal forester Herman Falk, who personally killed more than 100 bears.

The Swedish hunting laws of 1864 stated that anyone who had tracked a bear to its den had the sole rights to that animal, and made it an offence for anyone else to disturb the bear or try to kill it or frighten it away. There was a bounty on every bear killed, amounting to 40 Riksdaler, which was paid out of public funds. This bounty payment was discontinued in 1893, and from 1912 onwards the bear was granted full protection on all public land in Sweden. Norway banned most bear hunting in 1932, and in the same year bears were given full legal protection in Czechoslovakia.

Some bear hunting still goes on in Europe, but present day hunting is carefully regulated by strict close seasons and by imposing a quota of animals which may be shot. Unfortunately, illegal hunting still threatens the remaining Brown bears in the Alps. In the Karelia region of Russia there is an estimated population of 3,000 bears, and an annual cull of 150 animals is permitted. The future of the bear is much more precarious in the Alps and in Spain than in the north and east of Europe.

## Characteristics

The European Brown bear is a large animal, and a typical bear from Scandinavia will weigh up to 300 kg, although exceptional individuals have been found to weigh up to 350 kg. A mature male is considerably larger than the female, often by 50%–75%, and he may be twice as heavy. As the seasons of the year change, there are considerable variations in the bears' body weights. In autumn, before hibernation and after they have fed greedily on the rich foods of summer and autumn, they may weigh up to 70%–80% more than they do when they emerge from hibernation. In preparation for winter hibernation a bear may accumulate reserves of fat weighing as much as 100 kg. Male bears are not fully mature until they are ten years old, but females are considered fully grown by the age of six.

The bear has a short neck and relatively short legs. The ears, which are sometimes hidden in the long fur, are small and rounded, and the tail is so short that it often cannot be seen. The bear's muzzle is brownish-black and rather up-turned, and it is quite hairless and prominent. The comparatively small eyes are dark brown, while the feet are large and flat with hairless soles. The bear is a plantigrade, which means that it can walk upright on the whole sole of the foot. Each foot has five toes, each with a strong, curved claw.

The bear's fur is dark brown, but individuals can vary from near black to an almost golden brown. There are considerable variations between individual bears and between bears of different ages, which is helpful in identifying individual animals. The colouration also varies from one local population to another, and one example of this is the bears of eastern Europe, which have especially dark fur on their backs. Young bears often have a white collar, which may take many forms from a complete circular band to a pattern of large and small spots or flecks.

The Brown bear has a primitive, rather unspecialised tooth structure, which consists of 42 teeth. The molars and premolars are flat and blunt, and some of the premolars are often missing. The tooth pattern is:

$$\frac{3\ \ 1\ \ 4\ \ 2}{3\ \ 1\ \ 4\ \ 3} \text{ or } \frac{3\ \ 1\ \ 3\ \ 2}{3\ \ 1\ \ 2\ \ 3}$$

The bear is a easy animal to identify, but it is often very difficult to find. However, it leaves various clues which may lead you to it. Its excrement is often a large and shapeless pile, rather like cow dung, and sometimes it is in separate lumps in which can be seen the identifiable remains of food, such as leaves, berries and seeds, and the remains of insect material and hair. In snow and on damp ground bear activity can be seen clearly, for the animals leave large footprints with clearly marked heel prints, especially from their hind feet, and the five large, clawed toes leave unmistakeable marks. When a bear has been moving at its customary ambling pace the tracks look like two lines of parallel prints, but in a fast run or frightened gallop the bear leaves a pattern of tracks similar to those of other galloping animals. Bears also leave tell-tale scratch marks on tree trunks, which are easily identified by their size and height above the ground. Also, both males and females will rub themselves vigorously against tree trunks, leaving remains of hair. This occurs mainly when the bears are moulting their winter fur in spring and early summer.

## Reproduction

The male bear becomes sexually mature at the age of 4–5 years, and the female about a year earlier. However, the female may not give birth until she is 8–9 years old. Bears are promiscuous in their mating

habits: no permanent pairings take place between the sexes, and several females may have their home ranges within the territory of one dominant male bear.

Mating occurs in the period April–June, although in northern regions mating may not begin until May. The female is receptive to the male for about 2–4 periods during that time, and she may be followed by a male for several days, during which time the two may mate repeatedly each day, perhaps as many as ten times. After mating the male and female split up and go their separate ways.

Like some other mammals, the female bear has a delayed implantation of the foetus, and implantation normally takes place 2–3 months after fertilisation. The gestation period lasts for 7–9 months, but the foetus really only develops significantly during the final two months of pregnancy. The cubs are born in the period from December to February, and the female gives birth in her winter den. Young females may have only one cub, but more mature females usually have two, and triplets are not unknown. Record litters of up to five cubs have been recorded.

The cubs are very small, about 25 cm long and weighing 300–500 grams, and they are born blind and helpless. Their eyes open after 4–5 weeks, and the female and her cubs will emerge from the den when the young are about four months old, which is in the period March–May depending on the region, and by this time the cubs will weigh about 4 kg. The cubs will stay with their mother until they are three years old. They will suckle sporadically until they are 18 months old, and they will spend their second and third winters with their mother in the den. By their first winter the cubs will weigh four times as much as they did at birth. Because it takes so long for the cubs to mature and become independent, the mature female reproduces only every third or fourth year. In captivity Brown bears have been known to live for up to 50 years, but 35 must be considered the maximum for a wild bear. A lifespan of 35 years would give a female a maximum reproductive potential of 15–20 cubs, but only a few females actually give birth to that many. Their low rate of reproduction and the sedentary lives of the females make bear populations very vulnerable.

## Adaptations and behaviour

The bear has many special skills. It is an excellent swimmer; it can climb trees with ease from an early age; and it can run very fast, especially over short distances, when it can move at 40–50 km/hour. Its senses of smell and hearing are very acute, but its eyesight is relatively weak.

The bear can stand up on its hind legs, which it often does in order to pick up a scent or to listen to something suspicious or interesting. It is said that a full grown male bear can carry game as heavy as a reindeer with its forelegs.

Bears are mainly active by night, resting in their dens by day. Disturbance is known to affect bears' patterns of activity, and in undisturbed areas bears are active at dusk and dawn. During the almost unending daylight of the Nordic summer there is little difference between the bear's daytime and night-time activities.

The bear's habit of going into hibernation is well known. The den consists of a hole dug in the earth, sometimes under the roots of a fallen tree or in a cave among rocks. The hibernation den is lined with grasses, twigs, moss and dried vegetation. Hibernation is an adaptation to allow the bears to survive through the intense cold of winter and the seasonal lack of suitable food, for the bear is principally a vegetarian. Bears can be awakened from even their deepest hibernation sleep, when their body temperature has dropped by 4–5°C, and sometimes as much as 7°C. The heartbeat slows down to barely a quarter of the waking rate. As we have seen, the birth of the cubs takes place in the females' winter dens.

Old male bears are usually solitary wanderers, but females with cubs form a stable and sedentary unit, and the cubs remain with their mother for two years. Female cubs may remain with their mother even after she

*The bear has a thick skull, the largest among all European land predators. Like the wolf and fox, it has 42 teeth, one of the highest numbers among all predators. But often the bear loses some of its young molars and has only 38–40 teeth when mature. The canine teeth are especially large and strong. The rough, knobbly molars reflect the bear's general eating habits. The skull of an old male Brown bear weighs just under 2 kg, or 20 times more than the skull of an old male Red fox.*

has mated and given birth to a new litter, while the young males wander off on their own.

A mature male has a large territory, while the females have a much smaller home range. Bears often move along regular tracks and paths within their territories, and will wander for up to 10 km in a day. The distance travelled will depend upon the environment, the availability of food, and the density of the local bear population. Much more is known about the territories of the North American Brown bear, and the females have a home range of 15–90 $km^2$, while the male's territory may be as large as 1,500 $km^2$.

In a sample group of 500 bears aged two years and older, which were shot on the borders of Finland and Russia, 65% were found to be males. This area of Finnish Karelia is an important immigration area for Russian bears, and this abundance of males suggests a much higher degree of wandering among males than females.

Studies of the movements and lives of European Brown bears have been carried out with the aid of radio tags attached to individual bears. A radio-tagged Swedish female bear was found to move within an area of 120–130 $km^2$ over a period of three years, while another female remained within a range of about 200 $km^2$. Two Swedish males, by contrast, had territories of over 1,000 $km^2$. One radio-tagged bear in the Italian Alps used a territory of 56 $km^2$ in one valley, and these data from Europe tend to demonstrate that the European Brown bear behaves in a way that is very similar to that of the North American Brown bear.

## Habitat and diet

In Europe Brown bears live mainly in afforested and mountainous areas. In the Alps and the Carpathians bears live at altitudes up to 2,600 metres. In Scandinavia the bear is often found above the timberline. Bears normally wander in search of suitable environments to fulfill their basic requirements for food, daytime shelter and hibernation dens. Bears are reported to wander up to 80–100 km.

The bear's diet consists mainly of plants, ranging from dry roots and juicy herbs to berries and shoots. In preparation for winter it usually feeds greedily on various types of berries, including cloudberries, blueberries and crowberries.

Bears also like to eat insects, and signs of bear diggings may often be seen among anthills. The bear's fondness for honey is also well known. Large animals such as Reindeer, Elk and domestic cattle occasionally supplement their diet, and the bear can be dogged in pursuit and swift to attack. An animal which has been killed by a bear can be identified by the way in which the skin is torn off the parts of the carcase which the bear has eaten. A bear probably requires about 10 kg of food per day, but the precise requirements vary with the size of the bear, the type of food available, and the season of the year. Before hibernation, during which it lives only on its fat reserves, the bear is immensely gluttonous in its eating habits. It also hoards food and often buries whole carcases of game animals.

## Related species

Bears are found in all parts of the world, apart from Africa and Australia. The relationships between the various species and races has been the subject of a great deal of discussion and argument. Most zoologists now accept that there are seven species within the bear family *Ursidae*. The Eurasian Brown bear (*Ursus arctos arctos*) is the nominate form of the Brown bear subspecies, which also occurs in North America. About 20 different species and forms of bear which were described by earlier naturalists have now been combined into one single species, to which both the Grizzly bear (*Ursus arctos horribilis*) and the Kodiak bear (*U. a. middendorffi*) belong. The Kodiak bear is sometimes called the Giant Brown bear, and with males weighing up to 1,000 kg it is the largest of all the Brown bears. However, some scientists still regard the European Brown bear as an independent species with eleven different subspecies. If this reasoning is followed, then the Pyrenean bears must be referred to as the subspecies *U. a. pyrenicus*, the Italian Abruzzo bears become the subspecies *U. a. marsicanus*, and the northern, southern and Alpine bears must be referred to as *U. a. arctos*.

## Distribution

The Brown bear is found in Europe, Asia and North America. The European population is principally confined to Scandinavia, Finland and eastern Europe, and the existence of a wild population in the Pyrenees seems doubtful. In the Abruzzo National Park in Italy there are about 100 Dinarian bears, while in the valleys of the Italian Alps fewer than 10 individuals remain. There are about 300 bears in the Carpathians, and a small number exist in the mountains of Yugoslavia. In Scandinavia and Finland there are about 1,000 bears.

Russian Karelia has a large population, estimated at 3,000 animals, and it can therefore be seen that the bear is in difficulties in some areas while populations remain healthy and numerous in other areas. In some areas, such as the Beilawieza forest on the Polish-Russian border, bears have been reintroduced, but many other former parts of the bears' range are now unsuitable for such reintroductions.

In Sweden there was a veritable extermination campaign against the bear in the eighteenth and nineteenth centuries. By the eighteenth century the bear had been eradicated in most parts of southern Sweden, and the population remained buoyant only in the remoter northern parts of the country.

Swedish hunting legislation in 1864 encouraged the campaign of extermination by raising the bounty for bears from the 1808 figure of 12 Schillings to as much as 50 Riksdalers. In 1889 it was proposed in the Swedish parliament that the bounty payments should be stopped, and this was eventually done in 1893. This marked the beginings of policies of protection and a changing attitude towards the bear. In 1909 national parks were created, in which bears and other animals were given protection. The killing of bears was permitted only if human life or property was threatened, and this exemption was eagerly grasped by the Lapps and other dwellers in the high mountain areas. Until 1912 tracking and killing bears on someone else's land was permitted, and at the same time the right to shoot bears on government owned land without permission was also revoked. The hunting legislation of 1927 gave the bear more protection by designating it as "Crown game", which meant that dead bears became government property. But the bear has never been totally protected in Sweden.

Up to 1980 the bear was given complete protection in Sweden, with the exception of Härjedalen, Jämtland and certain areas of Lapland. In these areas there was a 14-day hunting season in Härjedalen, a one-month season in Lapland, and a two-month season in Jämtland, but since 1981 this has been replaced by a system of licensed hunting. In 1983 licences were issued for 40 animals, in 1984 for 42 animals and in 1987 licences were issued for up to 53 bears.

In 1960 the Norwegian population was estimated at 50 bears, with about 200 in Finland and 250 in Sweden. When the Swedish bear population was counted in 1952 the total number of animals was reckoned to be a little under 300. By 1957 that estimate has been reduced to about 250 bears, and after that the numbers of bears and their geographical distribution began to increase. In 1980 the estimated Swedish population was 600 bears, and the bear is now to be found in various parts of Sweden where it was formerly extinct.

Much the same situation has occurred in Finland, where the immigration of bears across the border with Russia has played a key role in the gradual increase in bear numbers. Between 1969 and 1981 some 682 bears were believed to have moved in, and during the same period at least 456 bears were killed. In the autumn of 1982 the bear population of Finland was estimated at 300–350 individuals, and further increases are believed to have occurred, so that the current Finnish population is probably about 500 bears.

The Norwegian population is currently put at about 200 individuals, and bears are still to be found in the Telemark region of southern Norway.

## Bear hunting

As has been mentioned, the bear is legal quarry in various parts of Europe. In a number of countries, regulated licensed hunting is permitted, which takes account of the numbers of bears present and their reproductive rate, and which reflects the growth of the bear population and its geographical distribution. The Nordic tradition of bear hunting is based on an ancient tradition of sport, perhaps more so than any other form of hunting. In bear hunting we can learn as much about ethnology and cultural history as about the hunt and its quarry. For the bear population of Scandinavia the bear hunts of the eighteenth and nineteenth centuries had a devastating impact, and people in the southern areas of Scandinavia will still talk about the circumstances in which the last bear was killed. Some especially large specimens have been preserved in museums. In the far north, however, the hunting tradition remains unbroken owing to the continued existence of the bear. This is an example of how the conservation of wildlife can be not only a matter of preserving our heritage of nature and wild creatures: it can also be a significant contribution to the preservation of a cultural heritage.

There is no longer any need to reduce the numbers of bears in Scandinavia, but conservation should take account of the value of bears for their pelts and for their meat. In this way a well managed and flourishing bear population should be of tangible value to local communities. Bears cause little damage, and under normal conditions when it is not provoked the bear is quite harmless to man. It is in the light of these facts that the Laplanders bear hunts should be considered, and also their attitudes and actions towards other predatory species. Some hunting traditions are kept alive by the traditional attitudes which are hostile to large predatory animals. Many predators have declined in numbers, and some have been exterminated, but the general view of the Lapps is that the numbers of reindeer which have fallen victim to predators has increased five-fold in recent years, and they regard this as evidence of the "failure of government policies towards predatory species". But the bear is not an important predator of the reindeer, and the Lapps' indignation is directed primarily at wolverines and the lynx, since the wolf has now been exterminated in these northern regions. The shooting of bears under licence is a lucrative source of revenue, but governments have an obligation to reimburse those who have suffered loss or damage by bears.

Recent assessments made by the Ministries of Forestry and Agriculture in Finland have taken a positive attitude towards the brown bear. They believe it is possible to allow the bear population to increase to 250–300 animals in reindeer grazing areas, and to a minimum of 700–800 bears in other parts of the country. If it can be done, this target would be well worth achieving.

# Wolf

## *Canis lupus*

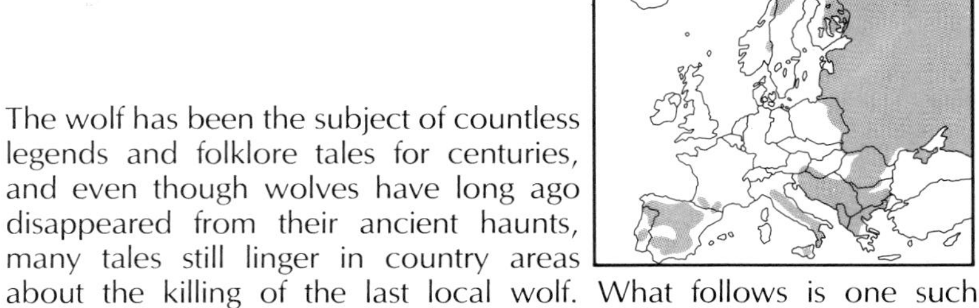

The wolf has been the subject of countless legends and folklore tales for centuries, and even though wolves have long ago disappeared from their ancient haunts, many tales still linger in country areas about the killing of the last local wolf. What follows is one such example from Sweden.

One day in February 1898, wolf tracks were seen crossing over the snow-covered Foljö fjord in the direction of Orust. It was midwinter on the west coast of Sweden, and the celebrated contemporary naturalist and hunter Gustaf Kolthoff and a companion tracked the lone wolf for a day, but failed to catch up with it. At Lilla Edet the wolf crossed the river Göte and later attacked several guard dogs. It probably spent the summer in the wild country east of the river valley, perhaps north-east of Lake Lygnern. On Christmas Eve of the same year a wolf was shot at Fjäras Bräcka at the southern tip of the lake, and there were no more wolves sighted that year along the west coast. It appears that this was the last wolf to be found along the west coast of Sweden during the nineteenth century.

The resident and wandering wolves in other parts of Sweden were hunted in the same relentless way, and a comprehensive country-wide campaign was organised to exterminate the wolves. Hunts in country districts could involve a line of beaters consisting of 4-600 men, and farmers were ordered to prepare wolf pits and to have wolf nets and other equipment ready.

The migratory movements of wolves had long been a familiar phenomenon. In 1854 a naturalist and sporting writer had written that "in February the wolves make their way to the sea coast to gorge themselves on newborn seals, since between Candlemas on the 2nd and Matthewmas on the 24th the seals give birth to their young". Other country folk used to say that each spring the wolves went down to the sea and lapped at the salt water, and it would therefore appear that when young wolves today migrate after breeding in southern Norway and Sweden, they are following ancient instincts and tradition.

As in ancient times, the activities of wolves are associated with the deaths of other animals, especially sheep. They are also known to attack guard dogs and hunting dogs; but man's attitudes towards the wolf are now more divided and ambivalent than ever before. Most people now adopt a positive attitude towards the presence of a wild wolf population in Scandinavia, but lively debates continue in those areas which have wolves, and unfortunately those who favour extermination have made their views obvious in practice. In the very year in which the Swedish parliament gave complete protection to wolves throughout Sweden, official permission was also given (after the law had been passed) for two wolves to be trapped in Lapland. During the 1980s the legal and illegal hunting of wolves has seriously reduced the size of the wild population, and have thereby jeopardised the species' survival.

In Norrland, the presence of the wolf population depends upon an influx of wolves from the east, but here also the existence of the wolf is compromised by illegal hunting. One of the last wolves which moved into this area in 1978 was illegally hunted with a snow scooter, and was killed in May 1981.

## Characteristics

The wolf is the world's largest member of the dog family. Within its European range the only other animal for which it could possibly be mistaken would be a large, light-grey Alsatian dog. However, the individual colouration of the wolf can vary between light greyish-white and greyish-brown with yellow-brown and yellow-grey variations. In the Arctic there are white Polar wolves, and black-coloured individuals can also be found among normal wolf populations.

The wolf's head is larger and wider than a dog's, and the chest is normally narrower. The ears stand erect and the eyes are slanted. The dark pupils of the eyes are round, and the iris is amber-coloured. In the dark the eyes shine yellowish-green when they reflect light. The wolf has an upturned nose, the tip of which is wide and dark, and the animal's lips are blackish. The legs are long and thin, and wolves have five toes on their forepaws, with the inside toe positioned high up, exactly like a dog's dew-claw, and four toes on its hind paws. The claws are large and the middle pad of the foot is large; like the pads of the toes it is hairless. The wolf's bushy tail usually hangs down, but it is held up when the animal is moving fast or if it is running away from danger. Adult wolves usually weigh between 35 and 55 kg, and very large females can reach weights of up to 55 kg too. But exceptionally large males can weigh up to 75 kg. A wolf is fully grown when it is 4–5 years old.

The wolf has 42 teeth, placed on the following pattern:

$$\frac{3\ \ 1\ \ 4\ \ 2}{3\ \ 1\ \ 4\ \ 3}$$

There is a carnivorous scissor tooth in each jaw, which is well developed with a serrated edge. The canine teeth are large and powerful. At five months old the cub's permanent teeth replace its milk teeth. The wolf's footmarks are large, the paw print measuring about 10 cm long and similar to that of a dog. The tracks are straight like those of a fox, with a distance of up to one metre between paces. Wolves can also swim well. Among its vocal noises are a harsh yelp and various rough snarling and barking noises. Both individuals and an entire wolf pack will howl from time to time, and this serves as a means of communication between the packs and individual wolves. The wolf's senses of sight, hearing and smell are all extremely well developed.

## Reproduction

The wolf is normally monogamous, and the pair stay together for a long time. Litters can comprise up to 10–12 cubs, but 4–6 is more usual. Older females give birth to larger litters, and the size also depends upon the availability of food. The female, also called the wolf bitch or the she-wolf, has ten nipples.

The cubs are born in a lair or den dug in loose soil or into a sandy bank, sometimes among uprooted trees and overhanging vegetation. The females are in season from the end of December until April, but the mating season is later in northern areas than in the south. The period of oestrus lasts for five days, and in the act of mating the pair of wolves "tie" like dogs for a period of up to 20–30 minutes. After a gestation period of 93 days the cubs are born, and this usually takes place in the period March–June. Newly born cubs weigh about 30–40 grams, but can weigh as much as 60–70 grams. They are born blind, and their eyes do not open until they are 10–12 days old. At this stage their fur is coal black, with a greyish head. The female remains with her cubs until they are weaned at the age of 6–8 weeks, and during this time the male wolf provides for his mate by bringing a share of his hunting spoils to the den. The young females from the previous year's litter also act as "helpers" and bring food. Later, the male, the female and "helpers" all provide the cubs with food, which they partially digest before regurgitating it. Shortly before they are one month old the cubs begin to venture to the mouth of the den and to inspect the immediate vicinity. By the time they are six months old they accompany their parents and the other pack members when they hunt. One mature and productive breeding pair of wolves forms the nucleus of a wolf pack. The young males are gradually able to go off on extended wanderings of their own, but the female cubs tend to stay with the family pack. Wolves are sexually mature at one year old, but they seldom breed until they are at least two years old, and the male cub is a little slower in maturing than the female. Wolves are said to be able to live to 12–16 years of age, and it would therefore be possible for a female wolf to produce 10–14 litters of cubs in a lifetime, which might amount to 60–100 cubs from one wolf pair.

## Distribution and subspecies

The wolf has a circumpolar distribution in the northern hemisphere. It is also found on many islands in the Arctic Ocean, including those in the northern Canadian archipelago and Greenland. Despite its widespread distribution and its many variations in colour and size, it is agreed that all wolves belong to one and the same single species — even though there are no fewer than 32 distinct subspecies.

The Iberian wolf is more reddish-brown than its other European relatives, and it has been classified as the subspecies *Canis lupus silicia*, while other continental wolf populations are included in the subspecies *C. l. crucigera*, and the northern woodland wolves of Eurasia are given the nominate form, *Canis lupus lupus*. The wolves on some of the Mediterranean islands have also been accorded subspecies status.

The forest and woodland wolves of Fennoscandia belong to the same subspecies as the Russian forest wolves that live in the

coniferous zone or taiga. This is helpful from the point of view of nature conservation, for it means that immigrant wolves from the east which help maintain the numbers of the Scandinavian wolves are actually members of the same subspecies, and no crossbreeding is involved. However, the wolves of the Russian and Siberian tundra belong to a different subspecies, *C. l. albus*.

## Habitat and diet

The wolf was once common and widely distributed across all of Europe, and it occurred in many types of habitat, including woodlands, moorland and grasslands, as well as mountainous areas. As a result of widespread and concerted extermination campaigns, which in the last century wiped out many of the southern European wolf populations, the wolf became confined to remote mountain areas and alpine terrain. For this reason it is often erroneously regarded as a mountain animal. In some areas it can occur at altitudes up to 2,500 metres.

The wolf is active both by day and night. It can travel as far as 70–80 km in a period of 24 hours, although longer distances are common. Its daylight life has been adjusted to take account of human movements and activity, as a result of constant persecution. Its diet consists of all kinds of mammals, and it mainly hunts large game such as roe deer, reindeer and elk. However, if there are many small rodents and rabbits available locally in summer the wolf will catch and eat large numbers of them. The wolf pack hunts within its own clearly defined territory.

In areas where there are many wild reindeer the wolf will feed himself almost entirely on them, especially in the winter when the wolves hunt in packs. As the present day Nordic wolf population is so small, most wolves live and hunt alone or in pairs, widely separated from one another. A single Scandinavian wolf can kill prey as large as an elk, and the hunt often involves a long lasting pursuit of the prey. Yet wolves are also capable of making sudden surprise attacks on their prey, and are able to run at speeds up to 65 km/hour over short distances. Circumstances will dictate the choice of prey and the method of hunting. In some areas wolves will attack grazing animals, especially sheep and reindeer, and hunting dogs may also be killed if they are working in a wolf's territory. Some wolves in southern Europe feed by scavenging on carrion and refuse. When tackling live prey the wolf concentrates on the hind legs of the fleeing animal, but the killing bite is on the throat, where the strong canine teeth are the principal weapon.

## A gregarious animal

Normally the wolf is a highly sociable animal. The structure of a wolf pack is determined by social characteristics such as sex and age, psychological and physical strength, and family relationships. A pack's territory is determined by external conditions such as the nature of the terrain and the availability of prey, as well as by the reproductive cycle. Wolves mark their territories by urinating and by howling, and they will defend their territories against neighbouring packs. In areas of North America where there is a high density of wolves, territories can be up to 300 $km^2$ in size, but they tend to vary according to the size of the pack and the availability of food.

The wolf pack is led and dominated by an older female, sometimes called the "lead wolf", and the males and young cubs attach themselves to her. The first year cubs are part of the pack, as well as the young wolves from previous years, especially the females.

*The wolf resembles a large German shepherd dog. Some breeds of dogs are descended from the wolf, others from the jackal. As with other canine animals, the wolf's mood is shown by how it holds its tail, ears and body, and draws up its lips. This makes it look fierce, and gives signals to members of its pack. The wolf seen here is strong and aggressive. Laid-back ears and a more drawn-up nose would indicate readiness to attack.*

Sometimes other wolves join this nuclear pack before the onset of winter. Disputes can lead to a member of the pack being expelled, especially young males and old, weak wolves.

Membership of the pack has positive advantages for the wolves' survival through the winter and for the pack's breeding success. Studies of rejected, lone wolves have shown that it is difficult for them to take care of themselves, and they often have to resort to eating carrion and small prey for their survival. Young female "helpers" bring food to the older females and to the cubs. Besides improving the cubs' survival chances, this behaviour is thought to be valuable experience in anticipation of the helpers' own future breeding and rearing of cubs. Living in packs has its advantages when it comes to tracking and encircling a large and fast-moving quarry animal, but the pack is also more exposed to its enemies than a cautious single wolf would be. When food is scarce there is intense competition. The right sized pack is ideal for hunting and for the successful defence of the dead prey from other groups of wolves and from antagonistic solitary wolves. Besides man, the wolf has few natural enemies.

## Wolf populations in Europe

The westernmost outposts of the wolf in Europe are found in eastern Europe, in the Balkans, and in the north-eastern parts of Fennoscandia. In addition there are isolated populations of wolves in Italy, in the Iberian peninsula, and in central parts of Norway and Sweden. The latter should probably be regarded as the outcome of occasional westward movements from northern Russia and Finland, and not as the remains of an earlier widespread Scandinavian population. During the last few decades this type of immigration has resulted in occasional breeding in both Norway and Sweden. The wolf population in the Baltic countries of Poland, Germany and Czechoslovakia are the results of colonisation from the east. In contrast to these, the isolated wolf populations of the Mediterranean countries are relict populations only.

The wolf was eliminated from most of Europe a long time ago: in England at the beginning of the sixteenth century; in Scotland in 1743; in Ireland in 1821; in Denmark in 1800; and in Germany in 1841. It is now extinct in 11 of Europe's 25 countries, not counting several very small states.

The present European wolf populations are low in numbers. In 1970 the Italian population was estimated at 100–150 animals, and the Spanish-Portuguese population at about the same numbers. The Finnish population varied from 20–30 animals, while the Norwegian-Swedish population was less than 10 animals. The eastern European-Balkan population of wolves is more numerous, and comprises more than 1,000 animals.

The wolf and its behaviour have been documented in many different ways. Where wolves have been common they have often spread fear and terror, which was then exaggerated as the tales were retold. In reality, the number of attacks on men have been few, and fatal attacks even fewer. Ancient literature and folklore often implied a connection between times of misfortune and an increase in the numbers of wolves. In England the year 1281 was described as a year "with large and devastating packs of wolves". The wolf population of Norway and Sweden was said to have increased in 1718, when the two countries were at war with one another, while in France the wolf population increased in the 1790s, in the years immediately following the French Revolution. In Estonia and Latvia 1822 was called a "wolf year", and a contemporary naturalist wrote that "war and internal disturbances at the time contributed to a great increase in wolves". Furthermore, the wolf increased in numbers and spread further west in Europe during the Second World War! The wolf is thought to have increased dramatically in numbers in European Russia at that time, and a population of 200,000 wolves was estimated at the end of the war. At that time wolves were also reported from Lüneburg Heath, a century after the species had been exterminated in Germany.

The wolf has been the cause of a great deal of local damage as a result of its attacks on domestic animals. In southern Europe sheep, goats and pigs on open pastures were especially affected, while in northern Europe domesticated reindeer were a common prey. The greatest losses were sustained when domestic animals were left alone to graze in the woodlands, and for this reason attempts were made to keep wolf numbers down, or to exterminate the animal. It is said that in France in 1797, following a government decree, 7,351 wolves were killed. These extermination efforts were so successful that it is difficult to believe that they could ever be repeated; and yet after the end of the Second World War 63,000 wolves were shot in European Russia.

## The wolf in Scandinavia

The wolf's former distribution included all the Nordic countries except Iceland. Inevitably, wolves came into constant conflict with man, and the first areas in which the wolf populations were severely depleted were in the well populated agricultural regions. The last Danish wolf was killed at the beginning of the last century, probably in 1813. A celebrated wolf hunt took place in southern Jutland in 1778, and no fewer than 12 wolves were shot on that one occasion.

In Sweden, Norway and Finland the wolf survived throughout the country until the end of the last century, and it has never been totally exterminated. One reason for this is steady immigration from Russia in the east.

Wolf populations appear always to have been characterised by fluctuations in numbers, just as there have frequently been extensive migrations. Linnaeus and others described the wolf as having been rare in Sweden and Norway at the beginning of the seventeenth century, and yet wolves were considered common in many areas in the sixteenth century and again in the eighteenth and early nineteenth centuries.

During the period 1827–39 a total of 6,790 wolves were shot in Sweden, and in the period 1851–60 a further 1,808 were killed. This means that an average of 540 wolves were killed annually up to the 1840s; 180 annually in the 1850s; around 40 in the 1870s; and 3–4 per year at the beginning of the 1950s. By then the remains of the Swedish wolf population had retreated into the most remote mountains.

Much the same pattern can be traced in Norway, and by the 1970s the total wolf population of Sweden and Norway was estimated at 10–20 animals at the most. There was evidence of wolves breeding on the borders of central Norway and Sweden in the 1980s, but the population probably never exceeded 20 animals. Some young wolves from this area have wandered widely, as far as southern Sweden, and a few have been shot or deliberately run over by cars, deliberately and quite illegally.

While the wolf in Sweden and Norway has been seriously depleted and effectively restricted to the north, where it continues to balance on the verge of extinction, it has enjoyed better fortunes in Finland. There an increase in wolf numbers was noted in the 1970s, from about 50 animals in 1976 to about 100 wolves in 1978. Some immigration occurs from Russia, where the numbers of wolves are known to have increased. However, the wolf is thought to have adapted itself to changing conditions, leading to an increased rate of breeding success and resulting in an increase in the population. The wolf is now completely protected in Norway and Sweden, while in Finland about 10–20 wolves are shot each year, compared to about 30 during the latter years of the 1970s. Compared to its Nordic neighbours, Finland has a vigorous wolf population, but nevertheless the Finnish Council for Natural Resources advocates an increase in the wolf population to about 400–500 animals.

# Lynx

## *Felis lynx*

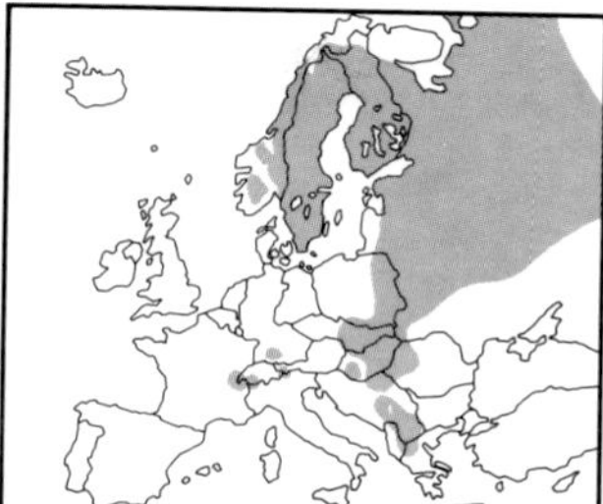

The lynx is the largest European representative of the cat family. It is characterised by its tufted ears, short tail and relatively long legs. The fur is soft and very thick. The basic colouration changes with the seasons, from a light yellow-white in winter to yellowish-brown in summer. Its underparts are always a light greyish-white. The back and legs are liberally sprinkled with dark spots, and there are a few spots across the light-coloured breast, but there are seldom any on the belly. There are many individual variations in the colour and number of spots. The rather snub-nosed face is framed by a long, beard-like growth of hair along the cheeks; and the lynx's eyes shine bright yellow. The tip of the tail is black. A mature lynx usually weighs between 20 and 30 kg, but some individuals have weighed up to 45 kg. The male lynx is considerably larger than the female.

Like most other felines, the lynx has a short, broad skull, which measures 129–144 mm in length and 100–111 mm wide. There are 28 teeth set in rather short jaws, and there is a reduced number of molars. The European wildcat (*Felis sylvestris*) has 30 teeth, and this can also occur woth the lynx, but both species have at least four fewer teeth than other European carnivores. The lynx's molars have serrated cutting egdes.

Like the domestic cat, the lynx leaves rounded footprints in snow or on damp ground. The claws are retractile and seldom leave an imprint. There are five toes on the forepaws and four on the hind paws.

The longest known lifespan of a lynx is 17 years, although most wild individuals do not attain this age. Man's persecution has been the principal threat to the lynx in historic times, although in recent years fox mange has severely depleted the population in Scandinavia.

### Reproduction

The male and female lynx meet only briefly for the purposes of mating. The mating season takes place between February and April, but it occurs at slightly different times in different parts of the lynx's distribution area. The male and female remain together for one or two weeks, but the female is in season and receptive to the male for only a few days. During this time the pair will mate repeatedly, and it is believed that the presence of the male stimulates ovulation in the female, thereby making fertilisation possible. During the mating season the female temporarily abandons any young which she may have, and she rejoins them when the male leaves her after mating. The female lynx is regarded as sexually mature when she is almost two years old, and the males are mature a year later.

The gestation period is just under 2½ months, and the yearling cubs stay with their mother until about six weeks before the new litter is born. A typical litter will comprise 2–4 young, but the numbers can vary. Litters of up to five cubs have been recorded. The cubs are blind at birth, and are born in dens in rocky cairns, natural caves and similar protected sites. The cubs' eyes are open at 16–17 days, and the newly born cub weights about 70 grams.

The young stay close to the den during the first two months of their lives while they are suckling their mother. Thereafter they begin to eat the regurgitated food the mother gives them, but they continue to supplement their diet of meat with milk until they are about six months old. They demonstrate their skill in climbing up and down trees, and the mother leaves them, often for quite long periods, while she hunts.

The cubs' development therefore proceeds slowly, and they stay close to her and live with her until her next litter is born. She provides them with food and teaches them how to hunt; and they begin to hunt successfully for themselves by about November, having significantly grown in size and weight during September and October. The cubs' early training is important, for they must learn not only to hunt prey of various sizes and types, but also to recognise potential prey and to kill it efficiently. During late winter, when they are about one year old, the young begin to lead fully independent lives.

### Related species

The relationship of the lynx to other feline species, and also its formal scientific naming, have been the subject of a good deal of controversy. This is because of the great variation to be found in the colouration, the pattern of spots and the animal's size. Some zoologists have included the three northernmost forms in one species, *Lynx lynx*, while others have regarded them as distinct species. The three forms in question are the pardel or Spanish lynx (*Lynx pardina*); which occurs in the Iberian peninsula; the northern lynx (*Lynx lynx*); and the North American lynx (*Lynx canadensis*).

The tradition in the Nordic countries during the early part of the last century was to divide the lynx into three types, depending on their colouration. These were the "wolf-lynx", the "cat-lynx", and the "fox-lynx". However, in 1847 Sven Nilsson of Sweden, a celebrated zoologist and an experienced hunter, expressed the view that "these three forms constitute only accidental forms of one and the same species". This observation has subsequently been proved correct.

### Diet

The lynx is believed to feed mainly on small mammals such as hares, rabbits, foxes and various woodland birds. They will also catch full grown Roe deer and their fawns, especially in winter. The most important prey species for the Spanish lynx is thought to be rabbits; for the North American lynx the Snowshoe rabbit is especially important; and hares account for about half of the diet of lynx in eastern Poland. In many areas the remaining half of the diet is believed to be composed of Roe deer and fawns. However, the availability of prey and the species which live in the locality determine what prey species are taken, and in matter of diet the lynx seems to be very adaptable and rather opportunist. Birds comprise only 1%–2% of their diet. In some areas lynx are known to hunt sheep, domestic cats and Reindeer calves, but these seldom comprise more than a small proportion of their diet. The lynx is believed to require about one kg of meat per day, and a single hare is more than adequate for the daily requirement of a full grown lynx. In some parts of Spain and Portugal the lynx feeds almost exclusively on Red-legged partridges, while in other parts of Europe it feeds almost exclusively on rabbits.

## Adaptations and behaviour

The lynx usually lives a solitary life, with the exception of the mating season and during the time when the female is followed by her young cubs. It is very much a nocturnal animal, spending the daylight hours lying up in a day lair, which may be inside a hollow tree but is more usually in a rocky outcrop where the animals has a commanding view of the surrounding countryside. From this type of place a lynx can spot any imminent danger or disturbance. From observations of lynxes in captivity it is known that they can detect movement as far as 4 km away, and can hear high-pitched sounds at up to a distance of 5 km. A lynx seldom uses the same day lair twice in succession, and they tend to roam over wide areas. The North American lynx is known to cover distances of up to 80 km in a 24-hour period. Territories vary in size: in northern Sweden territories are from 300 $km^2$ up to 1,000 $km^2$. In continental Europe the territory size is thought to be between 10 and 250 $km^2$. It is apparent that the size of the territory is partly influenced by the availability of prey species.

The lynx population is so small at present that there is little or no competition between individual animals for living space. Contact between individual animals is thought to be minimal, but each animal identifies its territory with scent markings. It deposits faeces on rises and hummocks on the ground, and sprays urine on rocks, tree stumps and tree trunks. Like many other felines, the lynx sharpens its claws by scratching on tree trunks and stumps.

The lynx's hunting methods have been the subject of much speculation. Lynx have often been described as animals which hunt in groups of adult animals working together in an organised pursuit and attack, but this is incorrect. Through repreated tracking of lynx in the snow it is possible to gain a good understanding of its hunting strategies, and other information has been gleaned from radio-tagged prey species and from examinations of lynx food remains. These studies all show that the lynx is a solitary hunter. The female goes out alone. leaving her young behind, and she pursues her quarry by stalking stealthily. Like many cat species she does not chase her prey but generally stalks it carefully and makes a surprise attack from close range. According to one study in Sweden, most attacks are made from a distance of not more than 20 metres. In winter the lynx often prefers to hunt Roe deer, seizing them by the throat and neck, and the puncture marks from the four strong canine teeth can often be seen clearly. The struggle is usually brief and the lynx kills quickly. Smaller mammals such as hares are usually killed with a single bite across the back, and there are no prey remains left. The lynx seldom buries any of its prey; and it has been known to drag dead Roe deer for distances of up to several hundred metres.

## Distribution and numbers

If the three lynx groups mentioned earlier are regarded as one species, the distribution of the lynx may be said to cover all of Europe, Asia and North America. Its spotted fur identifies the lynx as a woodland animal, although this does not prevent it occupying open country such as alpine areas bordering on forests of osier and alpine birch. In southern and central Europe the lynx can be found at altitudes up to 2,000 metres.

Intense persecution has greatly reduced the numbers and distribution of the lynx in Europe. In central and southern Europe it has been forced to retreat into remote mountain regions and reserves, and by around the middle of the last century the lynx was exterminated in Germany and Austria. The same was true of France, apart from a few animals remaining in the Pyrenees and the Alps in the early 1900s. Attempts have been made to reintroduce the lynx to its old haunts at various times, including an introduction to Germany in

1938, and later in France, Austria and Switzerland. Today the European lynx population probably comprises 2–3,000 animals. In the early 1980s the Swedish population alone was estimated at around 600 animals, having increased from an estimated 175 animals in 1952 and 2–300 in 1963. However, there has been a sharp decrease in the Swedish lynx population since the mid-1980s.

The majority of Swedish lynx are found in the north of the country, and during the 1960s and 1970s between 10 and 70 lynx were shot every year. In the 1985–86 hunting season only five were shot.

*European wildcat.*

The Finnish population is estimated at around 550 animals, and the lynx occurs in Finnish and Russian Karelia, straddling the border with Russia. The lynx has been protected in Finland since 1968, since when it has increased significantly in numbers. Some licensed hunting has been permitted, however, but the number of animals killed has been kept to not more than 10–20 per year. The Finnish Council of Natural Resources has determined that the population in Finland could be allowed to increase to a total of 900–1,000 individuals.

Lynx numbers have also increased in Norway, although the movement northwards has fallen off in recent years. The lynx exists across large areas of Norway, but it is only rarely found in the northern Finnmark area. Some 25–35 animals are shot annually.

## A European relative

The lynx has one wild feline relative in Europe, the European wildcat (*Felis sylvestris*). In general the wildcat looks like a large grey striped domestic cat, but it has longer hair and much thicker fur. The thick fur is especially prominent in the animal's big, bushy tail, which is distinctively ringed with alternating light and dark bands. The tip of the tail is dark, and the head is larger than that of a domestic cat, with more distinctive markings.

The European wildcat is closely related to the African wildcat (*Felis s. libyca*), and these two are now often regarded as local subspecies of the same species, and they are very similar in appearance. These two cats are often regarded as two racial groups among the species of small wildcats. Otherwise, three species are identified: the European, the African and the Asian wildcat (*F. s. ornata*). The African wildcat is regarded as the ancestor of the domestic cat, which is well known from images of four thousand years ago and from mummies in ancient Egyptian tombs and temples. The two races are so closely related, however, that the European wildcat and the domestic cat often interbreed. Over many years the European wildcats have undoubtedly acquired some genetic characteristics from the domestic cat, and vice versa. A study of the skulls of wildcats from different parts of Europe showed that 40%–70% of the populations were purebred.

The European wildcat is big and powerful. Its total length is rather more than one metre, and the maximum body length can be as much as 80 cm, to which can be added a further 35 cm of tail. Most individuals are smaller than this, however. The tail normally accounts for 40%–45% of the animal's total body length, and the males, at 11.5 kg, are significantly larger and more heavily built than the females, which weigh around 10 kg. A record weight of 14.8 kg has been recorded for a wildcat shot in central Europe, while the weight of the average Scottish male wildcat is less than 5 kg, with the females weighing barely 4 kg. The wildcat has a typical cat cranium with a snub nose, and it has a total of 30 teeth.

The European wildcat lives in several isolated populations across the continent, and there is another isolated northern population in Scotland. It is also found in large areas of eastern France and extending into Germany, where it also exists in several small, isolated populations. The largest continuous area which it occupies is in the Mediterranean countries, where it extends across Spain and Portugal, the Pyrenees, southern and central Italy and most of the Balkans and Yugoslavia, and into parts of Romania, Bulgaria and Czechoslovakia. It is also found on most of the large islands of the Mediterranean.

Historically, the wildcat existed after the Ice Age in Europe south of the Baltic and throughout European Russia and in Britain. However, it has never existed in the Scandinavian peninsula. In the Alps it will live at altitudes up to 1,500–2,000 metres, and in Scotland at up to

500 metres. As its Latin name implies, the wildcat is a creature of the woods and forests, where it preys mainly on birds and small mammals. In many areas the principal prey is small rodents, voles and mice. Wildcats roam over quite large areas, and they are mainly nocturnal in their activities. The male's home range may be as much as 175 hectares, but it is normally found to defend a territory of about 70 hectares. The females tend to be more sedentary in their habits.

Depending upon the food supply, the wildcat lives alone or in small groups. The females are in season for less than one week during the period February–March, and gestation lasts for 63–68 days, which means that the litters, which range in size from 1–8 young, are born mainly in April and June. They leave their mother's den when they are a little over one month old, and they suckle up to the age of four months, after which they quickly become independent.

Wildcats have been seriously persecuted across their entire distribution area, often due to the alleged damage they do to gamebird populations. The species is protected in many countries in central Europe, and there the numbers appear to be increasing. Increases have also been noted in Scotland, where the wildcat is still persecuted as it is regarded as a predator of grouse.

## Relatives of the lynx

The lynx is a cat, and therefore belongs to a rather exclusive group of animals. There are only 35 known cat species worldwide. The family, *Felidae*, is represented all over the world, with the exceptions of Australia and the inhospitable polar regions. Many of the larger species like the lion, tiger, leopard and jaguar are well known, but the smaller cats are less familiar.

The lynx shares many characteristics with its relatives, including the ability to retract its claws into a special pocket above the pads of its feet. Cats are known as digitigrades, which indicates that they walk on their toes. They have erect triangular ears, and they have black and white markings on the back of their ears, which is an important way of signalling in dim light, for cats are mainly nocturnal. Their eyes reflect brilliantly in the dark. Like most woodland cats, which are the dominant group within the cat family, the lynx is shy and cautious, and tends to lead a secretive life.

The lynx is included among the smaller cats, and is a member of their collective genus, *Felis*. *Felis* comprises the lynx's relatives, the desert lynx and the red lynx, as well as the European wildcat, the ocelot and the puma. There are only two wild cats in Europe, the lynx and the wildcat. Lynx are special in many ways, and one of these is that of all 35 cat species, the lynx is the only one that is bob-tailed.

*The lynx's skull has a typical shape, strongly domed on top, and with shortened jaws like those of other cats. There are normally 28 (30) teeth — two less than in domestic cats and European wildcats. Compared with weasels and canine animals, it has fewer molars. The total skull length is 129–144 mm, at least 28 mm more than the wildcat's.*

# Martens

The martens belong to the family *Mustelidae*, marten- and weasel-like animals which are distinguished by the elongated shape of the body and a medium-long tail. There are five toes on all four feet, unlike dogs and cats, which have only four toes on the hind leg. The marten's legs are comparatively short, and its movements are lithe and sinuous. The ears are small and rounded, and most species have anal scent glands. Both the Pine Marten and the Beech Marten occur in Europe, and there are five other species of the family *Martes* in North America and Asia.

# Pine Marten

## *Martes martes*

The European martens of both species belong to the nominate forms *M. m. martes* and *M. f. foinaf*, but some other populations with a different racial status live to the east. The well-known sable (*Martes zibellina*), so valued for its skin and now persecuted almost to extinction over large areas, is an Asiatic relative of the European martens.

It is easy to mistake the Pine Marten for its relative the Beech Marten. The Pine Marten's chest spot is smaller and coloured a yellowish-white, and it never extends as far down the animal's forelegs as it does on the Beech Marten. The fur of the Pine Marten is also a deeper, darker brown, and not greyish-brown like the beech marten's. The Pine Marten also has a darker black to greyish-brown nose, whereas that of the Beech Marten is greyish-pink. Both species have 38 teeth, but the Pine Marten's third molar, counted from the front, on the upper jaw, has a convex outer edge, whereas that of the Beech Marten is concave.

The soles of the Pine Marten's feet are richly covered with fur, and its ears are rounded, with a conspicuous light edge.

As an adult the Pine Marten weighs up to 1.8 kg. Generally the male is heavier, but the female may attain the size and weight of a small male. The male Beech Marten can weigh up to 2.3 kg, and the males are also generally heavier than the females.

The Pine Marten lives mainly in woodland, especially older coniferous forests and mixed woodlands, where the squirrel — its favourite prey — is most common. In Scandinavia the marten is the dominant predator of this type, while in mainland Europe it shares the same habitat with the Beech Marten.

The marten is said to avoid farms and human settlements, while the Beech Marten has adapted itself to them. This may be the outcome of competition between the two species, rather than an indication of the particular shyness of the Pine Marten. In Sweden it occasionally happens that a Pine Marten will make its home in a farm steading, and like the continental Beech Marten it has even been known to make its home in the attic of a house, and in outbuildings close to houses.

Both Pine Martens and Beech Martens are most active at dawn and dusk, and during the night, but the Beech Marten seems to be more active during the daytime. Both martens have a similar diet of small invertebrates and occasional berries and plants; but squirrels are the Pine Marten's main prey in northern parts of Europe, and it has been estimated that a Pine Marten will kill and eat an average of eight squirrels per month. Exactly how many it kills depends on the density of the local squirrel population. Young birds, rabbits and rodents are also important prey, and the Pine Marten has also been known to attack larger animals such as Roe deer.

### Reproduction

Both martens have a delayed implantation and development of the foetus, as do Stoats and Weasels. This means that their young are born 8–9½ months after mating has taken place. The mating period occurs in July–August, and both species may have a second mating period in January–February. The single annual litter comprises 2–5 young, but in years when squirrels are numerous and food abundant, martens can have up to seven young in one litter. The female marten chooses old squirrel dreys and holes in trees for her nesting place, and she also makes occasional use of bird nesting boxes and niches among craggy rocks.

The young are born blind and their eyes open at the age of about five weeks. They are dependent on their mother for a long time, and suckle for about three months. However, as early as two months old they will follow the mother on ventures out into the vicinity of the nest. They are sexually mature at the age of 2–3 years, and the Beech Marten usually matures earlier than the Pine Marten. Newborn Pine Martens weigh about 30 grams.

## Occurrence and subspecies

The Pine Marten is found in most parts of Europe. There are gaps in its European distribution, however, and these include Spain and Portugal south of the Pyrenees, the Balkans, south-west Finland and some Mediterranean islands, including Corsica, Cyprus and Crete. The Beech Marten was one of the mammals of southern Europe which did not have time to take advantage of the ancient land bridge between northern Europe and the Scandinavian peninsula, and they are therefore not found in Norway or Sweden. However, the Pine Marten did find its way to these countries, and it is now widespread from Lapland to the southern coasts. In some places it is a permanent inhabitant of the birch forests of the high mountain areas. Its numbers vary according to the food supply, and from time to time the pressures of sporting shooting and commercial hunting have had an effect on marten numbers. Several thousand martens are caught or shot every year, and in the hunting year of 1985–86 a total of 10,000 martens were taken by shooters and hunters.

Martens are thought to have benefited from the decline in red fox numbers due to rabies, and in Scandinavia their numbers have increased in recent years. Despite the fact that the marten is largely a tree-living species, there can be considerable competition for prey between martens and foxes.

*Beech Marten (left) and Pine Marten*

# Beech Marten

## *Martes foina*

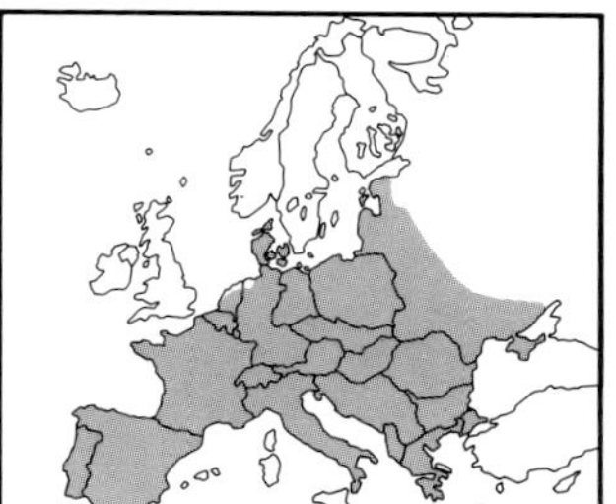

In bodily conformation the Beech Marten resembles its close relative the Pine Marten, but its fur has a thinner layer of shorter hair. Its white throat patch is larger and it extends down the forelegs, although the actual shape varies considerably from one individual to another. This animal is usually white and not yellowish like the Pine Marten. The pads of the feet have no hair, unlike the Pine Marten's, and although the tooth patterns are the same, the rear upper molars are shaped differently.

A Beech Marten will weigh 1.1–2.3 kg, the adult males being usually over 1.7 kg in weight, while the females are usually less than 1.5 kg. The Beech Marten is believed to be able to live for up to 15 years.

Beech Martens are mainly nocturnal, and they are rather drawn to human settlements, farms and villages; and they will even occur in large cities such as Berne and Berlin. They are also attracted to arable farmland where there are plantations of trees and rocky outcrops. In the Alps they will live at altitudes of up to 2,000 metres, and they are very skilful climbers. They have no difficulty in climbing into attics, lofts, dovecots and poultry houses, and they will take eggs, young birds and adult poultry, which makes them a great pest for the poultry farmer. They will also catch small rodents, birds and especially rats, around farmyards, houses and on rubbish dumps, and in most of continental Europe they are more common than Pine Martens.

## Reproduction

The Beech Marten appears to become sexually mature at two years old, and the females produce one litter of young annually thereafter. Mating takes place in the summer and the litters of 3–5 young are born in the period March–May. The mother, which has four teats, suckles her blind young until their eyes open at the age of about five weeks, and she continues to lead and feed them until they are about three months old.

Beech Martens are very widespread, and besides Europe they occur right across central Asia as far as Mongolia. They are absent from the British Isles, despite attempts to introduce them in the 1930s, and also from Fennoscandia, but are found widely elsewhere in Europe.

Their distribution extends further east than the Pine Marten's. Both species coexist in the Caucasus and parts of south-west Asia, and in the Alps the Pine Marten is found at heights of up to 1,800 metres, while Beech Martens occur at up to 2,000 metres.

In Denmark the Beech Marten is considered more common than the Pine Marten, and their numbers there have increased considerably this century. During the 1960s about 1,000 Beech Martens were shot or trapped annually, while in the 1980s the annual number had risen to about 6,000.

*The bull elk's antlers make an attractive trophy. This partly reflects the quality of the elk population. Royal bull elks, which are biggest, have the largest horns. The horns are judged in terms of the span between the outermost tines, the height and width of each horn, the number and average length of tines, and the circumference of the horn base. All are given points. To be classed as a gold medal head, 300 points are needed. The European record is held by a Swedish thirty-pointer scoring 414 C.I.C. points.*

# Elk

## *Alces alces*

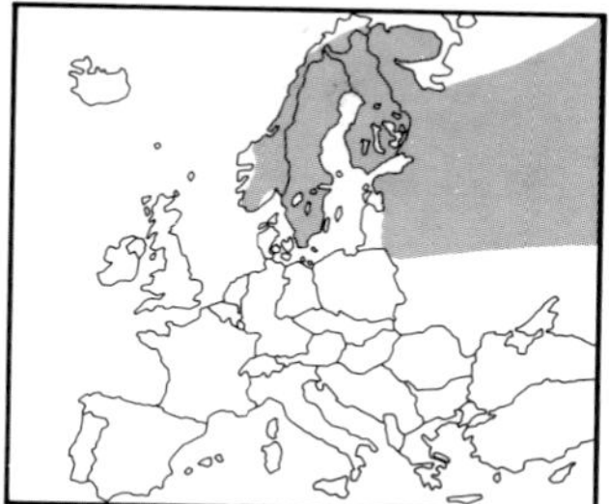

The Elk or Moose is the world's largest deer, and its great size and relatively dense populations have attracted man's interest ever since ancient times, when the Elk was an important quarry for primitive hunters. Despite changing social conditions and changing forms of land use, the Elk has maintained its status as a common and important animal. Attempts have been made to domesticate the Elk, but this has been as unsuccessful as attempts to tame African elephants. The Elk has also proved unsuitable for enclosing in parks, as is done with red deer and fallow deer. In 1590, when Elks still roamed freely over most of the country, the Swedish king, King John III, who had a great interest in deer, had a special Elk park created, in an unsuccessful attempt to build up a herd of park Elk.

At much the same time attempts were being made in Germany to reintroduce the Elk to deer parks in areas where it had once been common, before it was exterminated by over-hunting in medieval times, and these attempts continued in the seventeenth and eighteenth centuries.

The great Swedish naturalist Linnaeus was interested in these large deer, and recorded the fact that one of the Swedish provincial governors had a two-year-old tame Elk, and in his later lectures he stated that it was perfectly possible to tame an Elk if it was taken from the wild at not more than 14 days old. Any older and it was virtually impossible to tame. Peter Thunberg, a pupil of Linnaeus, recorded in 1798 that several Elk had been born and tamed in a park at Uppsala, and in 1779 the Royal Swedish Academy of Sciences had offered a substantial cash award to anyone who successfully domesticated the Elk and bred from it in captivity.

Elk were proverbial for being fast swimmers and also fast runners, and some Elk have been trained to pull sledges in both Scandinavia and Russia. King Karl IX of Sweden forbade the keeping of domesticated Elk in the 1600s, apparently for the bizarre reason that criminals fleeing from justice used them as mounts on which to escape, for their speed and endurance was far greater than that of a horse.

In the eighteenth century Elk were often caught in pits and traps, as the country people rarely had access to suitable firearms, and in summer Elk were even hunted with scythes and billhooks, and with wooden traps and hand-held spears. Whatever weapons came to hand were used to kill Elk, and this remained true until quite recent times. In 1936 a Swedish police survey of Elk hunters revealed that over 50% of them had no hunting licences, while the majority used shotguns rather than rifles. The shotgunner's tactics involved shooting at the Elk so as to smash its leg joints, after which the wounded animal could be approached and finished off with another shot. 73 hunters had killed 17 Elk in one season by this cruel method.

Today the countries of Scandinavia are models in wise and carefully controlled shooting of Elk. Hunting and culling are strictly governed by law, and the sportsmen take an important part in the decision making. In Sweden it is estimated that one of four sportsmen participate in Elk shooting, amounting to about 225,000 individuals, but in spite of this it is still difficult for them to achieve the annual cull targets which are usually in the order of 125–170,000 Elk. Research has shown that the populace of Sweden are 72% in favour of sporting shooting, and only 10% are definitely opposed to it. It is also clear that the sport of game shooting is not reserved only for those who are privileged by their wealth or social position. All the evidence points to the conclusion that Elk hunting is very much an activity within the reach of "the average Swede".

Throughout Europe the tradition has been that the hunting and shooting of Elk and similar big game animals has been a "royal" activity, what the Germans refer to as "die hohe Jagd", which had usually been reserved for the diversion of kings, noblemen and landed gentlemen. Today the pendulum has swung back towards the spirit of the Middle Ages, when it was generally held that "the Elk belongs to him who has killed it". However, this more democratic attitude has to be regulated by a system of laws and licensing arrangements.

### Characteristics

The Elk is distinguished from other European deer by the sheer size of its body, and also by the length of its legs. The head is large and elongated, and the muzzle is decidedly pendulous. There is a rough hump across the middle of the animal's shoulder blades. The males (bulls) carry antlers from September to December-January, and they have a long and hairy "beard" that hangs down more than 10 cm below the neck. The head of a cow Elk is more elongated than that of the bull; she lacks antlers and has only a hint of a beard. The forequarters of an Elk look disproportionately broad and powerful, while the back slopes down towards the rump, which is lower and narrower.

The bulls are by far the largest, weighing 20%–30% more than the cows; in Scandinavia a bull Elk may weigh up to 500 kg and more. In North America a bull Elk (or "Moose", as they are locally termed) can weigh several hundred kilos more, and weights of up to 800 kg are not unknown. The Elk's hair is shaggy and it is a greyish brown-black in colour, but the legs and the belly are a light greyish-white. The hair is about 10 cm long, but the mane of a mature bull may be 15–20 cm long. The tail is short, less than 10 cm long, and there is no rump patch, although cows and their female calves have a light-coloured ring around their genitalia. The Elk's eyes are brown, and in some regions Elk occasionally occur with reddish eyes.

Elk calves are reddish-brown during their first 3–4 months, and their fur does not become the typical greyish-brown until September. Unlike other young European deer, Elk calves do not have spotted coats. Calves have rounded but rather triangular shaped heads, and at birth they weigh 10–15 kg. By the time they are four weeks old they have quadrupled their birth weight, and by four months old they weigh ten times more than at birth. By its second autumn the young Elk will weigh close to 300 kg.

Elk droppings consist of rounded pellets, and an adult will deposit a pile of 100–300 pellets at a time. When the diet consists of soft and moist foods the pellets are soft and shiny, and they cling together in a clump. Calf droppings are smaller and rounder than those of the adults, and often have a soft and shiny surface.

### Antlers

During its first autumn the bull calf develops a pair of velvet-covered antlers which are only a few centimetres long. These continue to grow during the winter and spring, and by the time the Elk is a year old he has what can best be described as a pair of antlers shaped like bicycle handles, with one branch and one tine on each side. Well developed young bulls can have forked antlers, but this is more characteristic of two year olds, which may even have three tines on each antler.

The most powerful antlers are found on 6–12 year old bulls, and after this age the antlers begin to decline in size and quality. Ageing bulls have antlers which are described as "going back". Antlers are classified and described by the number of points, counting in even numbers. A bull Elk with two points on one antler and three on the other is described as a six-pointer, and if it is necessary to describe it so as to distinguish it from a bull with three points on each antler, it is described an an "uneven six-pointer". Pointed tines and forked antlers always branch off from the main beam of the antler, and mature bulls may have multiple points, where the points stem from an area of flattened or palmated antler. Bull Elk with palmated antlers are more common in northern Scandinavia than in the south.

The antlers begin to grow in April from the pedicle, a knob of bone on the skull, and the antler is fully developed by the end of summer. The growing antlers are covered in a downy-haired skin called "velvet", a membrane of skin which is full of blood vessels. The velvet dies and is rubbed off the hardening antlers in August–September, leaving the newly exposed antlers shining red from the tattered velvet and the blood vessels that had earlier nourished the growing antlers. The bulls rub their antlers against rough tree trunks and branches, and they become hardened and burnished, acquiring a deep brown colouration. The antlers are shed in the period December–March, and it is noticeable that the older bulls shed their antlers first, well before the younger ones. The appearance and size of the antlers depend upon the animal's age, but this is also influenced by the animal's genetic make-up and on the food supply which has been available to it while the antler has been growing in velvet. Antlers can weigh up to 20 kg, and their total length can be 1.6 metres, while those of the North American Moose can be as much as 1.97 metres long. In Scandinavia antlers of record length measure up to 1.5 metres.

## Teeth

A fully grown Moose has 32 teeth, with a structure:

$$\frac{0\ \ 0\ \ 3\ \ 3}{3\ \ 1\ \ 3\ \ 3}$$

but it can sometimes have an extra false molar or premolar in the upper jaw, which raises the total to 34 teeth. The calf has only four molars in both the upper and lower jaw. At one year old the calf sheds its milk teeth, and by 16 months old a complete change of teeth has taken place. During its second year the young animal can be distinguished from older Elk by the fact that the three front molars (premolars) are lighter than the three at the back.

An Elk is considered to live up to 20–25 years of age in the wild, although it is fully grown at the age of 4–6 years, and the cows mature somewhat earlier than the bulls.

The Moose normally appears to be a silent animal. It does make various noises, but these cannot normally be heard far off, and most of its sounds are like nasal snufflings.

## Adaptations and behaviour

The Elk's daily activities are largely determined by the amount of disturbance which takes place in its vicinity. Where Elk are free from human disturbance they can move and feed by day and by night, but in general the Elk is a nocturnal animal, which lies up in cover by day and becomes active at dusk and again at dawn. Elk often choose to rest out in open countryside, where they can easily detect danger while it is still a long way off. Elk often seek shelter by night near to their feeding areas, in grasslands, marshes and clear-felled woodland areas, sometimes moving into arable fields and often found along a woodland edge. They impressions made by the size and weight of their large bodies make their bedding places easy to identify.

Elk are generally solitary animals, but the cows are followed by their calves throughout the year, and the second year calves are replaced by the newborn ones. During the rutting period the bull joins first one and then another of these mother-and-offspring groups, for a short time in each case. In some northern parts of its range Elk will form winter herds of 20–30 animals, and herds can also form in areas of particularly good feeding. In southern Scandinavia it is common to find herds consisting of a few cows and their calves accompanied by one mature bull or several young males.

The cows are mainly sedentary, usually giving birth in the same place year after year. However, the bulls and young animals often wander over large areas of the countryside. In some northern areas they migrate seasonally, moving between winter and summer ranges. A typical migration is probably 50 km, but movements of up to 110 km have also been recorded.

It is known that young Elk usually return to their mothers' home ranges, but their wanderings may result in their acquiring new ranges of their own, to which they adapt quickly. Summer territories for mature Elk in northern Scandinavia can be from 800 to 3,000 hectares, and in the south territories may be of a similar size, but are usually much smaller. The bulls maintain large territories, which often include the territories of several females.

The Elk's sense of smell is usually regarded as its most acutely developed sense, but it also has very good hearing. The Elk's sight, however, is not so important, and it can have difficulty in seeing objects far off, or things which are not moving. Elk are colour-blind. When they sense danger they turn their heads towards the source of the threat, sometimes with the muzzle raised in an attempt to catch the scent of an intruder, and the ears are always swivelled forwards. When an Elk is feeling aggressive his fur bristles and his ears flick back and forth. The Elk attacks by lashing out with its forelegs in powerful kicks. Elk are fast runners and have been clocked at 60 km/hour. They are good swimmers and also seem to enjoy wading in water.

## Reproduction and rutting

The Elk rut occurs in September–October, and it begins earlier in the south than in the north. The gestation period is about eight months, and since the young females become sexually mature during their second autumn, the first breeding usually takes place at two years old. In more northerly areas the first calving may not take place until the third year, and it is assumed that all Elk breed annually between the ages of 5 and 11 years of age.

During the rutting season the bulls are active. They dig rutting wallows with their forelegs, mark them with urine and often wallow in them. They chase away smaller and weaker rivals, and court the females with snuffling sounds. At this time of the year the bull Elk may be seen with the cow and her previous year's calves, and this creates an idyllic illusion of a family group, for such gatherings are only very temporary.

The female is in season and able to be fertilised for only one day, but she comes into season every third week. Elk are polygamous. A surplus of females is not necessarily detrimental to the population, but if there is a large imbalance between the numbers of females and males it can cause some cows to come in season a second or a third time before they are mated, and this then results in the calves

*Bulls, cows and young Elk have very different head shapes. The bull has a thick head with a conspicuous mane and dewlap, while the cow's head is more slender. Calves and young Elk have short, almost triangular heads. The bulls can usually be identified individually by their horns' size, form, and number of tines. There is a distinction between beam antlers (shown above) and palmated antlers (below).*

eventually being born 3–6 weeks later than normal, and such a calf has a reduced chance of survival. In extreme cases a cow may not be mated, but barren cows are very rare. A low density of bull Elk is often brought about by the over-shooting of mature bulls, and this can result in a young or weak bull Elk mating with a number of cows and weakening the genetic characteristics of the species. This is partly compensated for by the fact that the breeding cows also make their own genetic contribution, and they tend to live longer than the bulls.

As a rule the young cows give birth to only one calf, but in older cows twins are usual, and triplets are not unknown. The Elk cow has four nipples. The abundance and quality of the food supply affects the Elks' sexual maturity, the timing of the rut, the number of calves and their weight and condition when they are born. Elk calves weigh 10–15 kg, and may weigh as much as 17 kg at birth. They are born with their eyes open and can stand after an hour or so of birth. Within two or three days they can follow their mothers without difficulty, and the vital bond between the cow and her calves is forged during the first day after birth. Disturbance at this critical time can result in the cow rejecting her calf or even kicking it to death. After the first day the risks of rejection are very slight. The calf is totally dependent on its mother's milk for 6–8 weeks, but they usually continue to suckle until they are about four months old. The yearlings depend on their mothers' guidance for the entire first year of their lives, and they are not rejected by their mothers until she gives birth the following year.

Cows normally give birth each year, and since they can live to 20 years of age and are still fertile at the age of 15–20 years, a female Elk in ideal circumstances may give birth to about 35 calves.

## Diet and nutritional needs

The adult Elk needs between 10 and 20 kg of food each day, depending on the nature of the food and the size of the animal. It is almost entirely herbivorous, and grazes both winter and summer on bushes and trees such as rowan, aspen, sallow and willows, as well as other deciduous trees and shrubs. During the spring and summer the shoots of the deciduous trees are laden with tender buds and leaves. During winter the shoots are not so tender, and the Elk's basic diet often consists of the previous year's growth of pine and juniper shoots. In summer Elk often forage for herbaceous plants along the edges of watercourses and in marshes, and it is not unusual for them to wade deep into the water. Sometimes they even duck their heads under water to feed! They are considered to be especially fond of the roots of water lillies and other water plants. In late summer and autumn Moose periodically invade cereal fields, and they are often attracted by apples in quiet orchards and gardens. Various kinds of low-growing bushes and shrubs, mainly blueberries and heather, are also part of their diet.

## Enemies, diseases and mortality

The Elk's greatest enemy is probably man, with the wolf coming second. By tracking wolves in Scandinavia it has been found that a single wolf is capable of killing an adult Elk, and that this is a common occurrence. In areas where wolves are numerous, prey of this large size is usually hunted by a pack of wolves, but in light of the present size of the Scandinavian wolf population, kills of this sort have a negligible effect on the numbers of Elk. Shooting remains the principal means by which Elk numbers are controlled. But in Sweden motor traffic accounts for the deaths of 5–6,000 Elk every year, and many more Elk also die of injuries after having been wounded by a shooter or hit by a car. The nuclear accident at Chernobyl is also believed to have hit some Elk populations very hard, the principal consequence being that the flesh of the animals was rendered unfit for consumption because of the levels of radiation and the accumulations of heavy metals in the liver and kidneys.

Of the natural disease to which the Elk may fall victim, about 1% of Scandinavian Moose are afflicted by the virus disease known as fibrous papillomatosis. The symptoms of this are greyish-black tumours on the skin, varying in size from that of a hazelnut to the size of an orange. Animals with a mild infection may recover, but otherwise the animal loses weight and dies. However, the flesh of Elk infected with this disease is considered safe to eat.

The Scandinavian Elk population has been affected in recent years by infestations of the so-called "brain worm", *pelaphostrongylus*, a threadworm which is about 5–6 cm long. The life cycle of this parasite includes a stage in which it occurs in snails, which the Elk eats along with damp vegetation. From the digestive tract the worms try to develop in the animal's tissue, and if they reach the brain or the nervous system they cause paralysis, emaciation and death.

## Distribution, subspecies and habitat

The Elk is a member of the holarctic fauna, which means that it occurs throughout the northern hemisphere in Europe, Asia and North America. Although the Elk does not vary very much in appearance across this wide distribution range, six subspecies have been identified, of which the giant Moose of Alaska (*Alces a. gigas*) is perhaps the best-known. Exceptionally large Alaskan Moose (as Elk are called in North America) may weigh as much as 1,000 km. Size is the main factor in identifying the various subspecies. The European Elk (*Alces a. alces*) is replaced in east Asia by the subspecies *Alces a. cameloides*. The other three subspecies and the giant Moose already mentioned are found in North America.

The Elk is a forest animal. It prefers young forests, especially those which are badly drained and where the trees grown in stagnant water, as well as in areas adjacent to marshes, rivers and bogland

rich in vegetation. Changes in timber extraction practices in northern Europe have benefited the Elk by providing large clear-felled areas where tender young leaves and secondary growth make good grazing ground. In some areas the Elk will move in summer to higher terrain, not above the timberline but up among areas of willow and mountain birch scrub. In winter Elk prefer young coniferous forests.

The Elk was formerly common over much of continental Europe, but it is now absent from areas south of the Baltic Sea. In Denmark there was a healthy population of Elk until the Neolithic age, and very occasionally an individual Elk may swim across the narrow sea passage from Sweden to Denmark.

Elk became extinct in many parts of Germany in the Middle Ages, but it survived in parts of Prussia and along the Polish border. In the mid-1930s the Elk population of eastern Prussia was estimated at about 1,000 animals, but these were shot out during the Second World War. The Elk was also rendered extinct in Poland, but it was reintroduced there in 1957. Occasionally a lone Elk may wander into Germany from the Baltic states and from Poland.

The Elk used to be very common in Scandinavia, but unfortunately the hunting legislation introduced in 1789 by King Gustav III gave greatly increased hunting rights to ordinary landowners. This had a devastating effect on a population of Elk which was already in some danger of extinction in certain areas, and in 1760 the last Elk in southern Sweden was recorded as having been shot. By the beginning of the last century there was only a small residual population of Elk left in northern Sweden, and the situation was much the same in Finland and Norway.

During the latter years of the last century and in more recent times the Scandinavian Elk population has made a most remarkable recovery. This was brought about first by protective legislation, but assisted by reduced competition from other animals for grazing in the woodlands, and by the decimation of predators such as the wolf. In

recent years modern forestry management procedures have made the mixed and coniferous forests more accessible and attractive to Elk, and a policy of shooting mainly calves has helped to preserve the highly productive mature cows.

The size of the Elk population has often been estimated by extrapolating from the numbers shot each season. In 1910 4,538 Elk were shot in Sweden, Norway and Finland. In 1982 the number has risen to 250,000!

The largest increase in the Elk population, amounting to a population explosion, took place in the period 1945–70. In 1945 the Swedish winter herd was estimated at 47,000 animals. In 1953 the numbers had almost doubled, at 90,000; in 1961 it was 118,000 animals; and in 1978 it was reckoned to be 250,000. By 1982 this had further increased to over 300,000 animals, and in 1986 no fewer than 126,000 Elk were shot as part of the controlled culling campaign.

In Finland the total population in the winter of 1950 was reckoned at 5,000 Elk, but by 1985 a total of 70,000 animals were shot under licence. The Norwegian population in the late 1960s was estimated at 35,000 Elk, but by 1985 no fewer than 25,000 animals were culled.

## Man and the Elk

The optimum size of the Elk population is related to the availability of suitable grazing land for it. Man's attitude towards the Elk is shaped by other considerations also, including the damage done to young trees by Elk and also the significant number of traffic accidents. In addition, as the Elk population of Sweden produces as much meat as sheep, cattle and reindeer put together, man is looking for high yields both financially and in terms of the animals' breeding rates. In Finland experiments were carried out based on a density of 4–7 Elk per 1,000 hectares; in the 1980s four Elk to the same area was the target, a number which is acceptable in terms of forestry and agriculture as well as from the viewpoint of traffic safety. In many parts of Sweden the Elk density exceeds this. In the Jönköping area of southern Sweden the winter population densities in 1982 and 1986 were 5.1 and 6.8 Elk per 1,000 hectares. In northern Sweden in the years 1982 and 1985 the winter densities of Elk were estimated at 6.7 and 5.0 above the timberline, and 11.4 below the treeline. These high densities may represent wandering Elk, but in the Royal Park area around Hudiksvall a winter density of 33.6 Elk per 1,000 hectares has been established, while the natural density in adjacent areas is 8.0. Experiments and studies of tagged Elk have shown that they will wander up to 50 km away in spring. The density and distribution of the Elk in winter is to some extent influenced by the animals' wanderings and by the availability of good feeding habitats.

## Hunting

Elk hunting and shooting cannot be described briefly. In Sweden, however, it is carried out under licence with a relatively long open season, and with a strictly predetermined quota of cows and calves to be shot. In addition there is what is called "general hunting" in smaller areas, where during a period of only a few days it is permitted to shoot one animal (the sex of which may be specified) and one calf. Elk hunting is permitted only in designated hunting areas.

*The bull Elk bears antlers for most of the year. Only for a short time, some months from later winter to early summer, does it lack antlers. The time for shedding horns and growing new ones, however, varies with the individual, its age, and the area. New antlers, skin-covered, stop growing in late summer or early autumn. The skin is then cleaned off, and the antler under it is often blood-red with traces of skin. Elk cows are antlerless.*

# Woodland grouse

The Capercaillie and the Black Grouse have long been important gamebird species throughout Europe, both for the country people who live close to heaths and forests and also to sportsmen from the towns and from other countries.

There has been a long tradition in northern Europe and Scandinavia of stalking the displaying males of both these species, and shooting them on their breeding grounds. However, this has generally been viewed by conservationists as potentially very damaging to the populations of both species, especially when it is combined with other environmental changes which have militated against these woodland grouse. The Capercaillie population has decreased over much of Europe, and in large areas of Germany the Capercaillie became extinct during the last century and early this century. Capercaillie in Britain began to disappear as early as the seventeenth century, and the last Capercaillies in Ireland and Scotland disappeared in the late eighteenth century. Throughout the species' European range a serious general decline was reported in the last century.

In mainland Europe the Capercaillie has been regarded by naturalists as a vulnerable species, which had originally been a creature of the old forests of northern and central Europe. These forests were chiefly composed of conifers interspersed with open clearings, marshes and boggy places, areas of exposed sand and gravel and with a rich understorey of shrubby growth. Gradually the Capercaillie has had to adapt to mixed and deciduous woodlands, and its habitats have been altered in various ways, especially by man's forestry activities. All these trends have given the Capercaillie an impoverished environment, with limited amounts of food and a reduced supply of the insects which are known to represent such an important element in the diet of young chicks. Capercaillie have therefore tended to be confined to remote and mountainous areas, where the old forest habitats have been least affected by modern changes. There have been many attempts to reintroduce Capercaillie to parts of Europe where they were once common, but most of these have been unsuccessful owing to the unsuitability of the habitat.

In the early part of the last century numbers of Swedish Capercaillie were introduced to parts of Germany, and also to Scotland. The Scottish experiment was highly successful, with introductions of adult birds to Perthshire in 1837–38. These were given full protection, and eggs from captive Capercaillie were placed in the nests of wild Black Grouse. Thereafter the Scottish population thrived and increased significantly in numbers and distribution.

The European range of the Black Grouse is similar to that of the Capercaillie, and it is to be found across the zone of coniferous forests which includes Norway, Sweden, Finland and northern Russia, with other more localised populations around the Baltic and in Poland. Black grouse also occur in the Alps and in Britain, although the species is absent from Ireland. Like the Capercaillie, the Black Grouse has been affected by environmental changes and also by shooting pressure in certain areas. This species has declined seriously across almost all its European range, and is now extinct in many areas where it was formerly common.

While the Capercaillie is chiefly a bird of the woods, the Black Grouse is a bird of more open countryside and of the woodland fringes. It is often to be found where upland forests meet areas of open heather moorland and damp upland pastures. Some recent upland afforestation has adversely affected the numbers of black grouse, as has intensive stocking of upland areas with sheep and hill cattle. There is not a great deal of competition between the Black Grouse and the Capercaillie, since the former prefers younger, open conifer woodlands and the moorland fringes, while the Capercaillie favours the depths of more mature woodlands.

Some of the most important information about the history and decline of Black Grouse populations has come from studies carried out in Denmark. Archaeologists have identified Black Grouse bones dating from 10,000 years ago in the Själland area of Sweden, and the species' may well have been present in Denmark and southern Sweden from those early times. There is further evidence from the Middle Ages and various written accounts from the seventeenth century onwards, In the 1800s Black Grouse were to be found in most parts of Jutland, but there were signs of a decrease as early as the 1830s. Initially this was attributed to the improvements in sporting firearms, but it seems likely that the gradual loss of heather moorland was the real reason for the beginnings of the decline.

In 1850 some 38% of Jutland consisted of heather moorland, but by the end of the century half of this had been lost to cultivation, and by 1950 only 9% of Jutland remained as heather moors. By 1967 this had further decreased to under 6%. An estimate in the autumn of 1940 said that the Danish population of Black Grouse numbered approximately 5,000 individuals. Two years later this was reckoned to have dropped to 2,400 birds, and two severe winters may have contributed to this serious decline. By 1966 the population was estimated at 1,100, and by the 1970s it was believed that the entire Danish population numbered not more than 100 individuals. This virtual extinction of the species has been attributed to the loss of habitat owing to moorland cultivation, combined with afforestation of other areas. The changed environment does not provide the foods which Black Grouse require, nor the abundance of insects which promotes good chick survival.

# Capercaillie

## *Tetrao urogallus*

The Capercaillie is the largest wild gamebird in Europe, and is regarded as one of the oldest of the gallinaceous species. The body weight of mature adults ranges from 2.4 kg–6.5 kg, and the males are about 30% larger than the females. The cock Capercaillie's wingspan is in the range 369–430 mm, while that of the female is in the range 283–310 mm. The accompanying illustrations show the appearance of both sexes. The young chicks are yellowish on the underside, flecked and speckled with brown on the head and back, and the tarsi are feathered down to the toes with yellowish-white down. As they fledge and mature the young birds resemble the female, but from the age of about four weeks it is possible to distinguish young males from young females by the fact that the males' bills are noticeably paler in colour. By the age of 6–8 weeks the young Capercaillies' plumage is richly flecked and speckled, and not dissimilar to that of the adult females, but the young males will not develop the subtle iridescence of the adult plumage until its second year.

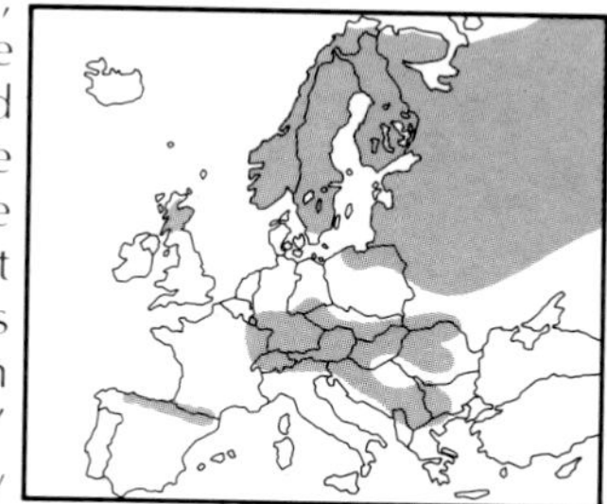

The annual moult occurs in the period March–July, by which time the summer and early autumn plumage emerges, to be followed

Black Grouse (left) and
Capercaillie, both cocks.

about two months later by the winter plumage. Yearling birds are known to moult slightly earlier than the older birds.

The Capercaillie can be a very vocal bird, and its best known repertoire of calls is heard at mating time. This comprises a series of clickings and sibilant hissing noises, and includes a loud and resonant "plop" like a cork being withdrawn from a bottle. Capercaillie males indulge in vigorous and dramatic mating displays, and some are remarkably fearless and casual about displaying close to busy roads, to houses and in front of passing people. The flight of a mature Capercaillie is deceptively fast, with powerful wingbeats and fast gliding movements.

## Reproduction

The mating displays of the cock Capercaillie involves a gathering of males at dawn at a traditional displaying area or "lekking ground". The birds come individually or in small groups from their night-time roosting places, and on clear, bright mornings the display may begin quite early. Conversely, dark skies and wet weather can mean the display does not begin until later, and it may not be quite so prolonged or dramatic. The females wait quietly around the fringes of the lekking arena, and once a female has mated briefly with one of the males she will slip away to her chosen nesting site, and thereafter the male has no further part to play in the breeding activity.

The eggs are usually laid in the period April–June, and a typical clutch will comprise 6–10 pale yellow eggs flecked with small, dark brown spots. The dimensions of the egg are 39–44 mm x 55–63 mm, and they weigh about 50 grams, which is similar to a typical chicken's egg. The incubation period lasts for 24–28 days and the newly hatched chicks weigh 35–40 grams. They grow quickly and are capable of short flights at the age of about two weeks, by which time they will weigh about 250 grams. Even at this early stage the males will weigh about 25% more than the females. The female takes sole charge of rearing and brooding her young, and the group will remain together until late autumn or early winter.

## Occurrence and diet

The Capercaillie is a member of the palaearctic group of bird species, and it occurs widely across northern Europe and Asia. In Asia there also occurs the Capercaillie's only close relative, the Short-billed Capercaillie (*Tetrao parvirostris*). There is also an eastern form of the Capercaillie, distinguished by the different shape of its rump, which is sometimes given the status of a separate species, known as *Tetrao urogalloides*. Others consider it is a subspecies of the short-billed Capercaillie. Various local populations of the Capercaillie have been referred to as subspecies, and opinions differ as to whether there are three, four or five of these. Among them are the Capercaillies of the southern mountains, such as the Pyrenees and the Cantabrian mountains, which are believed to have been isolated before the last Ice Age. The Capercaillie of Scandinavia are referred to as the subspecies *Tetrao u. urogallus*, while the slightly larger birds of central Europe are referred to as the subspecies *Tetrao urogallus major*. It has been argued that the Capercaillie of southern Sweden represent a separate subspecies, possibly the same as that which occurs in central Europe. However, Capercaillie vary considerably in size, and there is considerable variation throughout Europe, and it is probably unwise to use size alone as an indication of a distinct subspecies.

We have already noted the kind of habitat which Capercaillie prefer. The birds' food depends on what is available locally, and Capercaillie feed freely on a variety of vegetable matter. In the winter birds in coniferous woodlands eat mainly pine needles, and in spring it feeds on the fresh buds of deciduous trees, together with many kinds of fresh shoots and leaves, including the buds and shoots of coniferous species. In late summer and autumn Capercaillie feed on berries and also on a certain amount of animal matter, in the form of insects and insect larvae. They will eat many types of invertebrates and worms, and they will tear open anthills to eat the contents. Insects foods are an especially important source of protein for the growing chicks, while an adult Capercaillie needs to eat approximately 10% of its own body weight each day.

# Black Grouse

## *Lyrurus tetrix*

The appearance of the Black Grouse is familiar, especially that of the mature male in his iridescent purplish-black and white plumage. The female is rather similar to the female Capercaillie, but smaller and lacking her chestnut-coloured chest. A male Black Grouse has a wingspan of 248–291 mm., compared to a range of 224–244 mm for the female. Adult males weigh 0.8–1.8 kg, and are usually about 25% heavier than the females. The young birds resemble the female, but at the age of about 10 weeks the males grow dark, blueish-black feathering which is similar to that of the mature males. Both sexes have a dark-coloured bill, and the tarsi are feathered right down to the birds' toes, like other grouse species.

The mating behaviour of the Black Grouse is also quite well known, but it takes place in larger groups than that of the Capercaillie. It usually takes place on a low, flat place on moors or upland grassy areas not far from the treeline. The males fly into join the lek in the early morning, and the dominant males will mate fleetingly with a succession of females which wait around the perimeter of the lekking arena. There is often a second peak of lekking activity in the late afternoon and towards dusk, and the birds have also been known to display at night by the light of the moon. Black Grouse lekking activity usually takes place in April–May, and sometimes as early as March.

### Reproduction

The females each select their nesting sites, which are in a depression or scrape in the ground, often in the shelter of low shrubby growth or other vegetation. After mating with one of the cock birds near the lekking ground, the female goes off to lay her clutch of 6–10 pale yellow eggs, which are flecked with dark brown and reddish spots and marks. The eggs measure 47–52 mm x 33–37 mm and typically weigh 33–37 grams. The female incubates the eggs for 24–28 days, and she looks after her brood of young well into late autumn and early winter. The chicks grow quickly and are fledged by the age of about four weeks. When first hatched, they weigh about 24 grams, but by the time they are two weeks old their weight will have increased to about 100 grams. By six weeks old they will weigh about 800 grams, as much as some small adults. By September the birds of the year will only weigh slightly less than the mature adults, but it is not until November–December that they attain their full adult weight.

## Occurrence and diet

The Black Grouse, like the Capercaillie, is a bird of the palaearctic, occurring widely across northern Europe and Asia. Its only closely-related species is the Caucasian Black Grouse (*Lyrurus mlokosiewiczi*) which is confined to the Caucasus region. This local species is seemingly a relict of the birds which existed before the last Ice Age, when the birds were forced to move further south. Smaller distinct local populations also exist elsewhere, and the Black Grouse of Great Britain are usually referred to as a distinct subspecies, *L. t. britannicus*. A total of five (and possibly six) different local subspecies of the Black Grouse are recognised.

The Black Grouse's diet consists chiefly of vegetable matter. Like the Capercaillie, it searches for its food on the ground and also on low trees and bushes, the latter mainly in winter. Among the plant food they eat, the most common is a selection of buds and shoots from deciduous trees and bushes, and they will also eat fresh shoots from pine trees. In autumn Black Grouse will feed on berries of various kinds. When the chicks are young, however, they are heavily dependent upon insects and small invertebrate food items, which represent a vital source of proteins for the growing birds, without which the rate of chick survival is very poor. The adults will also eat insects when they are available, which is mainly in summer.

*The Hazel Grouse is a true woodland bird. It belongs to the same family —* Tetraonidae, *the Woodland Grouse — as the Capercaillie, Black Grouse and Ptarmigans. This is a separate group of 16 species, in contrast to the 183 species of "field hens" such as pheasant, partridge and quail. Woodland Grouse occur only in the Northern Hemisphere, in Europe, Asia and North America. The Hazel Grouse, unlike some of its relatives, is monogamous. How long the pair bond lasts is not well known, and it seems to weaken as the young grow. But often the cock follows his family while the young are small. As is normal, the young have a close bond with their mother.*

# Hazel grouse

## *Bonasia bonasia*

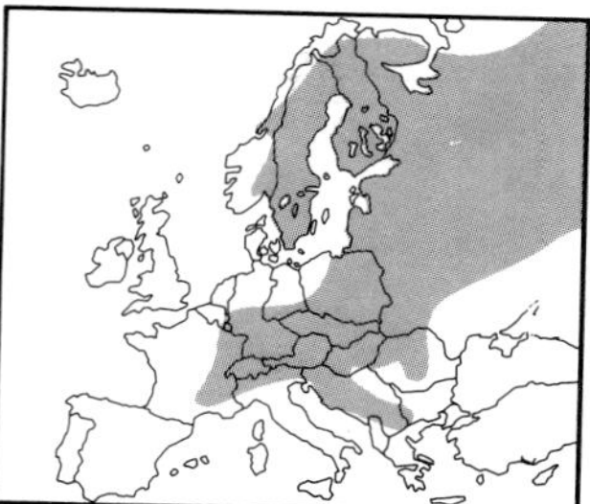

The Hazel Grouse is one of the least well-known gamebirds of Europe, despite the fact that it is widely distributed across northern and north-eastern Europe and also in the Alps. The various vernacular names for this species in Nordic countries apparently derive from the old word *japr*, which means brown. The English and German names tend to indicate that this is a bird which is hazel-coloured, or which can be found among hazel bushes, while the French name emphasises that this is a bird of the woodlands. The Italians, however, call it *Fracolino di monte*, which indicates that the bird is found in mountain habitats. This selection of European names give us some hints about the bird's appearance and where it can be found.

### Characteristics

The Hazel Grouse is classed among the so-called woodland grouse, and it is quite closely related to the Capercaillie, the Black Grouse and the Ptarmigans. Like these species, the Hazel Grouse has feathered tarsi but its toes are quite bare.

At present there are four or five genetically distinct subspecies which are recognised by ornithologists, although there are more than that number of local populations. The subspecies *B. b. bonasia occurs* in Scandinavia, north-west Russia and Poland, and this is a relatively light-coloured bird with a grey back. The Hazel Grouse of central Europe is referred to as the subspecies *B. b. rupestris*, which is an altogether browner bird. In Russia there occurs the subspecies *B. b. griseonota*, a lighter-coloured bird which tends to live further to the north. Within the eastern parts of this bird's range, towards the Pacific coast, there are further distinctive populations.

The Hazel Grouse is the size of a domestic pigeon, and the sexes are similar in size. The typical body weight lies in the range 315–500 grams, and the wingspan is 170–184 mm. The male can be distinguished from the female by a black spot on the throat and a more conspicuous rust-coloured spot along the front edge of the wing. Both sexes can raise their crown feathers into a small erect nape tuft or crest, and they do this as part of their courtship activity and also when they are alarmed. They also cock up their tails. Hazel Grouse communicate by melodic whistles or "peeps", especially during the breeding season, and the male's display call can be rendered phonetically as *tsitsi tseri tsi, tsi tsui*.

The young birds resemble the female, for they lack the mature male's throat spot, and there are many other similarities. The downy chicks are coloured reddish-brown, with black eye bands.

Like other woodland grouse, the Hazel Grouse is a ground-living bird but it will also perch in the lower branches of trees. They are shy and elusive birds, and can run quite fast. They fly with a vibrant wingbeat, and Hazel Grouse are active by day. By night they roost in trees, and in winter they are capable of digging down into snow holes, like the Ptarmigan.

The Hazel Grouse is a bird of the woods, and can be found in both coniferous and mixed forests, but it prefers woodlands with a variety of tree species of different heights and ages, and it is fond of woodland which has a fairly dense understorey of shrubby growth.

Its diet consists of the buds of deciduous trees, berries, tree moss and seeds of various kinds. It will also eat insects, insect larvae and various invertebrates.

### Reproduction

The Hazel Grouse is monogamous and is often to be found in pairs, even outside the breeding season, and under optimum conditions the territory of a pair will comprise about 10 hectares. The female takes sole responsibility for incubating the clutch of 8–10 eggs, which are reddish-yellow and sparsely flecked with small brown dots.

The female lays one clutch of eggs per year, and these measure 37–41 mm x 27–29.5 mm, and typically weigh about 20 grams. The eggs are laid in April or May, at the rate of two eggs every three days, and the nest is a simple scrape or depression in the ground. Incubation takes 22–27 days and the female takes sole responsibility for the chicks, which will fledge within about two weeks of hatching. Up to this point they are almost entirely dependent upon insects, invertebrates and worms of various kinds for their food. The availability of this particular kind of high-protein insect and animal food will affect the survival rates of the newly hatched chicks.

The young birds are quite independent by the time they are a month old, but the family tends to remain together until late in the autumn. The birds will pair up shortly after this, and they are able to breed in their second year.

At hatching the chicks weigh about 11–12 grams, and when they fledge they will weigh 25–40 grams. At the age of one month they will weigh about one quarter of their adult weight.

### Distribution and numbers

The Hazel Grouse belongs to the Siberian group of birds and animals, which means that it occurs primarily in the large coniferous zone of northern Europe and Asia. The numbers of Hazel Grouse are lower than that of other woodland grouse species, and in the 1970s the Swedish population was estimated at around 150,000 breeding pairs. In continental Europe the population has declined seriously during this century, and in many areas it has disappeared altogether. However, it can still be found in many mountain areas, and in the Carpathian mountains and in the Alps this species occurs at altitudes up to 1,800–1,900 metres above sea level.

Despite its small size and declining numbers, the Hazel Grouse is still a popular sporting bird in large areas of its range. It is usually hunted by sportsmen using a special whistle to imitate its call, and some northern sportsmen also use a spitz-type dog to locate the birds perched in trees and bark to indicate when they have found them. In Sweden 23,000 Hazel Grouse were shot in the 1985–86 shooting season, which represented a decrease of 30% on the previous season's bag.

# Arctic hare

## *Lepus timidus*

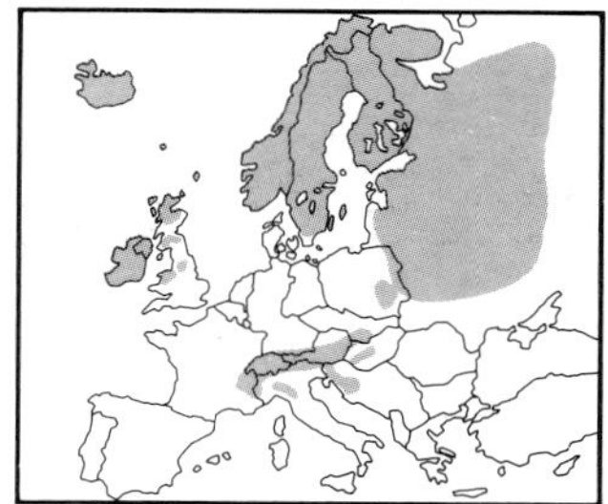

There is an old Nordic legend told about this hare, which relates how hares would move into pastures where cattle grazed and would proceed to milk the cows. These were known as "milk hares" and they became extremely fat as well as developing a white meat, similar to veal or chicken. The "milk hare" was always described as being black in colour, and melanistic hares with an unusually large amount of black pigment have been shot in various parts of Scandinavia, and a number of mounted specimens are preserved in museums. However, this does not quite fit in with the legend, which went on to say that any hunter who succeeded in shooting a milk hare would see the dead animal transform itself into a puddle of milk and a pile of sticks! The Milk hare or the "Troll hare" was believed to be in league with the powers of darkness and evil, and it was said that after drinking the cows' milk the hares would regurgitate it into a bowl or a saucer for a witch.

Some early writings about Arctic hares stated that these animals were especially attracted by the sounds of a drum beating, and that they would drum on the ground with their front feet in such a way as to create a resonant drumming sound. This was regarded as rather similar to the way in which rabbits will thump on the ground with their hind feet as a contact sound and to give a warning to others.

The idea of a hare with delicate white meat like chicken or veal, instead of the usually dark and rich flesh of most hares, led to attempts to interbreed hares and rabbits. One of those who tried to do this was the celebrated French scientist Broca, and this may reflect the fact that the French as a nation have always been very fond of eating rabbit.

The so-called Milk hare was reputed to be extremely difficult to shoot, which fits in well with its reputation as a creature of the night in league with evil powers. It was believed that hunting dogs would die if they touched these witches' familiars, and it was thought that they could only be killed with specially prepared bullets.

Hares play a most prominent part in the folklore of many European nations, and in the medieval fairy tales of Germany and also of Scotland there figured the purely fanciful character of the "Horned hare", which was said to have a horn in the centre of its forehead like a unicorn. However, in art it was more often depicted with two horns, rather similar in appearance to those of a Chamois and occasionally of a Roebuck. Nowadays the "Milk hares" are recognised as nothing more than a melanistic form, and the "Horned hares" are purely a fable, but they were at one time given their own Latin name, *Lepus cornutus*. To play along with the beliefs of folklore, many European taxidermists created deliberate forgeries — animals with the body of a hare and the antlers of a Roebuck. This idea also spread from Europe to North America, where it is perpetuated in the legend of the "Jackalope" — a hare with antelope horns — which was said to exist in Wyoming, which is where the Prong-horned antelope is to be found.

In many countries the hare is associated with the celebration of Easter. However, it has never been credited with the ability to lay eggs, apart from one Swedish taxidermist's creation which purported to be a cross between a hare and a Capercaillie!

## Characteristics

The Arctic hare, also called the Blue hare or the Changeable hare, is rather similar to the Brown hare. However, the Arctic hare is smaller, has shorter ears, and its tail is completely white in summer and winter, when the tip of the Brown hare's tail turns black. The Arctic hare lacks the chestnut-coloured areas that appear on the chest of the Brown hare and frame its white belly. The Arctic hares of northern Scandinavia, northwards from the lakes of central Sweden, have an entirely white winter pelage, and only the very tips of the ears remain black. In southern populations of this animal the tips of the ears are blueish-grey. The Arctic hares of Iceland are completely white in winter, as are those which occur in the Alps, and the same is true of the Blue hares of Scotland and the Irish hare, although both the latter may retain their summer coats in certain instances. In particular, the Irish hare often only develops a partially white pelage in severely cold winters, and in mild winters it may remain quite reddish-brown with little or no white. The Arctic hare does not occur in any other parts of continental Europe.

It has been discovered that the Arctic hare's annual moulting is influenced and regulated by the amount of daylight and by the temperature. An interesting experiment was carried out by introducing Arctic hares to the Faeroe Islands. These hares came from a stock which was normally quite white in winter, but after 35 years on the islands 95% of the descendants of those hares were found not to change their fur colour. This change must have taken place as an adaptation to the local conditions in the Faeroe Islands, which normally have no snow cover in the winter.

The sexes look very much alike, and the maximum weight is in the range 3.8–5.8 kg, with the lowest weight for a mature adult hare being around 2 kg. A typical average weight would be in the range 3–3.5 kg. The Arctic hare has 28 teeth, and its tooth structure is the same as that of the rabbit. Hybrids between Arctic hares and Brown hares have been found in Finland, southern Sweden and in the Baltic states, but hybrids with rabbits have never been found in the wild state.

During the day the Arctic hare usually lies concealed in its form, which comprises a shallow depression which is often situated at the base of a low-growing fir tree or in a rocky cranny. Hares often have several forms within their territories, but sometimes the same form is used day after day. At twilight the hares become active and go out in search of food, and in the breeding season courtship may take place at night. The hare's nocturnal movements can cover quite considerable distances, up to several kilometres and as much as 50 km in a single night. A typical territory is 20–30 hectares in size, and they often overlap, but they may also be very much larger than this. Arctic hares are usually solitary animals, but it is common to see several together during the breeding season.

The Arctic hare can swim, but this is an ability which it rarely uses unless it is hard pressed. It can also dig short burrows of one or two metres in length, and this is particularly common in the eastern parts of its range. Arctic hares will also dig shelter holes for themselves in deep snow, and after heavy snowfalls or in stormy weather it may remain there for several days, crouched down and well protected in its underground form.

The hare stamps its hind feet when it is alarmed and sometimes screams shrilly when in pain, for example when it has been wounded by a shot. Otherwise it is mainly a silent creature. Old accounts of hare hunting describe how hares will retrace their steps and make long jumps and sideways leaps to try and shake off their pursuers. Hare tracks in woods may seem to go helter-skelter in all directions, but hares will often follow well-defined paths and tracks when they are covering longer distances. Tracks sometimes run along ditches, by the side of stone walls and along paths and rides in woodland, and hares will often run along roadways. Holes in stone walls and also gateways are frequently used by hares as crossing points from one area to another.

The Arctic hare inhabits woodlands, in addition to open areas of heathland and moorland. It prefers the fringes of forests and woods which have open glades and marshy corners, and generally avoids open cultivated land, unless this is adjacent to other woodland or moorland habitat. In many lowland and agricultural areas of Europe the Arctic hare is absent, and its place is taken by the larger and more prolific Brown hare.

The Arctic hare likes to have access to young deciduous trees, tender twigs and the fresh shoots of willow, birch, rowan and sallow. Aspen is a favourite food of the Arctic hare in many areas, and hares may be encouraged by felling young aspen in winter. The hares will first devour the tender twigs, and will then turn their attention to the bark on larger branches and on the trunk of the tree. Hares will also eat heather, blaeberry plants and juniper twigs, and during the summer they will feed freely on all sorts of grasses and herbaceous plants. Its diet varies depending upon what is locally available, and in Scotland 90% of the hares' winter food consists of heather shoots.

In winter it is not difficult to follow a hare's movements, for it leaves clear tracks in the snow and also leaves a trail of round, dry pellets of excrement as it goes. During one night's wanderings a hare may deposit 200–500 such pellets, and each day it requires about 400–500 grams of food. While it lies in its form the hare excretes a soft type of excrement, which is then eaten. This form of digestion, which is known as refection, enables the hare to extract the maximum nutrition from what it has eaten by passing it through its digestive tract a second time.

## Reproduction

The Arctic hare is capable of breeding when it is less than one year old, and mating usually takes place at the age of 9–10 months. The male will usually mate with several females, and each female may also have a succession of male partners.

Two to three litters of young are born each year, and a typical litter will comprise 2–5 young, called leverets. In quite exceptional cases as many as seven young may be born, but the Arctic hare is not as prolific a breeder as the Brown hare, and the young leverets are born into a much harsher environment where mortality is heavy. Only

20% of leverets survive their first autumn, and Scottish studies have shown that only 14% of hares may survive to the age of one year. In northern parts of its range it is common for hares to have only one litter of young per year.

The mating season varies with the location. In the Alps it takes place in the period March–June, while in Scotland it extends over the period from January–September. In Scandinavia the breeding season lasts from February–June, and sometimes into July. Mating takes place again immediately after the birth of a litter of young, but not during the period of gestation, which lasts for some 45–52 days. When a female is in season she may be courted by more than one male.

Unlike the young of the rabbit, the leverets of the Arctic hare and the Brown hare are born well furred and with their eyes open. The fur is quite well developed and the leverets are soon on their feet and able to move about quite freely. Some Arctic hare leverets have an elongated white blaze, similar to that which can occur in some Brown hares. After about three weeks of suckling their mother, the young become more independent and able to take care of themselves. The mother's milk is especially rich and nutritious, with a much higher fat content than cow's milk. At birth the leverets weigh about 125 grams and they grow very quickly, but on average they still weigh about 10% less than an adult hare during their first winter. They attain sexual maturity from the age of nine months and onwards.

## Distribution and subspecies

The Arctic hare occurs from the Atlantic to the Pacific Ocean in a wide belt across Europe and Asia. It is also found on various islands, including Britain and Ireland, Iceland and in the Japanese archipelago. In North America there occurs the closely-related Polar hare, which is sometimes regarded as a separate species and sometimes as a subspecies of the Arctic hare. It is also to be found in Greenland and it is sometimes referred to as *Lepus timidus groenlandicus*. Another hare, *Lepus americanus*, the Snowshoe hare, replaces the Arctic hare in similar habitats in North America, and it enjoys a similar status as a game animal.

The Arctic hare appears in several different sizes and pelages within its wide area of distribution. The Polar hares in Greenland and North America, *Lepus timidus arcticus*, are white throughout the year and are considerably larger (2.7–5.8 kg) than the European Arctic hare.

The hare found in the Alps, *L. t. varronis*, is now considered a distinct subspecies, as also are the Scandinavian Mountain hare (*L. t. timidus*) and the "Mohare" (*L. t. canescens*). The Alpine hare is somewhat smaller and tends only to be found above 1,300 metres above sea level. The Mountain hare has snow-white winter pelage, while the "Mohare" is grey with brown patches on the forelegs and the head. The Mountain hare is found in northern Scandinavia, north of the Norrland border, and in Finland, while the "Mohare" is found south of that line. A wide border zone exists containing mixed forms and, to make matters more complicated, hares have been introduced into various areas.

## Lifespan, diseases and enemies

Hares are believed to live to the age of 9–10 years, and some for as long as 13 years. However, in some years the mortality rate for adult hares can be as high as 75%. The fox is usually the commonest cause of death, and Golden eagles also account for a number of hares. Goshawks will also take occasional hares. From time to time bacterial and parasitic diseases can wipe out whole populations of young hares.

Bacteria can be transmitted from voles and lemmings, and from infected hares by mosquitos and ticks, and the hares die a few days after being infected. They tend to become apathetic and suffer from breathing difficulties and diarrhoea. Hunting and shooting also account for considerable numbers of hares. In Finland 544,000 Arctic hares were shot during the 1984–85 hunting season, while a further 127,000 were shot in Norway and 200,000 in Sweden.

## Scandinavian hare hunting

The hare is one of the most popular Scandinavian game animals, and there is a long tradition of hare hunting and shooting. An estimated 100–200,000 hares are shot annually in Sweden alone. There is an old saying that "no-one knows which way the hare will go", such are its tendencies to turn and double back on its tracks and to move in circles and twists when it is pursued. These antics have long been of special interest to those sportsmen who hunt hares with scenting hounds, which follow their quarry not by sight but by scent.

Traditionally, sportsmen have tended to draw a distinction between the "Mohare" of the woodlands and the wilder country-side and the so-called "Garden hare" which frequents cultivated land, farms and meadows, although both are forms of the Arctic hare.

The Brown hare, sometimes referred to as "the German hare", should not be confused with the various forms of the Arctic hare, and it was not present in Scandinavia until well into the nineteenth century. The Arctic hare, however, was found throughout the Scandinavian peninsula from the far south to the Finnish border, and it was not until the beginning of the present century that the introduction of the Brown hare to Scandinavia began to make its mark.

Many different breeds of dog are used in hunting and shooting hares. Long-legged harriers are the traditional choice, although many modern sportsmen prefer to shoot over dachshunds and German pointers. It is important that a good dog for hares should be staunch to that quarry, and it should not be sidetracked into chasing other animals and deer such as Roe.

Hare hunters have always made use of the fact that hares are territorial, and will rarely stray out of their familiar areas. This fact is exploited by the sportsman, who will go to seek hares in places they are known to frequent. Hares were hunted with dogs long before reliable modern sporting guns were available, and the aim was often to drive the hares into nets which had been set to intercept them.

Once a hare has been started from cover and the chase is underway, the sportsman has no influence over the direction which the hunt will take. However, the aim is to ensure that the hare passes within shot of one or more standing guns, and these will position themselves to as to make the best use of known hare tracks and of other open spaces such as forest rides and clearings.

# Woodcock

## *Scolopax rusticola*

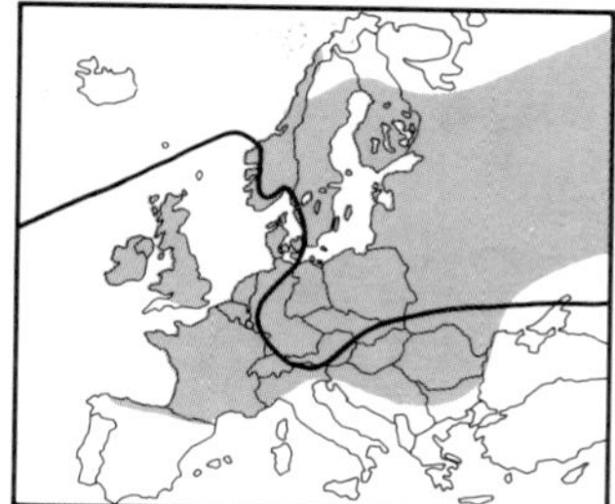

The Woodcock's Latin name is linked to a whole family of wading birds, the *Scolopacidae*, which comprises no fewer than 81 species of wading birds world-wide. This family includes not only familiar species such as Curlews, Snipes, Redshanks and Sandpipers, but also some unusual species such as the Ruddy Turnstone and the Ruff. All of these are distinguished by their relatively long legs and bills. The various Woodcock species and subspecies are the only ones to combine a very long bill with relatively short legs. Five species of Woodcock are recognised world-wide, including Eurasia and America, and wherever they occur Woodcock tend to be very highly regarded as sporting birds.

In some of the earliest European writings about gamebirds, we find references to the Woodcock's desirability as a fine table bird, tender and finely flavoured. More modern books on game and game cookery tend to refer to the tradition of cooking Woodcock without eviscerating them, and of serving the individual bird on a slice of toasted bread.

Among European nations, the French are especially fond of shooting and eating Woodcock (and Snipe), and there are specialist societies and organisations dedicated to the hunting and gastronomic enjoyment of these species. Shot Woodcock may be allowed to hang for a few days, and they are cooked with their entrails still in place. The birds are cooked individually and browned in butter with a good red wine added to the natural juices. In Scandinavian countries, where Woodcock breed in large numbers in spring and early summer, the numbers shot in autumn are comparatively small, because these northern Woodcock are very migratory in their habits, and most move southwards and south-westwards to winter in southern and western Europe. In Denmark there is a small breeding population of Woodcock, but quite large numbers of birds are shot on migration.

## Characteristics

The Woodcock's bill is about 75 mm long, and has a slightly bulbous tip. The eyes are very large and dark, and they are set high up on the sides of the bird's head, giving it an almost all-round field of view. The sexes are externally similar, and have bold cross-barring on the crown of the head, and vivid white tips on the underside of the main tail feathers. Typically, an adult bird will weigh 245–390 grams. The wingspan lies within the range 185–201 mm.

The position of the Woodcock's eyes also gives this bird the ability to see almost all round without moving its head, and its field of vision remains large and clear even when the bird is in a head-down feeding attitude. The eyes also give the Woodcock a very clear view overhead, which is important in the breeding season, when males make display flights over the females which rest on the ground in woodland.

Woodcock prefer deciduous and mixed forests, but they also live and breed in coniferous woodlands. They especially favour woodlands which are quite open, with rides and roadways, marshy clearings and open glades, and they are mainly attracted to medium-sized and large forests, rather than to small copses. They also like areas of countryside where expanses of woodland are broken up by grassland and pastures.

Woodcock feed on many types of invertebrate animals, especially worms and insect larvae, which the bird locates by probing in soft soils and ooze. The slightly bulbous tip of the bill is full of highly sensitive nerve endings to detect underground prey, and like the snipe, the upper mandible is flexible and capable of grasping food items like forceps. Woodcock will also eat insects, especially in the larval and pupa stages, and other creatures such as millipedes and small molluscs, but earthworms remain their principal food.

The Woodcock is active mainly at twilight and during the night. The well-known spring display or "roding" flight is carried out at dawn and again at dusk by the male Woodcock only. This begins as soon as the birds have returned to their breeding grounds, and can be seen from early March to as late as July. The displaying birds fly slowly and steadily around a circular or triangular course, following regular flightpaths and passing just above the woodland canopy. These flights take the birds over prime nesting habitat and advertise the males' presence to breeding females waiting in the woods. The slow and prominent flight is rather bat-like or owl-like, and the displaying male utters a distinctive call, which consists of a series of deep, frog-like croaking notes followed by a high-pitched *twisick*. Sexually-receptive females attract displaying males down to join them on the forest floor, and there is a brief period of display on the ground before mating takes place.

Dominant male Woodcock are territorial, and will chase off intruding males within their displaying area, and a male's displaying area may overlap with the breeding territories of a number of females.

## Reproduction

Female Woodcock may begin breeding at one year old, while the majority of males are believed not to breed until their second year, and form a non-breeding reservoir of young birds in the meantime. The males are described as "successively polygamous", which indicates that a dominant breeding male will mate with a succession of females during the somewhat prolonged breeding season. The male bird only stays with his mate for a short time after mating, and as soon as egg-laying has begun he will leave her and recommence displaying.

A typical clutch will consist of four rounded-oval eggs, yellowish in colour and flecked with large reddish-brown spots. The nest consists of a shallow depression in the ground, and when the eggs are unattended the nest and its contents may be very conspicuous. The dimensions of the egg are within the range 40–48 mm x 31–36 mm, and typically weighs about 26 grams. The newly hatched chick will weigh about 17 grams, and the females takes sole responsibility for incubating the eggs and rearing the young. Incubation lasts for 20–24 days. Nests with eggs may be found in Britain and Ireland as early as the end of March, while in the northern parts of the Woodcock's breeding zone and in the mountains of central Europe the peak egg-laying time may be in June–July. The chicks are fledged and quite fully grown at about five weeks old, and some are able to make short flights as early as ten days old. The female Woodcock produces one brood of young each year, and although double-brooding is suspected it has never been proven.

For many generations there has been controversy as to whether the mother bird can carry her young in flight, and it now seems certain that this is occasionally done, with the adult bird grasping her young individually between her thighs.

Like many birds, Woodcock have a rather high rate of mortality in their first year, and this is estimated at 60%. Mortality among the mature birds has been put as high as 50% per year. Shooting accounts for a significant proportion of Woodcock mortality, but these birds are also vulnerable to severe winter weather.

## Distribution and numbers

The Woodcock belongs to the palaearctic group of birds and therefore occurs across northern and central Europe and Asia. In mountainous areas Woodcock occur at considerable altitudes, up to 3,700 metres in the Himalayas and 1,700 metres in the Alps. Despite the Woodcocks' extensive range, no locally distinctive races or sub-species have been identified.

Although there are some sedentary populations, the Woodcock is migratory over much of its range, especially in spring and autumn. Migration takes place mainly at night, and the northern European birds begin to move southwards and south-westwards from late September and October onwards. However, it is not unusual to find a few Woodcock remaining in southern Sweden and south-western Norway in winter. Generally, however, the coming of winter ice and snow drive the Woodcock away to find milder winter quarters in southern and western Europe.

The wintering areas of European Woodcock depend upon the availability of frost-free habitat, and the vast majority of birds winter to the south of the 1°C January isotherm. Marginal areas such as northern Germany and Denmark may have wintering Woodcock in a mild winter but few or none in a hard winter.

Woodcock which have wintered to the south migrate northwards from March onwards, reaching their most northerly breeding haunts by June. In Sweden a breeding density of 0.5 birds per $km^2$ of woodland has been estimated. Woodcock breed widely in Britain and Ireland, although not in the far south-west of either island, and there is clear evidence that the numbers and distribution of the British and Irish breeding population have increased steadily over the past 100 years and more. This in turn has been part of a general increase in the extent and numbers of Woodcock breeding in continental Europe this century, and it is assumed this increase has taken place as a consequence of habitat improvement, especially the afforestation of areas with damp soils.

## Woodcock shooting

In northern and central Europe Woodcock have traditionally been hunted in spring, and the appearance of the first Woodcock of the year was regarded as a welcome harbinger of spring. Woodcock were shot on their display flights by sportsmen waiting below in forest clearings and along the woodland edges, to intercept the male birds on their dawn and dusk roding flights. Nowadays, Woodcock shooting in spring is prohibited in many European countries.

In earlier times, however, the first Woodcock of spring were highly prized, and sometimes they were presented to royalty. Records from the mid-nineteenth century tell us that Swedish sportsmen would send the first Woodcock of spring to the royal kitchens in Stockholm. In Denmark the old sportsmen would hunt Woodcock with pointing dogs among the sand dunes and along the coast, where migrant Woodcock would stop to rest on their journey northwards.

Otherwise, most spring hunting was directed at the roding birds, and one trick of the hunters was to lure the birds within range by throwing a cap or a hat in the air. This simulated the fluttering jump of a female Woodcock trying to attract a displaying male, and occasionally the males were seen to drop right down and alight on the woodland floor when lured in this way.

In southern and western Europe, Woodcock shooting has long been a popular winter sport, and many sportsmen shoot Woodcock by using a bird-dog to locate and point the birds in cover, or use spaniels to beat out areas of likely Woodcock undergrowth. In Great Britain it is estimated that a high proportion of Woodcock are shot incidentally, in the course of pheasant shooting days.

*The Woodcock's plumage reveals its nocturnal habits. By day — when it usually rests, incubates eggs or cares for the young — the bird needs to have camouflaging colours and patterns. Yet at night, for activity on the ground and for signals between flying birds, a valuable aid is the outer, light-reflecting tail-ends. (The outermost primary feather, small but stiff, was once used by painters for watercolours.) When flushed in woodland, the bird gives a red-brown and plump impression as it flies off quickly between the trees.*

# Animals of the mountains

The harsh conditions of the mountains and of the high Arctic tundra demand special adaptations in the animal species which live there. For example, on Spitzbergen the animals have just three months in summer to prepare themselves for a winter which lasts for nine months. Many animals barely survive, and the number of species is limited. Fluctuations in the food supply result in massively increased mortality or in vastly increased breeding success. The size of many animal populations fluctuates, with ups and downs periodically over a number of years.

For some species the summer is not always long enough for breeding. This can be an especial problem for geese: a late spring with a delay in the spring migration and a late arrival at the breeding grounds leaves too little time for egg-laying, incubation and fledging of the young before the weather makes it essential for the birds to fly south again.

A number of species try to avoid this precarious situation or to take advantage of temporary resources by migrating seasonally; the migrations of reindeer are a well-known example. Curiously enough, none of the animals of the high mountains or of the tundra hibernate, neither the largest nor the smallest. Some, such as the wolverine, build up a food supply. The arctic fox survives the severe winter by eating the remains of reindeer and seals, and the prey remains left behind by wolves, wolverines and polar bears. Others build up food reserves in their bodies. In autumn the small reindeer of Spitzbergen have fat reserves of 20–25 kg, corresponding to one-fifth of their body weight, and by the end of the winter this has all been used up. There are also differences between the northern and southern populations of the same species. Rock ptarmigan and willow ptarmigan of the arctic subspecies both build up reserves of fat, while the southern subpecies do not. The same difference occurs between the reindeer of Spitzbergen and those further south, and between the alpine reindeer of southern Norway and the Scandinavian forest reindeer.

Many animals exhibit both external and internal adaptations to the harsh environment. Most of the animals have small ears, which usually lie protected in the fur and are covered with thick hair. The soles of the feet are also often hairy. Although the arctic hare's ears are long, they are proportionately shorter than those of its southern relatives. Many alpine and polar animals have a white winter fur which protects them from loss of body heat and also provides camouflage. The white hairs of the polar bear's fur and the grey hairs of the reindeer are filled with tiny air cavities and act as insulators. With the exception of the lemming, very few small vertebrates live the year round in these areas. However, the lemming makes use of pockets of air and passageways under the snow to remain active throughout the winter.

It is generally thought that species living in more northern, colder regions tend to be larger in size than the same species living in southern latitudes, although this is not an invariable rule. A large body benefits the conservation of warmth as it radiates proportionately less heat than does a smaller body. A large body, however, requires more nourishment, and the balance between these two characteristics has been solved in various ways by different species. On average, polar bears are larger than the bears of more southerly latitudes, just as the arctic hare is larger than the southern hares. Northern reindeer, however, are smaller than their southern relatives, and this is also true of the bean goose. The Brent goose, the barnacle goose and the Bewick's swan are all smaller than the more southern breeding geese and swans.

# Arctic fox

## *Alopex lagopus*

The Arctic fox is considered to be a very close relative of the common Red fox, and both are descendants from a common ancestor. The ancestors of the Arctic fox adapted to arctic conditions by developing a dense, thick fur with additional underfur, and a complete change of fur before the winter. The winter-white Arctic foxes are protected both by their fur and by effective insulation, as the hairs are filled with air cavities. In this way they are capable of enduring much colder conditions than their red fox cousins, and they have less heat loss and are therefore able to conserve energy. The pads of the feet are thickly covered with hair, and both this and its ears, which barely protrude from the head, are adaptations to the winter climate. It is probably only when there is severe wind-chill that the arctic fox suffers from the cold.

In winter they roam widely in search of carrion to eat, but they are mainly dependent on lemmings and other small mammals for their existence. They build up a considerable store of fat before the winter, but like the Polar bear they do not really go into hibernation, even if they are snowed in for days or lie protected in burrows and hollows. In the mountain areas of Scandinavia the thick winter coat of the Arctic fox has been the cause of its threatened existence from hunting and trapping for the fur trade, and the Arctic fox has been a protected species in Sweden since 1928. In other areas of its circumpolar range the Arctic fox is still a highly regarded sporting trophy because of its fur. For the Eskimos of Canada the fox is an important source of income. In Europe, however, it is mainly raised on fur farms, although it is still hunted and trapped for its pelt in Russia.

### Characteristics

The Arctic fox and the Red fox are similar in shape, but the Arctic fox is smaller, and the ears are noticeably smaller and rounded. The soles of the feet are covered with hair, and 95% of Swedish Arctic foxes are

completely white. They occur in two different colour forms — one which is white in winter and one which is grey-brown, known as the "blue fox". Both have dark grey-brown summer pelage, and the belly, neck and throat are white. They can weigh up to 8 kg, and their body weights are at their maximum in autumn, just before the onset of winter. The sexes look alike, and young of both colour types can be found in the same litter.

The Arctic fox is less shy than the red fox, and this is especially noticeable when it emigrates from its home range. Wild foxes come so close to man and can be so fearless that they may be mistaken for escapees from a fur farm. Like the Red fox, the Arctic fox can occasionally contract rabies. Otherwise the commonest cause of death is lack of food and resulting death by starvation. Mortality is especially high among the young animals, particularly during the winters when starvation takes a big toll when there are food shortages, and many foxes freeze to death. This is often the case in northern Canada. In captivity the Arctic fox can live to be 14 years old.

Most canine animals will howl, and so does the Arctic fox. Its most common call is a barking howl not unlike that of the Red fox, and this is heard mainly during the mating season.

## Reproduction

The Arctic fox mates during the spring months of March and April, when it is still winter in the far north. A gestation period of 51–54 days means that the cubs are born in May–June, a period when the arctic region is once again able to afford ample supplies of food. Late litters in July–August are usually second litters, and these only occur in those years when lemmings are very abundant. The fox cubs are born hairy but blind, and they weigh 250–300 grams. The fur is dark grey-brown. Both parents take part in the rearing of the cubs, and both bring food to the den. The cubs are born in a den which is dug by the parents in well-drained gravel or sand banks, and the cubs remain in the den until they are a month old, at which time they are still suckling the vixen. They are several weeks old and their eyes have opened when the gradual weaning process takes place, and they begin to eat the same food as the adults. The female either regurgitates food for the cubs or carries it back to the den for them.

Both the size of the litter and the survival of the young depend upon the food supply. A normal sized litter comprises 5–6 cubs, but litters numbering as many as 12–14 cubs have occurred. Since some females may have two litters in good lemming years, the reproductive capacity of the species is high. The young are sexually mature as early as one year old.

In years when the population is high and breeding conditions are good, several pairs of Arctic foxes have been reported to settle down together in a breeding colony, rather like badgers. Because the Arctic fox's breeding potential is influenced by the lemming population, it is cyclical, and increases and decreases in the numbers of cubs occur about one year after the corresponding variations in the lemming population. In the high mountains of Scandinavia this means peaks of breeding every three or four years. In 1967 an increase in breeding was recorded. 1970 was a record year and 1974 another. 1978 was also a good year for lemmings and small rodents, during which the Arctic fox population had a very high breeding success rate. In 1980, when there was a scarcity of rodents, no breeding was recorded. This was also the case with the Snowy owl, another Arctic species which is heavily dependent on the lemming population. The numbers of Arctic foxes and Snowy owls fluctuate in similar ways and for the same reasons.

## Habitat and diet

The Arctic fox, as its name suggests, is an animal of the arctic regions, where it lives on the tundra and in open country. It is often found along the coast, and it can sometimes be found in the fringes of cold and windswept forests. In the Scandinavian mountains it occurs further south, where it lives in summer on the bare mountains among the lichen and willows. At times it will also live in the high birch forests.

The population density varies with the location and the season, but the food supply is the critical factor. The same is true of the size of the animal's range, which is the distance which a fox will travel from its den. The territory may be as large as 30–35 km$^2$ in years with little food, but it can be as small as 8 km$^2$ in years when food is abundant.

In many ways the Arctic fox is an opportunist feeder, a characteristic it shares with the Red fox of southern and western Europe, but it is also a specialist whose breeding potential and winter survival are determined mainly by the supply of small rodents and the carrion left behind by other, larger predators. In addition, the Arctic fox will eat baby birds and birds eggs, while in the autumn berries supplement its diet, and in winter an occasional hare or ptarmigan may be caught. In some areas the Arctic fox has become a specialist in taking the chicks of upland birds when they leave the nest. Some northern geese, such as the Barnacle goose, do not inhabit flatland areas, but they breeds near cliffs and crags where the Arctic fox also occurs. When the goslings leave the nesting ledges the foxes take some of them as prey.

## Distribution

The Arctic fox has a circumpolar distribution, and it is found throughout a wide zone across the Arctic tundras, including Spitzbergen, Jan Mayen Land, Bear Island and in Iceland. In the Scandinavian mountain areas the Arctic fox is found much further south.

The young foxes leave the family group when they are one year old, and they often wander far to the south of their normal breeding areas. Large scale emigrations often take place when there has been a good breeding season, but this was perhaps more common in the past, when the Arctic fox was more numerous in the high mountains of Scandinavia. Arctic foxes have even been seen in the most southerly parts of Sweden! In the winter of 1831–32 some Arctic foxes reached as far south as Falsterbo and Skanor, "and killed a great many geese and lambs". Large scale emigrations apparently took place on a cyclic pattern, and old records show that the years 1831–32, 1841 and 1860 were years of great fox movements. One of the very largest seems to have taken place in the winter of 1831–32.

## The arctic fox in Norway and Sweden

The popularity of the Arctic fox as a quarry for skin trappers and hunters caused the species to come under intense pressures. It was shot, but trapping was the main way of catching it, and this remains the commonest way of taking Arctic foxes in Canada, where Eskimos and trappers make a living from the fur trade. In the last century the shooting and hunting of the Arctic fox in the mountains of Scandinavia does not appear to have had a devastating effect, and at the end of the last century the fox was still common in all mountainous areas. However, by 1928 its existence was considered to be in such jeopardy that it was fully protected in Sweden, although hunting continued in Norway. It was widely believed that this gave rise to a situation whereby illegally shot foxes were taken across the border from Sweden into Norway, where they were sold. Later protection was provided in Norway, and the arctic fox has enjoyed complete protection in both countries ever since.

One important reason for the decline of the Arctic fox was the decimation of the larger predatory mammals in the high Arctic, and especially the wolf. The remains of reindeer killed by wolves were an important source of food for Arctic foxes, and the animals would deliberately follow packs of wolves and wait in the vicinity of a kill until it was safe for them to move in and eat what the wolves had left behind.

As a fur-bearing animal the Arctic fox has always been a highly prized animal both for sportsmen and for commercial shooters and trappers throughout its range. Fur fashions change from time to time, but Arctic fox fur has been in fashion for a very long time.

Paradoxically, this has been somewhat to the benefit of the species, for the breeding of captive Arctic foxes has helped to reduce the hunting pressures on the wild populations. In Russia the Arctic fox has also been introduced to areas where it did not formerly exist. The pelts of farmed foxes, like those of farmed mink, are considered to be better and more consistent in quality. Farmed foxes can be found in the "blue fox" colouration, as well as the all-white variety.

Occasionally farmed foxes escape, which causes confusion for zoologists trying to study and count the wild populations. It is possible to determine with some accuracy if a dead animal is a farmed fox or a wild one, by examining its skull, but this is very difficult to do among free-living foxes. Both wild and farmed foxes will aproach very close to man, seemingly without fear, and the absence of fur farms in the immediate vicinity cannot be regarded as proof that the foxes are truly wild. Escapees can cover large distances in a very short time. An all-white eleven-month-old Arctic fox, a female, which had been kept as a family pet near Gothenburg ran away on 12th April 1984 and was found three days later about 100 km away, having negotiated many obstacles including one major river.

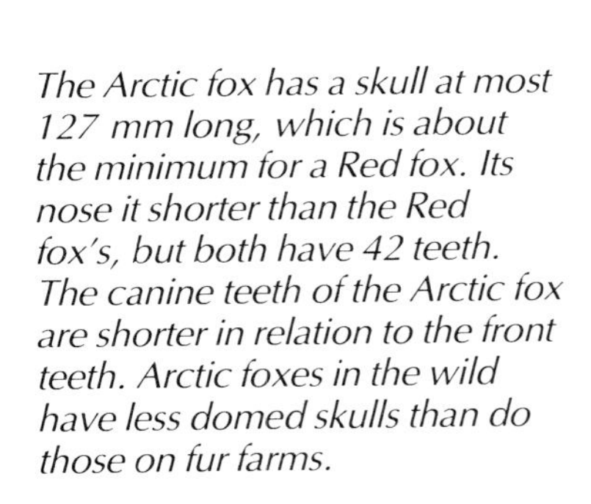

*The Arctic fox has a skull at most 127 mm long, which is about the minimum for a Red fox. Its nose it shorter than the Red fox's, but both have 42 teeth. The canine teeth of the Arctic fox are shorter in relation to the front teeth. Arctic foxes in the wild have less domed skulls than do those on fur farms.*

# Wolverine

## *Gulo gulo*

The wolverine is an animal of the far north, and it lives close to the timberline, and often above it. Few people have seen a wolverine in the wild, apart from Lapps and others who live in the mountainous far north, and this unfamiliarity may explain why the wolverine has such a strange tradition in animal folklore. The wolverine is sometimes referred to in English as the "glutton", and the German name "Vielfrass" also implies that the wolverine has a gluttonous appetite. Wolverines were at one time thought to be so fond of eating that their normal bowel processes could not act quickly enough. It was therefore alleged that the glutton would squeeze his body tightly between two trees growing closely together and push himself through the narrow opening, thereby squeezing himself like a tube of toothpaste. This extraordinary behaviour was depicted in various old engravings and woodcuts!

The wolverine is an interesting animal. Few people outside professional zoological circles would be able to attribute it to its correct family of species, however familiar they might be with the wolverine's appearance in pictures or in mounted museum specimens. It looks not unlike a small bear, but it actually belongs to the same family as the stoat and the weasel. It is also related to the badger, and shares a number of its characteristics. All these creatures are marten-like animals, and the wolverine is actually the largest member of this family which lives on land, the sea-otter being the largest of the entire group.

### Characteristics

The wolverine has dark brown fur with relatively long guard-hairs or bristles, and a dense underfur. The short tail is rather bushy There is a broad U-shaped band of yellowish-white which runs across the animal's flanks and around the back of its body. It is darkest on its underbelly. The ears are small and rounded, with yellowish-brown edges, and the legs are short. The sole of the foot is large and the toes are equipped with large and powerful claws. The body weight of a full grown adult wolverine is in the range 16–28 kg, and the male weighs 50% more than the female. The wolverine has a large skull with a short nose, and its total length is 13.3–15.4 cm. The eyes are rather small.

The wolverine moves with clumsy, half-jumping movements and it is not particularly fast. Its tracks run diagonally in the direction of movement, and the footprint shows five toes pointing forwards, and a clearly defined heel. A wolverine will swim when it has to, and its powerful claws enable it to climb trees. It usually lives a solitary life except during the mating season and when it has dependent offspring.

The wolverine is a clumsy hunter that feeds mainly on carrion, but it is also capable of killing live prey as large as a reindeer. It is a strong and persistent predator, and it will dismember large prey and drag away the pieces, some of which may be several times its own weight. Meat is hidden in caches under the snow or in loose, damp soil. Wolverines will also occasionally take foxes, ptarmigans and hares, especially if it can catch them by surprise, and it will also catch lemmings. Wolverines are not thought to feed on berries or similar vegetable matter.

The wolverine kills by biting the throat or fastening its powerful jaws across the victim's neck and shoulders. In areas where the wolverine is

*The wolverine's skull is thick, with a short total length and nose section. Its 38 teeth, and the areas for fastening its robust jaw muscles, are very strong. These are an adaptation for biting powerfully and crushing big bones.*

numerous and where fur-bearing animals are caught in traps, it will investigate a trap line and spring the traps, much to the trapper's irritation. This animal has acutely developed senses of smell and hearing and it relies principally on these when it is hunting. Its sense of sight is not particularly well developed.

The wolverine is usually active by night, and at dusk and dawn, and it is more active during the twilight of the brief northern summer nights than by day.

Male wolverines have larger territories than females, and a male's range may overlap that of several females. These territories are identified and marked with urine, excrement and scent secreted by the animal's glands. A wolverine will often urinate on the remains of a carcase which it does not intend to finish eating.

## Reproduction

The wolverine is sexually mature at the age or two or three years, after which it breeds every second or third year. Wolverines in captivity have lived for up to 15–18 years, and under optimal conditions a wild female may be able to produce 6–8 litters of young during her lifetime. The males mate with a number of females, and the males' genes are therefore spread through more litters.

Female wolverines maintain litters of differing sizes, ranging from 10–20 $km^2$ up to several hundred square kilometres. During the mating season, which runs from April to August, the males will mate with one or more females, and as with many marten-like animals the development of the foetus is delayed. The young, two or three in number and occasionally four, are born in February and March, and full development of the foetus may take up to eleven months. The young are usually born in a burrow deep in a snow bank or in a crevice or small cave in rocks.

The young are born blind, with yellowish-white fur, and their eyes open at the age of 2–3 weeks. At birth they weigh about 100 grams and are 12–13 cm long. They suckle for the first 2–3 months and remain with the mother until they are one year old. Young wolverines grow rapidly and by the age of 3–4 months are almost as large as their mother.

## Distribution

The wolverine is found in high mountain ranges and in the border zone between the tundra and the northernmost forests, but it can also be found in the coniferous forests around the Arctic regions of Finland, Norway and Sweden. It also inhabits the northern islands of Canada and up to the northernmost regions of Greenland, but it does not occur in Iceland, on Spitzbergen or other northern islands of Europe or Asia. The wolverines of North America were long regarded as a distinct species, for they are larger than the Eurasian animals and are lighter in colour. However, the Eurasian wolverine (*Gulo g. gulo*) and the North American (*Gulo g. luscus*) are now regarded as subspecies of the same animal.

In Scandinavia the wolverine now occurs only in the mountains down as far as northern Dalarna in Sweden and to Nord Tröndelag in Norway. Finland has wolverines in the north-east and in the forests of Karelia, but their total numbers everywhere have been greatly reduced owing to human persecution. At the beginning of the 1970s the total population of Norway, Sweden and Finland was estimated at 225 individuals, and since that time the numbers appear to have remained constant. Occasionally wolverines stray south of their usual haunts, and occasional sightings have taken place well to the south of the species' normal range. In Sweden the wolverine has been protected since 1969.

# Reindeer

## *Rangifer tarandus*

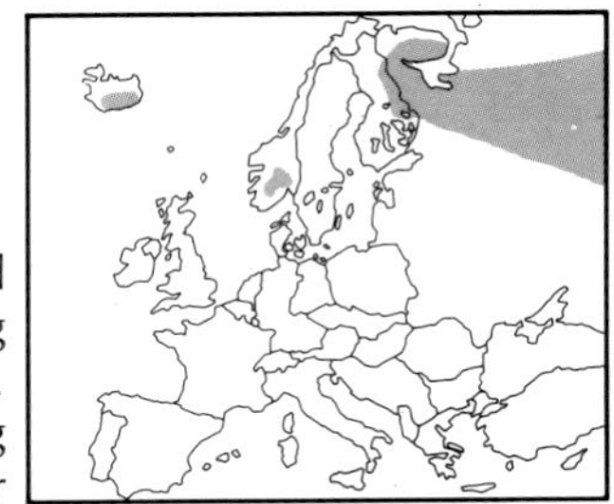

From time to time conservationists and sportsmen discuss the effects of hunting upon the size of wild game populations. Among most small game species hunting has virtually no significant effects, for other factors play a much more important role. With larger game species, however, experience has shown that man's hunting activities have been of considerable significance in the past, and continue to be important today.

Such was especially true in the past for the European bison and the wild ox, two large mammal species which have been almost completely wiped out. The reindeer has not been unaffected, too. Many predatory animals and especially the "big four" — lynx, wolverine, wolf and bear — have been seriously depleted by man. In addition to the direct effects of hunting and persecution, the animals' habitat and behaviour have also made certain species especially vulnerable. The beaver was seriously depleted in numbers and finally exterminated over most of Europe, and yet the fox, despite intense persecution over the centuries, has been able to thrive. Important factors have included the pressures of the human population, conflicts between animals and commercial forms of land-use, and the availability of improved weapons and traps. Man's attitudes towards animals, whether it be fear, appreciation or indifference, have also played a part.

The reindeer, like the wild ox and the wild boar, has had the misfortune to compete for living space and food with domestic livestock. At the turn of the nineteenth century, when the Swedish Laplanders began keeping herds of domesticated reindeer, the wild reindeer became threatened. In approximately 75 years man succeeded in exterminating the wild reindeer in Sweden, and it is believed that the last wild Swedish animal was shot around 1880.

### Characteristics

The reindeer is, of course, a member of the deer family, but it is exceptional in that males and females both bear antlers, and both sexes shed their antlers annually, just like other deer species. However, there is a striking difference between the size of the males' antlers and those of the females.

In general, reindeer bulls are the largest, weighing up to 50%–60% more than the cows. Bulls can weigh up to 160 kg, and cows up to 100 kg. If we consider the North American caribou a subspecies of the European reindeer, the maximum weights can go up to 270 kg for a bull and 157 kg for a cow.

The basic colouration of the reindeer's pelage is greyish-brown, with a pale belly and a greyish-white rump patch. The neck and withers are light, often greyish-white, and the delineation between the back and the belly is marked by a broad band, often black in colour. The winter coat is lighter in colour than the summer fur, but there are great individual variations.

Two subspecies of the northern European reindeer are recognised. One is the alpine reindeer, which is the ancestor of the domesticated reindeer, while the other is the larger forest reindeer.

The reindeer has canine teeth in its upper jaw, although these are absent in most species of European deer. Its cloven hooves are broad and in winter they are covered with an extra growth of hair that extends slightly up the animal's leg.

## Reproduction

The female reindeer become sexually mature at the age of 16–18 months. The rut occurs in September–October, which is when the small all-male herds of alpine reindeer bulls break up. The older bulls gather harems of females which they defend against mating attempts by other mature bulls and also younger animals. An old bull's ability to hold his harem of females will depend upon the number of animals in the harem. Among alpine reindeer, it is unusual for a bull to be able to hold more than 20–40 cows. When the dominant bull mates with an individual cow, the young bulls in the vicinity often try to take advantage of his preoccupation. The rut is a very demanding time for the mature bulls, and even though they will have built up a large reserve of fat during the summer months, this will have been completely used up by the time the rutting activity has finished. Forest reindeer bulls lead chiefly solitary lives in summer, but they too gather harems of cows during the rut.

The reindeer cow has a gestation period of eight months. Before winter begins alpine reindeer bulls, cows and young gather in herds numbering several hundred animals, and occasionally several thousand. Herds of forest reindeer, however, usually number about 40. These communities break up in spring, before the first calves are born, and this takes place in May or June. Cows give birth in special calving areas, and the incidence of twins is high, and triplets are rare. Most cows give birth within a period of 1–2 weeks, and calving takes barely 20 minutes, with the young calf able to stand and walk after a further 20 minutes. Two days later the calf is able to follow its mother, and the calves are suckled up to the age of two months. However, the youngsters remain with their mothers for a year, until it is time for the next crop of calves to be born. The yearlings will continue to grow until they are 2–3 years old.

At birth the calf weighs 5–8 kg, but the calves of the smaller Spitzbergen subspecies (*Rangifer t. platyrynchus*) weigh only 2.5–3.5 kg. Cows with calves rejoin the herd when the calves are only a few days old, and they are joined by yearlings of both sexes, while the bulls form separate summer herds.

## Antlers

As we have noted, the reindeer is unique among European deer because the females also carry antlers. But while the cow's antlers may weigh only a few kg, those of the bulls may weigh up to 11–12 kg. The calves develop small skin-covered button-like growths as early as their first winter, but the first set of proper antlers does not develop until their second autumn.

The old bulls cast their antlers as early as December–January, and sometimes even in November, while the young bulls cast theirs in February–March. The cows shed their antlers just before calving, in April–May. New antler growth, covered with a downy membrane called "velvet", begins immediately after the old antlers have been shed, and the new antlers are hard and free of velvet by August–September.

Because the reindeer of North America, northern Europe and Asia were at one time considered to be separate species, but are now regarded as subspecies of the same animal, it is worth considering the size of the antlers. The largest of all belong to the North American forest caribou (*R. t. caribou*), which has an antler span up to 1.52 metres, while the small tundra caribou (*R. t. groenlandicus*) has a span of about 1.12 metres. The maximum span of the European alpine reindeer (*R. t. tarandus*) is 1.2 metres.

## Teeth

The reindeer's tooth structure is similar to that of other deer, in that it has no incisors in the upper jaw, but, like the red deer, reindeer have canine teeth. The tooth structure is

$$\frac{0 \quad 1 \quad 3 \quad 3}{3 \quad 1 \quad 3 \quad 3}$$

## Adaptation and behaviour

The reindeer occurs in areas where winter temperatures can drop as low as −50°C. It is often compelled to live in conditions where the snow is a metre deep, and life in such a harsh environment is only possible through the development of specialised behaviour and adaptations. The reindeer is well suited to life in snow and extreme cold, for it has an extremely warm winter pelt which prevents heat loss and excludes the cold. Its winter pelage is shaggier and more compact than its summer coat. The individual hairs contain small air cells to give good insulation, and the hair is very thin, which allows it to grow very densely. It is difficult for water to penetrate this very thick growth of hair. In winter even the reindeer's muzzle and nostrils are covered with hair, but the thinner summer pelage begins to emerge in April–May, while the animals assume their winter coats again by September.

Reindeer swim well, and the insulating layers of air in the coat give the animal buoyancy in the water. Reindeer often need to swim when they have to cross rivers and lakes on migration, but the extent of their migratory movements vary in distance and direction. In Scandinavia the wild alpine reindeer of Norway's southern alpine areas are comparatively sedentary.

The reindeer is a highly social animal, living in large herds. In June 1969 an aerial survey was made of the alpine reindeer living in part of Norway, and the aerial photographs showed more than 21,000 reindeer living in 33 herds, the two largest of which numbered more than 4,000 animals each. However, the herd formation among the forest reindeer of Finland (*R. t. fennicus*) is very different. One population of 540 animals was divided into 37% solitary animals,

The significance of deer's antlers has been much discussed. At one time they were regarded as nothing more than decorations, of little significance. It was also believed that while the antlers were growing and covered in "velvet" with its abundant supply of blood vessels, this acted as a means of regulating the animals' body temperature, with surplus heat being radiated through this thin membrane of skin. The presence of these blood vessels is clearly seen when the velvet begins to peel off, revealing the antler bone reddened with blood. In most deer species the velvet is shed in autumn. In the majority of deer species the females have no antlers, which invites the question-how do they maintain a temperature balance? It is evident that reindeer are troubled by heat, and on sunny days they often look for patches of remaining snow, where they will stretch out to cool themselves. Both sexes of reindeer are antlered, of course.

9.7% living in pairs, and 52% living in groups. In winter the average size of a group was only 44 animals, even though there were 250 reindeer in the largest group. The average size of groups in summer was only 2.5 animals.

Besides the fact that the bull's antlers are larger than those of the cow, the male reindeer also has the largest antlers in relation to his body weight of all deer species. It is known that colour and size are important signals in the animal world, and it has also been shown that the ability to interpret such signals improves with age, so it is presumed they must be partly instinctive and partly learned. Reindeer use their antlers in combat between bulls, during the rut and when defending the harem. The largest and oldest bulls, which are usually the most powerful, also have the biggest antlers. Animals which lose their antlers also lose their position in the ranking order of the herd. A young bull can successfully challenge an old bull that has shed its antlers. The annual shedding and regrowth of the antlers also allows the animals to overcome the disadvantages which may result from a set of antlers which are damaged or misshapen.

## Population structure and numbers

Because reindeer tend to live in herds they are relatively unevenly distributed, with high densities in some areas and very low densities in others. There is no pattern of distribution similar to that of, for example, roe deer or moose.

At the end of the 1800s there was considerable concern for the future of the European reindeer. Laplanders had started what was a virtual eradication campaign against the wild reindeer, which competed for space and food with their domesticated herds. In Sweden the last wild reindeer were exterminated in the 1880s. In Norway the remnants of the population were confined to the southern mountains, and in Finland the numbers of wild reindeer dwindled and eventually disappeared completely. In Sweden and Norway it was mainly the alpine reindeer which suffered, while in Finland it was the forest reindeer.

From the remains of a large population which had formerly ranged continuously from the high mountains of southern Norway, eastwards across Finland and into Russia, considerable recovery was achieved through positive conservation measures. By the 1970s it was estimated that there were 60,000 alpine reindeer in southern Norway and in the Kola peninsula of Russia; while the Russian-Finnish population of forest reindeer was estimated at a little more than 5,000 animals. The small Spitzbergen subspecies is estimated at 6–10,000 animals.

In Fennoscandia the alpine reindeer exists only in southern Norway. In 1968 the winter population comprised 35–40,000 animals. In 1911 the Norwegian population had been estimated at only 10,000 animals, but by 1930 this had dwindled to only 2–3,000. In 1954, after the instigation of conservation measures, the pre-calving season population had increased to 12,000 animals, and at the same time in 1960 there were 20,000 reindeer. By 1965 there were 25,000. Since that time there have been attempts to manage the reindeer numbers by culling, because over-grazing has led to food shortages and a decline in the breeding abilities and the general physical condition of the reindeer.

## Distribution and subspecies

The reindeer is a member of the Arctic group of fauna, and it has a circumpolar distribution across Europe, Asia and North America.

The relationships between the different reindeer populations has been the subject of much discussion and study. At one time three distinct species were recognised, each with its own subspecies. Nowadays, however, all reindeer are deemed to belong to one species, which is subdivided into more than 20 subspecies. Generally speaking, two ecological types can be distinguished — the tundra and alpine reindeer, which live mainly in open country; and the forest reindeer which live in woodlands and afforested areas.

Based on this grouping, three different subspecies have been identified in North America. Farthest north, in Greenland, are relatives of the Eurasian reindeer (*Rangifer tarandus*). The tundra area of northern Canada is inhabited by the tundra or "barren grounds" caribou (*Rangifer arcticus*). The woodlands to the south of the tundra are the home of the woodland caribou (*Rangifer caribou*); and the main difference between these subspecies is their size. The smallest are found in the north, where the Greenland reindeer's maximum weight is about 135 kg; the barren grounds caribou reach a maximum of 180 kg; while the woodland caribou can reach a body weight of 270 kg. The size of the antlers also differs in the same proportions.

According to another theory there are five completely different subspecies. Besides the Greenland mountain reindeer (*R. t. tarandus*), of European origin, and the woodland caribou (*R. t. caribou*), the reindeer of the Arctic archipelago north-west of Baffin Island are included in the subspecies Peary's caribou (*R. t. peary*), and they can be distinguished from the barren grounds caribou, which according to this classification are named *R. t. groenlandicus*. Finally, there is the Alaskan animal called Grant's caribou, *R. t. granti*. The differences between these subspecies include size and migration habits.

However, current thinking is that these classifications are all to be regarded as subspecies of one and the same single species, *Rangifer tarandus*. This means they belong to the same species as the reindeer of Europe and northern Asia, and in northern Europe two subspecies are recognised. One is the alpine reindeer (*R. t. tarandus*), which was the first reindeer to be given a scientific description by Linnaeus in 1758. The other is the forest reindeer, *R. t. fennicus*, which was described as recently as 1909.

In general the reindeer has a widespread distribution in northern Europe, as have a number of other species that now occur in the northernmost parts of the Old World. Before and during the last Ice Age reindeer existed in central Europe and the British Isles, but after the ice retreated the reindeer was one of the first mammals to move northwards. Pieces of reindeer bone which have been discovered in Denmark and southern Sweden were found in peat bogs with the remains of the contemporary vegetation. which included alpine willow and dwarf birch, which indicates that an arctic climate prevailed. As the tundra gradually emerged the reindeer pushed north. About 10–11,000 years ago, when the ice lay over the plains of central Sweden, reindeer inhabited the lands to the south of it, and bones from southern Sweden have been dated back to 8–11,000 years ago. From southern Sweden reindeer went on to colonise Norway. It is probable that a second wave of reindeer came from the east and colonised westwards, and these were the forest reindeer that followed the expansion of the Siberian taiga and the spread of the coniferous tree zone.

These two groups of reindeer have different habitat requirements, and they also differ in size. The forest reindeer are the larger, and they are also better adapted to life in forests under deep snow-cover, for it has longer legs than the alpine reindeer.

In the 18th century European alpine reindeer were successfully introduced into Greenland and Iceland, and attempts have also been made to introduce domesticated alpine reindeer into parts of mainland Europe, but without success. A large reindeer herd was kept for some time near Viborg in Jutland, and at its height the herd numbered about 1,200 animals, but they were eventually wiped out by disease.

In more recent times Siberian domesticated reindeer, originating from the wild reindeer of the tundra, were introduced to the Kola peninsula of Russia, where they now predominate. In 1952 domesticated reindeer were introduced to part of the Cairngorm mountains in Scotland.

## Habitat and diet

The alpine and barren grounds reindeer prefer open terrain, an environment which seems to offer little shelter or food to a large herbivore. These reindeer are essentially grazing animals, and in summer they eat grass, herbs and lichen, while in winter they survive mainly on the reindeer moss which they scrape out from under the snow. They also eat twigs and shoots from bushes and trees, and any grass they can uncover.

The forest reindeer remain in the woodlands all year, and their main foods are leaves and shoots of bushes and low trees. They also graze on herbs and grasses, and they like to browse in marshy areas, and close to streams and rivers where the undergrowth is rich. In winter they eat the reindeer moss which they kick out from under the snow. If the snow-cover is very deep they eat "beard moss" (*Usnea barbata*), which grows on trees; and also small shoots and twigs. It is believed that their sense of smell helps them to locate lichen under the snow more readily than alpine reindeer can.

## Man, diseases and enemies

As we have mentioned, man has affected the numbers and distribution of wild reindeer in different ways — by hunting to extinction as well as by successful introductions. In many areas the reindeer is highly regarded as a game animal, and traditionally many Indian and Eskimo communities in Canada are dependent on the herds of wild caribou. In Europe the domesticated reindeer is equally important for the Lapps, while the wild reindeer is essentially a quarry for the hunter and the sportsman.

The wild reindeer is hunted in Europe, principally in Norway, and according to Norwegian statistics the number of reindeer shot annually is between 10,000 and 13,000. In many parts of its range the reindeer is susceptible to predators, and the wolf is the principal enemy, although both lynx and wolverine will regularly attack reindeer. However, the latter usually claim calves and young animals.

# Ptarmigans

## Willow ptarmigan — *Lagopus lagopus*

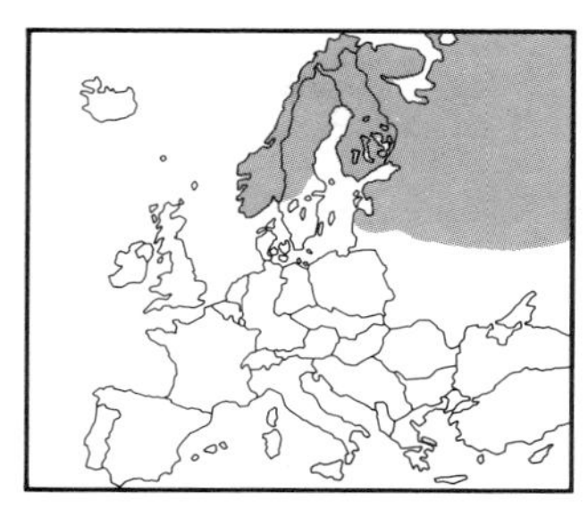

## Rock ptarmigan — *Lagopus mutus*

The ptarmigans are members of the grouse family, and they are rather similar in appearance and habits. Two species, the Willow ptarmigan and the Rock ptarmigan, are found in Europe. (The Red grouse of Great Britain and Ireland are a local subspecies of the willow ptarmigan, sometimes known as the willow grouse, and are discussed in an earlier chapter.) The term "snow ptarmigan" is occasionally used, but chiefly by restaurants and poulterers, and it refers to either of these ptarmigan species when they are in their white winter plumage.

The family *Tetraonidae*, to which the grouse belong, is represented by 17 different species in northern Europe, Asia and America. The capercaillie is the largest of these, and the hazel grouse is the smallest. Grouse are comparatively sedentary by nature, and this has led to the development of distinctive local characteristics. The rock ptarmigan has been identified in no fewer than 23 different subspecies, and the willow ptarmigan in 16.

Grouse are highly esteemed as game-birds. For most European sportsmen the willow ptarmigan is considered more significant, as it lives in terrain which is more accessible and easier to hunt than the high and rocky haunts of the rock ptarmigan. This is especially the case when shooting involves the use of shotguns and bird dogs such as pointers and setters. The rock ptarmigan is also less inclined to squat and lie still, blending with its surroundings, which makes dogging more difficult. In winter professional hunters shoot the rock ptarmigan with smallbore rifles, and do not use dogs. Snares are also occasionally used.

The annual bag of both species shot annually in Sweden has been estimated at 40,000, while in Norway some 675,000 birds are shot annually. In Finland the number has been estimated at 100,000. In the past there was a very strong tradition of professional grouse shooting in the Scandinavian countries, and in the 1890s there were instances of individual hunters accounting for up to 3,000 grouse in a single winter. In 1893 the records show that 30,000 grouse were

*Willow ptarmigan, female in summer plumage (left) and male in autumn plumage.*

exported through the customs office at Funäsdalen. However, there has never been any tradition of professional shooting of grouse in the British Isles, although sporting shooting can command very high prices and contributes significantly to the revenue of many upland moors.

## Characteristics

The winter plumage of both the ptarmigan species is very similar. Both are completely white, with feathered toes and tarsi, and the outer tail feathers remain tipped with black. Above the eye there is a bright red wattle, and the male rock ptarmigan has a prominent black eye stripe from eye to bill, which may also occur in older females. The rock ptarmigan has a more delicate bill and smaller claws than the willow ptarmigan. In summer plumage both species have white primary feathers, and these are prominent when the birds are in flight. However, the body feathering of these two species is quite different. The willow ptarmigan is reddish-brown with pale crescent-shaped scallops at the tips of the feathers, especially on the bird's back, wings and flanks. The rock ptarmigan is greyish-brown with gold-brown patterning, especially on the flanks and the trailing edge of the wings. The throat and breast of the male bird is grey to greyish-black with lighter greyish-white scallops. The females are brownish to golden brown with similar greyish-white bordering, and both sexes have light greyish-white throats and white bellies.

These grouse have a complicated moulting pattern. The white winter plumage changes slowly to summer plumage, and there is a gradual change back again to the winter plumage. In addition, the rock ptarmigan has a breeding plumage which is grey to greyish-black, becoming more brownish towards the autumn. The claws are shed annually. Both species are mottled with white in their transitional plumages in spring and autumn. The young birds' primaries are dark and change to white when they are about one month old.

The wingspan of a willow ptarmigan is about 193–211 mm, with that of the female falling within the range 188–196 mm. A fully grown bird will weigh 440–730 grams, and the males are heavier than the females. The wingspan of the male rock ptarmigan is in the range 189–213 mm, with the female's measuring 184–206 mm. Weights are in the range 347–610 grams, and in this species the male is also the heavier bird. Generally, the willow ptarmigan is a slightly larger bird than the rock ptarmigan, and it is believed that both species attain their maximum annual body weights in the period September–November. It has also been noted that the northern populations of both species are heavier than the southern populations, and it is believed that this is an adaptation for maintaining body heat in winter, for a larger volume of body weight releases proportionally less heat than a smaller body.

The willow ptarmigan makes a long cackling, chuckling call, while that of the rock ptarmigan is brief and harsher. The chicks of both species utter a high-pitched peeping sound. In the long summer days of Nordic lands grouse are active both by day and night.

In spring these grouse are found in pairs, and later when the chicks have hatched they will all remain together as a family. In autumn several families may join together to form larger autumn and winter flocks or packs. By this stage the young birds are largely independent, and they do not necessarily join the same flocks as their parents or siblings. The winter flocks are composed of birds of both sexes and all ages, and in rare cases they may number up to several hundred birds. In spring these flocks disperse and the breeding pairs spread out, each in its own territory.

If a flock of grouse are disturbed they burst into flight with a fast wing action, which then turns into a flight which involves gliding fast and making bursts of wing-beating, and the birds may scatter over a wide area.

## Reproduction

Like many other gallinaceous birds, these grouse species produce only one clutch of eggs per year under ideal conditions. If eggs are lost or nests destroyed, however, most birds will lay again. Grouse pair for the breeding season, and they are ready to breed in their second year of life. In northern areas the male claim their territories in March–April, and defend them vigorously against other neighbouring birds, and territorial males make prominent territorial flights and displays on the ground.

The eggs are laid in May–June, but in the case of the rock ptarmigan this may be delayed to as late as July. It has been discovered that the amount of daylight, and therefore the length of the day, influences the development of the sexual organs and triggers the onset of the breeding season. The willow ptarmigan typically lays a clutch of 8–12 eggs of a yellowish colour, richly speckled with black and dark brown markings. The dimensions are 40–47.4 mm x 29–32.5 mm, and they weigh approximately 21 grams. The rock ptarmigan lays a smaller clutch of eggs, usually 6–10 in number and similar in appearance to those of the willow ptarmigan, except that the blackish-brown markings are larger and sparser. The dimensions are 39–45 mm x 28–31.5 mm, and the egg weighs about 22 grams.

The eggs are laid at one day intervals, and the nest is a simple hollow in the ground, sometimes in the open and sometimes in low vegetation. Both species incubate for 21–24 days, and incubation is the sole responsibility of the female, although the males remain nearby. The newly hatched chicks weigh about 14–15 grams.

As with the young of other gallinaceous birds, the newly hatched young are well developed and active, and they can leave the nest and follow the adults without delay. Their eyes are open, they can hear and move about freely, but they tend to stay close to the mother. The mother and her young make contact mainly by sounds rather than by sight, and if the chicks are cold their calls will induce the female to gather them together and brood them. Brooding continues for about three weeks, and the male bird remains nearby and watchful. If danger threatens he may pretend to be wounded and try to lure a potential predator away from the female and her chicks. The chicks can make short flights at the age of two weeks, and their tail feathers begin to show when they are only two or three days old. The young stay in the care of the female well into the autumn.

## Habitat and diet

The willow ptarmigan and the rock ptarmigan are both birds of the open countryside. The willow ptarmigan lives mainly on heather moorland, on boggy ground and in forested areas where there are stretches of open heather with dense ground cover. Above all it occurs in the birch and willow forests close to the mountains. The rock ptarmigan is a bird of the higher mountains in Scandinavia and in the Alps. It also occurs in the birch and willow areas, but it prefers bare mountain slopes, often on rocky ridges and areas strewn with boulders. In winter it often descends to the birch forests on the lower slopes.

In Finland it has been found that the willow ptarmigan moves between four different habitats during the year, although this does not involve long distance migrations. In the Finnish woodlands the willow ptarmigan principally lives in areas of pine trees and heather where there is a good ground cover of blueberry bushes in September–October. In winter, from November to May, it lives mainly in areas of dense undergrowth close to rivers, brooks and lakes. In Finnish Lapland it winters in the mountainous birch forests, but there also it prefers areas near water.

Studies of the winter diet of willow ptarmigan along the coasts of northern Finland have shown that the principal winter food is the twigs and shoots of green willow and common birch, and willow ptarmigan were estimated to eat 90% of the available food of this type. It has been calculated that a willow ptarmigan requires about 30 metres of willow to fulfill its calorific requirements for one cold winter's day. The willow ptarmigan is active by day during the winter, for it must feed without delay during the short hours of daylight on winter days. Its preferred willow is usually supplemented by dwarf birch.

The population densities of these two species vary. In Sweden it is estimated that about five pairs of willow ptarmigans nest in each square kilometre of mountain forests, but that only one tenth as many nest in the coniferous forests. The rock ptarmigan is thought to have an average density of 1.5 pairs per km$^2$. In the Alps it may be as much as five pairs per km$^2$ in suitable habitat.

The rock ptarmigan lives in areas where average summer temperatures are a few degrees colder than in the typical habitat of the willow ptarmigan.

The newly hatched young are completely dependent upon insects for food during the first week or so of their lives. As they mature, plant foods become a more important component in their diet, and by the age of 2–3 months they have almost completely adapted to the same diet as their parents.

The willow ptarmigan eats the soft parts and buds of branches and herbaceous plants, seeds and in the summer insects and small invertebrates. In autumn they feed almost exclusively on berries, while the winter diet consists of birch and willow buds. The rock ptarmigan eats the soft parts of mountain plants, insects and small invertebrates in summer, and at this season it finds most of its food on the ground. In winter it shares the willow ptarmigan's taste for low-growing willow bushes and birches, and entire flocks sometimes feed together.

## Adaptations and behaviour

Both species of ptarmigan are known for their habit of burying themselves in the snow when the weather is very severe. Their reserve layers of fat make it possible for them to survive during quite long periods, during which they lie still and consequently use up little energy. Their feathers and the snow insulate them from the cold, and these are two of our hardiest bird species.

Grouse are rather sedentary birds, but they make periodic and rather short migrations according to the weather and the availability of food. Rock ptarmigans will move in winter to lower and more sheltered areas, where food is more readily available.

Grouse numbers and the density of pairs in the breeding season are known to vary periodically, and sportsmen and naturalists have shown great interest in the reasons for this. Some research has concentrated on looking for internal, self-regulating causes in these species, while other studies have looked at the significance of external factors. It has been noticed that there is a connection between the birds' behaviour before breeding and the number of breeding pairs. It has also been noticed that there is a reservoir of non-breeding birds — a floating population, that compensates for birds that disappear at the beginning of the breeding season, and such birds also took the place of birds that were removed for experimental reasons. It is also believed that the young grouse can become more aggressive if they are hatched from clutches laid by hen birds with nutritional deficiencies, and because of this aggression they later claim larger nesting territories, resulting in reduced pair-density. The numbers of grouse goes up and down in cycles with intervals of 8–10 years between the high-density years.

Availability of food and the weather conditions can also affect grouse numbers, but these factors are usually responsible for irregular and short-term fluctuations of grouse numbers. It has also been established that there is a connection between the numbers of game-birds and the abundance of small rodent populations. It is thought that an abundance of small mammals in rodent years reduces the level of predation on grouse, grouse nests and eggs by crows, raptors and predatory mammals. Studies in Norway have shown that certain species of predators can take over from one another. When crows were completely exterminated on one island, it was found that stoats increased and took their place as the principal predator of eggs and chicks.

## Distribution and numbers

Both the willow and the rock ptarmigan are holarctic birds, which occur across northern Europe, Asia and North America. The European distribution for both species is much the same, but there are certain areas where only one species is to be found. The rock ptarmigan occurs in the mountains of Scandinavia and in the northern tundra, and it also occurs in Iceland, on the islands of the far north, and in the Alps. The willow ptarmigan is not found in Iceland, but elsewhere it occurs in the far north as well as significantly further south than the rock ptarmigan. In Finland Russia it breeds in the coniferous forests, and in the 1800s the willow ptarmigan was to be found far south of its present range, in East Prussia and in the Baltic states, where it could be found up to the 1900s.

In Sweden the rock ptarmigan is found only in high mountains, but the willow ptarmigan is also to be found in afforested land as far south as the Gulf of Bothnia. In the mid-1970s the Scottish population of the rock ptarmigan (known in Britain simply as "the ptarmigan") was estimated at around 10,000 breeding pairs on the mountains of the central, western and northern highlands. The British population of red grouse, which is the British Isles subspecies of the willow ptarmigan, has been estimated at around 500,000 breeding pairs.

## Grouse hybrids

Members of the grouse family will hybridise readily, and there have been many recorded instances of rock ptarmigan and willow ptarmigan interbreeding in areas where the two species coexist side by side. These two species have also been known to hybridise with black grouse and hazel grouse.

*Male rock ptarmigan in summer (left), winter, and spring (right) plumage*

## Chamois

### *Rupicapra rupicapra*

The short nose, the triangular shape of the head and the hooked horns combine to give the chamois a rather alert appearance. The horns are short, only 32 cm at most, and they point straight up from the skull, with the tips curling backwards to form the characteristic hooks. Both the male, or buck, and the female, or doe, have horns; but those of the female are slightly smaller than the male's. The horns of the young, or kids, begin to develop when they are three months old.

The chamois has a goat-like appearance. The hairs of its coat are coarse and lie in dense layers, although they are not particularly long. The bucks have a ridge of longer hairs along the line of the backbone.

The chamois has a compact and muscular body, and a mature buck will weigh 35-50 kg, while a mature female will weigh 25–42 kg. The body is coloured a solid blackish-brown, but this turns lighter brown in winter. The head has a light greyish-white mask with a wide, dark band running from the ears over the eyes to the nose, which is covered with hair. The short tail is not more than 15 cm long and is black. The chamois is fully grown at the age of four years, and the longest recorded lifespan is 25 years.

The chamois can be divided into an Alpine race and a Pyrenean race. Those which live in the Pyrenees are slightly different in colour, for they are more reddish-brown in summer and dark brown in winter, and the light-coloured areas on the throat and head are more yellowish than in the Alpine chamois.

### Reproduction

The chamois becomes sexually mature at the age of 18 months, but it is believed that the majority of them do not breed until they are 3–4 years old. The rut takes place between October and December and gestation lasts for 25–27 weeks, resulting in the birth of the kids during the period April–June.

During the rut the buck utters a husky bleating call as he rounds up a harem of breeding does, for the bucks are polygamous. Both sexes have glands in the neck behind the horns, and the buck marks his territory with this glandular secretion. The bucks wander widely before the rutting season begins, with the gathering of a harem of does and the establishment of a territory. Bucks display for territory and status by raising the hairs along the ridge of the spine and holding them erect, while bucks of equal status will collide head-on in fights.

Before the young are born the chamois move up to the higher parts of the mountain, and the pregnant females leave the harem in preparation for the birth, which is usually a single kid if this is its first birth, but often there are twins and even triplets if the mother is older. The kids are born well-developed and are able to follow their mother immediately after they are born. A newly born chamois kid will weigh 2–2.5 kg. They grow quickly and are suckled until they are about six months old, but they begin to eat grass and herbs from the age of about two weeks. Some kids remain with their mothers until they are 18 months old, by which time they are mature and capable of breeding.

## Herds

Chamois usually live in small herds numbering 10–20 animals, but in winter they can band together into local herds of as many as 100 animals. Like the moufflon, the relationships within the herd are flexible, and individual chamois may leave one herd and join another. Old bucks may live as solitary animals, and the harems, guarded by the dominant buck, can represent well unified groups even though they are only temporary. The remaining herds at rutting time will comprise old and young animals. The mothers take their kids to join a herd a few months after they are born.

## Habitat and diet

The principal habitat of the chamois is high mountains. They live at higher altitudes in summer than in winter, and they may range from 800 metres up to 3,000 metres above sea level. They occur most commonly in rocky, forested terrain between 1,500 and 2,500 metres, and their preferred habitat consists of rocky outcrops, steep slopes and cliffs.

Chamois can jump distances of up to 7–8 metres. They have powerful legs with large hooves which provide them with a good grip on the rocks because the sole of the hoof is soft. The chamois can move across steep cliffs where there are ledges only 10 cm or more in width. Chamois feed during the daytime and at night they seek shelter and sleep. The major factors in their diet are grasses and herbs, but chamois will also nibble at the twigs and branches of bushes, and at low-growing brushwood. They graze chiefly after dawn and before dusk.

## Numbers and distribution

The chamois' natural distribution areas are the mountain massifs of central and southern Europe. Because chamois confine themselves to craggy and inaccessible parts of the mountains, they have not been as threatened by man as the moufflon and the ibex. The largest numbers and greatest local concentrations of chamois are to be found in the Alps, especially at the eastern end of the range. However, chamois have been introduced to many other places, including the south of Germany. Several thousand now exist there, mainly in the Black Forest.

The family *Bovidae* includes the ox as well as sheep and goats. There are several small members of this family in Europe, the best known of which are the chamois, the ibex and the moufflon. The saiga antelope (*Saiga tatarica*) also deserves to be mentioned here. It occurs on the steppes of southern Russia and can now be found almost as far west as the Black Sea. What these species have in common is that they are all mammals indigenous to Europe. The musk ox, described elsewhere in this book, is also a member of this group of animals, although its appearance suggests a closer relationship to the ox than to sheep and goats.

The species in this group of animals can be distinguished from deer by the fact that both sexes bear horns, with the exception of the female moufflon which often lacks horns. The horns are curved but not branching, and they are not shed annually but grow throughout the animal's life. All have short tails.

If we disregard their size, the shape of the skull and the horns distinguish this group from their bovine relative the ox. In addition, the moufflon, ibex and chamois only have two teats. Members of the family *Bovidae* lack incisors and canine teeth in their upper jaws, and their 32 teeth are arranged according to the formula:

$$\frac{0\ \ 0\ \ 3\ \ 3}{3\ \ 1\ \ 3\ \ 3}$$

The chamois, ibex and moufflon are all adapted to life at high altitudes and in extreme conditions. They feed on all kinds of vegetation, and all are agile climbers even at a very young age. The number of red blood cells is high, the heart muscles are especially well developed, and both the pulse rate and the body temperature are high.

The mature males gather the females when they come in season, and they guard their harems with care. To enable them to do this they build up large reserves of fat before the autumn rut, and during the autumn and winter mating period these master bucks lose up to 25%–30% of their body weight.

Within their various distribution areas, each of these animals lives at a different elevation. The saiga antelope lives on the steppe, while the others are animals of the mountains. The moufflon was previously a creature of the deciduous forests, and of the moors and the *maquis*, a low scrub of evergreen aromatic plants covering rolling hills and the lower slopes of the mountains. The chamois lives mainly in the high mountains, and in summer it grazes the high areas of its range, from 1,500 metres upwards. The ibex lives in even higher regions, above the timberline, and it is said to live at altitudes between 2,300 metres and 3,200 metres in both summer and winter. All three species prefer broken rocky terrain and slopes littered with large boulders.

All three species have been greatly reduced in number and have suffered at the hands of man. However, there have been some successful introductions in new areas, especially this century, and this has replaced much of what had been lost elsewhere.

To an even greater extent the larger bovine animals — the wild ox, the European bison (as well as the wild boar and the reindeer) — have met with serious persecution when they have competed with man's domestic animals for food and space. However, the moufflon can be crossbred with the domestic sheep, and hybrids now exist in several areas.

# Ibex

## *Capra ibex*

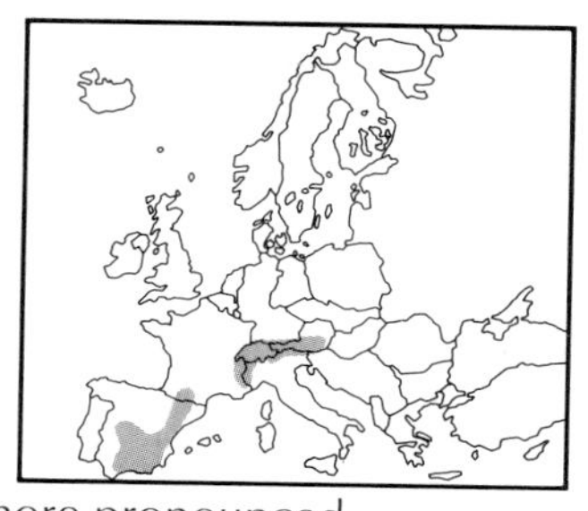

The ibex is a large mountain goat. The buck can weigh up to 124 kg, and the female up to 55 kg, which makes it twice the size of the chamois and the moufflon. The difference in size between the males and the females is also much more pronounced.

The ibex coat is a relatively uniform greyish-brown in summer and blackish-brown in winter, and there is a well-defined grey-white area on the belly. It has a light-coloured rump patch and a short tail (not more than 15 cm), which is dark on the upper side and greyish- white on the underside. The hocks of the hind legs are greyish-white, and it has a dark "billygoat" beard. The ears are relatively small.

The horns of the ibex are different from those of chamois, moufflon and other wild, horned animals of Europe. Both sexes have long, unbranched, sabre-like horns, and those of the buck may reach a length of one metre, while those of the ewe can be 30 cm long. The maximum weight is 5 kg. The front edge of the horn has thick, semicircular, transverse ridges which increase in number according to the age and growth of the horns. The first buttons of horn emerge on the kids of both sexes at 3–6 months old, and they continue to grow thereafter for 10–12 years. The longest recorded lifespan of an ibex is 30 years.

Three species of ibex occur in Europe: the Spanish ibex, the Caucasian ibex and the wild goat (*Capra aegagrus*). There are very few of each species, which are restricted to their home ranges in the Pyrenees and the Caucasian mountains, and parts of the Greek islands.

## Reproduction

Ibex live in small herds of 10–20 animals, consisting mainly of females and young animals. The mature bucks are either solitary or form separate male herds. During the rut in December–January the bucks engage in fierce combat, colliding head to head, and the sound of clashing horns can be heard a long way off. The most powerful bucks take over the herds of females when they are in season at mating time. The female is only in season for 2–4 days. Ibex are polygamous. The incubation period lasts 22–23 weeks, which means that lambing usually takes place in June–July (occasionally in May). Normally there is only one kid, but twins may also occur, and the newly born kid weighs about 2–3.5 kg. The food supply in winter determines not only whether there will be twins or only a single young, but also whether the foetus will develop or be reabsorbed.

The kids are able to follow their mothers within a few hours of birth, when they can hear, see and smell. As with so many other cloven-hoofed animals, the female leaves the herd before the birth and rejoins it about a month later. The kids are suckled until they are 6–12 months old, and they become sexually mature at four years old.

## Habitat and diet

Ibex inhabit the highest and most remote parts of the mountains. They move elegantly and unhesitatingly along cliffs and sheer slopes, surpassing even the chamois in agility and sure-footedness. Only a few hours after their birth, the kids are able to climb with their mothers.

In both summer and winter they can be found at altitudes between 2,300 metres and 3,200 metres, and they are often to be seen above the timberline. They graze on Alpine meadows and on slopes, where their diet consists principally of grass. In winter when there is snow-cover they eat twigs and shoots from bushes, shrubs and low-growing trees. They also search for lichens that are scratched out from under the snow or scraped from trees. Ibex are active by daylight, but are also said to move about and feed on moonlit nights.

## Herds

Ibex live in herds, although an exception to this is the old bucks, which tend to live solitary lives. Usually the mature bucks form all-male herds consisting of 20–30 animals, and within these groups there is constant skirmishing and sparring, which determines the animals' ranking order. As some animals leave the herd and others join it, the males' interrelationships are constantly being tested.

## Distribution and numbers

Of the three mountain animals — moufflon, chamois and ibex — the ibex has suffered most from man's hunting and other activities. But, paradoxically, the last remnants of the European ibex were saved from extinction by the dedication and enthusiasm of sportsmen. As early as the sixteenth century the ibex had gone from most of the eastern Alps. By 1706 it was exterminated in southern Germany, and by the 1720s it had been wiped out in Austria. By the 1840s the Swiss population had also vanished, and the only remaining ibex were in the Graian Alps, in north-west Italy.

This remnant population survived in what had previously been a royal hunting reserve called Gran Paradiso, and in 1879 the ibex population was estimated at about 600 animals. This area was declared a national park in 1921, but already by 1914 good conservation and management had caused the population to increase to about 3,000 animals. Numbers dropped during the two world wars, but by 1958 there were some 3–4,000 ibex living in this region. From there, ibex have been reintroduced to many parts of their former range in the Alps, and the present day wild population of ibex has been estimated at around 20,000 animals, plus those which are preserved in zoos and enclosed reserves.

Outside Europe the ibex has a wide distribution. It exists in many of the mountain ranges of Asia and Africa; it occurs in large numbers in the Himalayas, in the Caucasus and in the mountains of Afghanistan and Iran. Ibex are also found in arid mountain areas, such as the Sudan and Ethiopia, in Arabia and in the Sinai peninsula.

Despite this wide distribution, with a number of relatively isolated local populations, only four subspecies are recognised: the *ibex* subspecies in Europe; *severtzovi* in the Caucasus; *nubiana* in north-east Africa; and *sibirica* in the mountains of Asia.

# Moufflon

## *Ovis ammon*

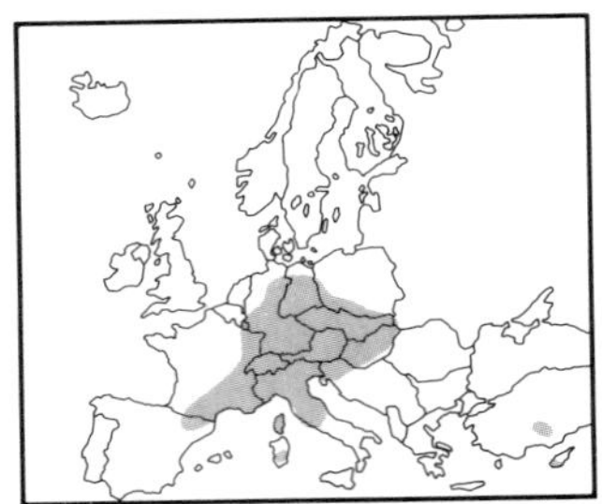

The moufflon is a relatively small wild sheep with a compact body. Its legs are comparatively long and the fur is brownish, with light greyish-white "stockings" on the legs, with a blackish-brown border at the top. The abdomen and the rump patch are the same light colour. The ram often has a big, greyish-white area on each flank, connected by a broad band across the back, and rams also have a well-developed mane on the throat and chest. The tail is dark on the upper side and greyish-white underneath. The rams' horns can be as long as 85 cm and can weigh up to 6 kg. The horns of the females are only about 10 cm long, and the females of the Sardinian population lack any horns. The ram's horns begin to show at the age of four months, and they continue to grow in length and thickness up to the age of 8–9 years. In Cyprus the rams often have spiral horns. The body weight can be up to 55 kg, and rams weigh about 30% more than ewes. The normal weight of a fully grown ram is 35–40 kg, and a typical ewe will weigh 25–30 kg. They are fully mature at the age of 4 years, and the longest known lifespan for wild moufflon is 15–20 years.

### Reproduction

The moufflon ewe comes in season for the first time when she is about 8–9 months old, and the rams become sexually mature at 18 months. Gestation takes 21–22 weeks, and since the rut takes place in October–December, the lambs are born in March–May. First-time mothers usually have only one lamb but older ewes often have twins, and the newborn lamb weighs about 2.5 kg. The young are very well developed and can stand upright after only a few hours, when they can hear, see and smell and follow their mothers readily. After about 14 days they start to eat grasses and herbs, but they continue to suckle for up to five months. By the time they are six months old they are almost entirely independent.

### Herds

Moufflon form herds of 20–50 animals in winter, but in summer the young animals and the mature rams form their own herds, while ewes with lambs live separately. In autumn old ewes with lambs form herds which are pursued and guarded by one or more mature rams. An old ram often leads the herd in winter. Old moufflon rams are usually solitary animals. The herds seem to be loosely formed, with individual animals belonging first to one herd and then to another. Old ewes often form small herds of up to ten animals, led by one old ewe. The ram herds follow these female herds carefully but at a distance.

### Habitat and diet

Like most species of wild sheep, the moufflon is found in relatively open terrain, where the herds can wander and the members of the herd will graze together. But moufflon also live in dense deciduous and mixed forests with cliffs and steep slopes. In mountainous regions the moufflon can occur above the timberline, but they are usually found at lower altitudes. Moufflon are excellent climbers.

In the Mediterranean region its favourite habitat is the shrubby *maquis*. It seems likely that this adaptable sheep has been forced to live in higher and wilder mountain areas than it would normally have preferred.

The moufflon is a ground grazer, eating mainly grass, herbs, brushwood and mosses. It also likes young twigs from bushes, but it does not strip the bark. In some areas it will feed in cultivated fields. Moufflon are active both by day and night, but in many places it feeds mainly at dusk and dawn, and through the night, perhaps as a result of disturbance.

### Distribution

The moufflon has been hunted relentlessly within its European range, and 8–10,000 years ago this comprised most of mainland Europe. The northern limits of its range ran through Poland, Czechoslovakia and France, but moufflon have never occurred in the British Isles. The species has been forced into the remote mountains of Corsica and Sardinia where it has survived, and in 1956 these island populations were estimated at 250 animals in Corsica and 3–4,000 in Sardinia.

Many wild moufflon have been introduced to Europe. They were reintroduced to Germany in 1902–3, and by 1938–9 the German population was estimated at not less than 7,500 animals. This population was almost completely eradicated during the Second World War, but by 1953 a combination of careful conservation and introductions resulted in a population of 1,850 animals. In 1968 the population was estimated at 4,000 animals, and in the hunting season of 1969–70, 900 animals were shot in West Germany.

The introduction of moufflon is generally regarded as non-controversial, for these animals are sedentary and rarely leave the woodlands. They cause little or no damage to trees, and only rarely do they raid arable fields. Their numbers can increase very rapidly, as the West German experience has demonstrated.

It is therefore not surprising that moufflon are now to be found in many parts of Europe, in large and small populations. The largest concentrations of moufflon are believed to be in Germany, Czechoslovakia and Austria, which account for about 65% of the European total. In Scandinavia moufflon exist in a number of wildlife reserve areas, and occur on some of the islands. Wild moufflon are also found in Denmark.

The Corsican and Sardinian moufflon have often been conisdered to be genetically distinct from the moufflon of Cyprus, and they are sometimes given the status of a separate species, for the Cyprus moufflon is known as *Ovis ophion*. The moufflon which were introduced to various parts of Europe were originally from Corsican and Sardinian stock, and they are now referred to as *Ovis musimon*. The moufflon is closely related to domestic sheep, with which it will interbreed.

## Animals of the mountains

The European mountain regions occupy a relatively small part of the continent, and only 1% of the land surface is above an altitude of 2,000 metres. This is comparable to the area of land which is covered by lakes. The Alps and their outlying peaks account for a large proportion of the European mountain area.

If mountains are high enough, they have the same climatic and vegetation zones which are to be found by moving northwards to higher latitudes. Glaciers and treeless areas are to be found in the high Alps, while green deciduous forests occur below the foothills, with coniferous zones and Alpine meadows occurring in between.

Because of the lie of the land, some mountainous areas are often very inaccessible. Furthermore, mountain conditions such as snow cover and sub-zero temperatures also vary, both from season to season and from place to place. All these factors influence the composition of the fauna and the environment. The inaccessibility of some European mountain areas has been the salvation of the brown bear, to take only one example. It remains to be seen if such species can continue to exist despite greatly improved modern communications and in the face of a widespread disregard for their welfare.

Compared to lowland areas, the mountain regions of southern Europe have their own special selection of resident animals and birds. The Alpine marmot (*Marmota marmota*), the Alpine fir vole (*Pitmys multiplex*) and the snow vole (*Micotus nivalis*) all occur here, well above the timberline and all are to be found at altitudes above 2,000 metres, even though they will live happily at much lower levels. This is also true of the rare mountain vole (*Dolomys milleri*). The snow vole is the highest living of all European mammals, and in the Alps it has been found at up to 4,000 metres.

There is an isolated population of arctic hares in the Alps, surrounded by the brown hares which are common across the lowlands of Europe. This is an important relict population which has successfully survived since the last Ice Age, and the alpine hare has been seen at altitudes up to 3,500 metres. A number of bird species also belong to the relict populations of Alpine fauna. The rock ptarmigan is perhaps the most obvious example, and it is perhaps the bird equivalent of the arctic hare. The only other place in Europe where these two are found is in the far north. For the thousands of years during which the rock ptarmigan and the arctic hare have lived isolated from their northern cousins, they have developed certain distinctive characteristics. The rock ptarmigan lives at heights up to 2,800 metres. Among the specialised birds of the Alps the following should be noted: the Alpina accentor (*Prunella coloris*); the wall creeper (*Trichodroma muraria*); and the snow finch (*Montifringilla nivalis*).

All these species are sedentary residents, although some migrate between the higher and lower parts of the Alps depending on the season and the weather. The hare is seldom found lower than 1,300 metres in summer, but in winter it descends to as low as 600 metres. The smaller mammals live beneath the snow cover, with the exception of the alpine marmot, which can weigh up to 8 kg, and which goes into hibernation.

# Europe's changing fauna

Only one of Europe's indigenous mammals — the wild ox — has been rendered totally extinct. On the other hand, no fewer than 32 mammals with origins outside Europe have been successfully introduced. Only one bird species — the great auk — has become extinct, while there have been 13 successful introductions of non-European bird species.

Even when they are not threatened to the point of extinction, many European species are under threat. This is principally due to environmental changes, but in a few cases man's hostility or intolerance has been to blame. Hunting for meat has seldom made any significant impact on animal and bird populations; more significant is the way in which certain large animal species, especially those living close to human settlements, have been driven away from their accustomed habitats.

The European bison became extinct as a wild species owing to the disruption of two world wars, and it was only thanks to the efforts of zoos that the species survived and was eventually reintroduced to parts of its original home range. The wild reindeer was forced away from much of its natural range because it was seen to be competing for grazing with domesticated livestock. The wild ox was wiped out for the same reason. The great bustard, Europe's largest land bird, has declined sharply in numbers and has disappeared from most of its former range across the grasslands and farmland of Europe. Illegal shooting and trapping threatens the remaining populations of wolves and bears in Europe, while environmental pollution is a threat to the survival of the otter in much of Europe.

Some of the introduced fauna of Europe has been able to find environmental niches, and this has enabled them to flourish. The pheasant and the Canada goose are two species that have made valuable additions to the range of European sporting species, without competing with other native species. The opposite is true of the American grey squirrel. For certain other species such as the mink, the racoon and the muskrat, the future is uncertain. Other introduced species such as the rabbit have become serious agricultural and horticultural pests, while others are considered to be potential carriers of various diseases. One case in point is the racoon, which is believed to be a carrier of rabies. In parts of Scandinavia the growth of the populations of introduced whitetail deer have caused concern for the well-being of the indigenous elk.

Regardless of whether or not a species enjoys man's support and encouragement, the scope and quality of the environment is of fundamental importance for its continued existence and future development. It may seem an easy matter to introduce and establish a foreign animal species in Europe, but it is exceptionally difficult to exterminate an introduced species which has proved undesirable. Proof of this can be seen in the cases of the mink and the muskrat.

It is therefore wise not to introduce any non-native animals. On the other hand, game conservation and management must be accepted and encouraged to support those indigenous species which have become threatened by man's activities. Wildlife conservation for the management and control of wild game species is likewise justified.

Environmental harmony is the crucial factor in promoting populations of animals large and small. It is as important for the polar bear as for the rock ptarmigan; for those that have increased in numbers as well as those that have declined.

# Muskrat

## *Ondatra zibethica*

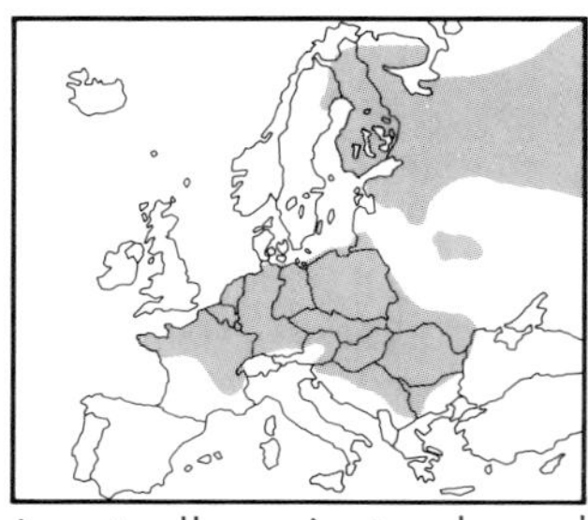

When the first trappers and hunters explored the interior of America and Canada they came across many species of animals whose furs became the basis of their livelihoods. One such species was the muskrat, which despite its name is actually a giant vole and closely related to the small voles which we find in woodlands and on farmland.

The word "musk" is a reference to the often powerful scent which this species secretes in glands in its groin, although this does not appear to have deterred the hunters and trappers. The muskrat was mainly caught in traps which were placed in such a way that when they were sprung, the trap and the trapped animal fell together into the water and the muskrat was drowned. During the last century several million muskrats were trapped each year for their pelts, which were sold at local pelt markets and were then sent on to furriers throughout the world. The muskrat's rapid rate of breeding has sustained its numbers despite the pressures of large-scale trapping.

Like the beaver, the muskrat is a rodent. The shape of its body is not dissimilar to that of the beaver, but the two animals are not closely related. Unlike the tail of the beaver, the muskrat's tail is flattened at the sides, and it is hairless and has a scaled surface. The basic colour of the muskrat is brown, sometimes lighter, with a tendency to beige, sometimes darker and with a hint of reddish-brown.

Like the beaver, its underfur is very soft and dense, with stiff and shiny guard-hairs. The underfur is lighter, an almost yellowish-beige, and the guard-hairs are often greyish-brown to dark brown, and the legs and tail are darker. The hind feet have five toes with a rim of coarse, stiff hairs, and the feet are webbed. The "thumb" of the forefeet is rudimentary, and the muskrat's whiskers are black. The outer ears are small and partly hidden in the fur, while the eyes are brown and small, like those of most mice and voles.

The muskrat has a typical rodent skull, and it has only 16 teeth, which is four fewer than the beaver. The body length, not including the tail, is 30–40 cm and the tail measures a further 20–25 cm. At most it can weigh a little over 2 kg.

### Reproduction

The muskrat can breed very prolifically. The litters are large, comprising up to 10–12 young, and at birth these are small, blind and hairless, weighing about 20 grams each. Their eyes open after about 11 days. The female muskrat has eight teats, and she is capable of bearing as many as five litters of young each year, although this will vary with the geographical location. In Scandinavia there are usually no more than two litters, and in mainland Europe the average is 3–4 litters. In North America, its native home, muskrats may produce up to five litters annually, but 2–3 is more usual. Muskrats become capable of breeding at or around the age of one year, and they often form monogamous breeding pairs, although where they are numerous they will often be promiscuous. The young are cared for by the female alone, and they suckle for up to three weeks, after which the mother leaves them. The gestation period is about one month, but its precise length can vary. It is believed that there is a substantial surplus of males at birth, about four males for every three females.

## Senses and communication

The muskrat has the usual range of senses shared by all mammals, but since it is nocturnal by nature its senses of smell and hearing are especially well developed. The males mark their territorial boundaries with secretions of musk, and the animals use a variety of sounds to communicate with each other, although they are normally silent. Males, however, are said to scream occasionally, a sharp and penetrating shriek, like that of a hare. When they are excited or conscious of danger they make a whistling call. They have long and sensitive whiskers, useful aids in moving by night through narrow runs and paths.

## Adaptations and behaviour

The muskrat reveals several adaptations to aquatic life. Its fur is water repellent and acts as an effective insulator, while the webbing between the toes of the hind feet and also the flat tail make swimming easier. A muskrat can dive and remain submerged for up to 20 minutes without difficulty. Ice-cover can make life difficult in winter, but like the seal, the muskrat breathes through holes in the ice which it gnaws using its sharp incisor teeth. These holes are plugged with pieces of vegetation, probably to prevent them from freezing up, and to make them readily available in case of emergency. On land the muskrat moves slowly, at a maximum of 5 km/hour. Like many other species of voles and mice they use their forefeet to hold their food. They sometimes live in small groups, in which they can defend themselves against other groups or individual intruders.

In undisturbed areas where conditions are ideal, muskrats can achieve very high local densities. With 2–3 litters a year, as is usual in southern and central Europe, one pair can produce up to 35 young in a year, although the average is about 20. By the end of summer there may be some populations with very high densities, but these are usually decimated in autumn and winter. This species is said to live for up to four years in the wild, but the average lifespan is very much shorter. When local populations become very dense, some of the animals will emigrate.

Several other aspects of behaviour can be seen when the population density increases. There is much more aggressive behaviour, as well as an increase in deaths caused by disease, starvation and competition for the available food. The breeding rate drops, and some females may only have one litter of young. These phenomena do not all occur at once; one can replace the other, and periodically they will each play a predominant role. Muskrat numbers can fall in both thriving and weak local populations when the habitat is changed by drainage or when there is prolonged ice cover. A series of unusually hard winters prevented the spread of the muskrat in northern Scandinavia. But it is after periods of decline in muskrat numbers that the species' very rapid reproductive rate comes into effect.

## Habitats and food

Watercourses of various kinds provide muskrats with the environments they like. They are attracted to lakes and running water, and they prefer areas of shallow water where there is a good supply of both aquatic vegetation and reeds. Like other rodents, the muskrat is principally a plant eater, and its diet comprises floating plants like water lillies and pondweed, and various reeds and rushes. Floating islands of mud with watercress, wild iris and bullrushes are also favoured. They will also eat molluscs, crawfish, aquatic insect larvae and small fish, young voles and baby birds when these items are available to them.

## Origins and distribution in Europe

The muskrat is a native of North America, but it has been introduced into various places in Europe. These introduced animals are considered to belong to the subspecies *Ondrata z. zibethica*. The first introduction is believed to have taken place in 1905, when four pairs were released near Prague. In a little over 30 years the species spread over 200,000 km$^2$ of central Europe, and in 50 years it had spread over 11 million km$^2$, including parts of Russia, Poland, Czechoslovakia, Romania, Bulgaria, Hungary, Austria, Switzerland, Yugoslavia, West Germany, Holland, Belgium and France, as well as Finland and Sweden. This rapid spread was assisted by the releasing of further pairs, and the European population is now very numerous. About two million muskrats are killed annually.

In its original environment in North America, some muskrat populations have been reduced to one-third of their normal numbers, but by the following autumn the numbers have returned to normal, such is their capacity for rapid reproduction.

The muskrat was introduced into Finland in 1922, where it spread rapidly and now occurs throughout the entire country. Up to 160,000 animals have been shot in a single year. The muskrat arrived in Sweden by natural immigration from Finland, but it has also been released. The muskrat has not yet found its way to Norway, Denmark or Iceland.

## Relations with man

The muskrat was originally regarded as an animal of considerable commercial interest. It had long been raised in captivity on fur-farms, and the wild population was considered as a supplementary stock. In the U.S.S.R. the muskrat is protected for that reason, as there are some areas that have wild populations which can produce millions of pelts. In these areas special dams and canals have been built, along with platforms for nesting and feeding areas, and in this way the muskrat is easier to hunt. However, in most European countries the muskrat is regarded as a pest and it is the target for various extermination campaigns. In Sweden, for example, the muskrat can be hunted all through the year, and Great Britain, uniquely among European countries, has actually been successful in exterminating its muskrat population. In 1929 a number of muskrat farms were started up, from which some animals either escaped or were released. It then became prohibited to keep this species in captivity, and by 1939 the feral population of muskrats in Britain had been completely wiped out.

# Grey squirrel

## *Sciurus carolinensis*

The Grey squirrel is a common animal in eastern North America, where there are a good many related species, more or less similar to it. There are approximately 250 species of squirrels world-wide, of which some 20% are found in North America, while in Europe there are only 2.5%. They grey squirrel was introduced into Great Britain in the 1830s, although the squirrels which are now found in much of the British Isles are descended from grey squirrels introduced to Woburn in Bedfordshire in 1876–89. At Woburn the Dukes of Bedfordshire introduced many non-European species, and either released them into the wild or kept them in fenced-in areas. (Père David's deer is one species which was saved from extinction by being kept in an enclosed and protected British herd, and some specimens of this species were eventually returned from there to their original range in China.) Other introduction also took place in Britain, and these releases together with subsequent natural spread have led to the species' present existence as a common mammal in much of England, central Scotland and parts of Ireland. However, it is not found elsewhere in Europe.

In appearance, the grey squirrel is not dissimilar from the European red squirrel except that the fur is grey, a solid grey in winter and grey tinged with brown in summer. The ears are small and round, and their fur covering changes from white in winter to greyish-brown in summer. The tail is very bushy and somewhat flattened, with white hairs along the sides, and the belly is light grey. The grey squirrel is also rather larger than the native red squirrel, and its adult weight is between 340 and 800 grams, nearly double that of its European relative. It achieves its highest weight in the autumn, as it builds up reserves of fat before the onset of winter, and then its weight begins to drop. Melanistic or black-pigmented specimens are not uncommon. The male and female grey squirrel are similar in appearance and size.

Like the European red squirrel, the grey squirrel is active by day. It has a preference for deciduous and mixed woodlands, but it is not absent from coniferous forests. It lives mostly in the trees, but descends to the ground from time to time. Much of its diet comes from plants and trees — nuts, buds, fruit and berries. It also eats small animals and occasionally young birds. It will tear the bark off trees to get at the nutritious moist layers which lie underneath, a characteristic which does not endear the species to foresters. Squirrels store nuts and acorns in caches in the ground, and will often carry these items for 20–30 metres to hide them. The grey squirrel has also become a common sight in many urban and suburban areas, where it will scavenge for food and steal bird food put out on bird-tables.

In Britain the litters are born in spring and summer, in the periods January–March and May–June. The nest, or drey, is usually placed in a hole in a tree. Like the European squirrel, the younger females give birth to one litter a year, while older females will produce two litters in a year. A typical litter comprises 3–5 young, but up to eight young have been recorded in a single litter. The gestation period is approximately 44 days, and the young are born blind and naked, weighing up to 27 grams. They are suckled by the mother until they are two months old, by which time they will weigh 100–200 grams. They are fully grown at the age of six months, and the young females become capable of breeding at one year old. The males are not sexually mature until they are 18 months old.

The males range more widely that the females, and in spring and autumn the young animals emigrate from the family territories in which they grew up. In North America large-scale emigrations of young have been observed. The population density of grey squirrels varies from year to year, but it usually lies within the range 5–50 individuals per hectare in North America, and in Britain density ranges of 1–13 individuals per hectare have been found. Only a very few wild grey squirrels, barely 1%–2%, live to be older than six years, although in captivity these animals can live for up to 15–20 years.

# Raccoon dog

## *Nyctereutes procyonoides*

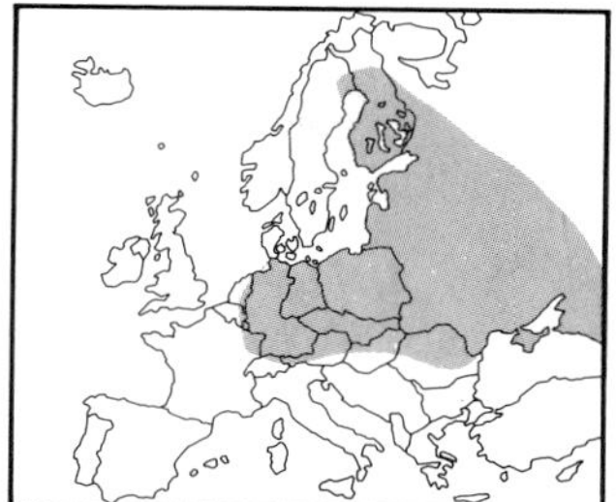

The name "raccoon dog" implies a cross between a species of marten and a dog. However, a raccoon has more in common with a bear than a marten, even though it is included as a member of the extended family of the dogs, the *Canidae*. In appearance it brings to mind a long-legged raccoon, but it has a bushy tail like a fox, and it lacks any ring markings on the tail. The hair on the neck and head is long and bushy, while the colour of the fur is not dissimilar to that of a badger. The fur is soft and light, with attractive guard-hairs, and these characteristics have led to the popularity of the raccoon dog as a hunter's quarry, as a farm-bred animal and as a candidate for various releasing experiments. It often looks plump, especially when it is in its thick winter coat, and when its fat reserves of up to half the animal's body weight contribute to the impression of plumpness. Raccoon dogs weigh about 10 km. Like the raccoon and the badger, the habits of the raccoon dog are mainly nocturnal, and this is reflected in its black and white face mask. The raccoon dog utters a call which is a combination of a whine and the meowing of a cat.

The natural range of the raccoon dog is in the Far East, in the countries of China, Korea, Japan and Vietnam. However, it has also been introduced into European Russia as a fur-bearing animal. At one time its fur was sold as "Japanese fox". In the easternmost part of its range the raccoon dog has been hunted both for its pelt and for meat. The first introduction into Russia took place in 1936, but this was followed by several more releases. Thousands of individuals were released at hundreds of locations from the 1930s onwards. Apart from Russia, raccoon dogs have been kept on fur farms in Poland, Sweden and Finland, and there have been various escapes from these captive populations. Feral raccoon dogs in Russia have undertaken a natural westwards spread, and by the 1940s they had reached Finland and northern Sweden from Russian Karelia. Only a single individual has been found in Norway. Flourishing feral populations have become established in Poland and Finland, and the species' range is spreading steadily westwards each year. The western limits of its range now run through East Germany, Poland, Czechoslovakia, Hungary and Romania. In 1962 the raccoon dog made an appearance in Westphalia in West Germany, and thus met the feral range of the raccoon, which was introduced earlier and has now spread throughout West Germany.

The raccoon dog feeds mainly on small mammals, birds and their eggs, frogs and various insects. It is omnivorous and will eat both plant food and carrion. In some places, research has established that raccoon dogs can cause considerable damage to other species of wildlife in a number of ways, including the destruction of the nests of duck, gallinaceous birds and other ground-nesting species. As a result the raccoon dog is widely regarded as a pest species.

When the young pups leave the den, raccoon dogs often hunt in family groups or packs, spreading out across the countryside and then reassembling. In eastern Asia they search for food along the margins of rivers and lakes, and in those areas it is said to have a special fondness for fish. After a successful quest for food, the raccoon dog often lies up and rests. Like the raccoon, the raccoon dog is mainly nocturnal, and this can lead to confusions of identification in those areas where both species occur. Another similarity is that both species favour lowland habitats and areas of woodland close to rivers, lakes and marshes. The raccoon dog is unique among members of the dog family in that it sometimes hibernates right through the winter, which is perhaps not surprising in view of the large fat reserves which it accumulates in autumn.

Raccoon dogs are thought to live in pairs, but generally very little is known about their family lives. Both parents are known to take care of the large litters of 7–8 young, which are born in a hole or hollow, either natural or excavated by one of the adults. The pups become relatively independent by the age of six months, but the family ties are not dissolved until the next year's litter is born. The density of local populations can vary from year to year, depending upon local conditions, and it is influenced by the establishment of new breeding colonies. However, where the species is carefully protected the breeding density is usually quite high. Individual territory size varies, and in Japan the average is known to be very much smaller than in Europe. Questions also arise about the future impact of this species upon indigenous European species of wildlife, when the raccoon dog is considered both as a potential predator and a competitor for food and space.

In Sweden the raccoon dog has been declared an undesirable introduction, and it may be hunted throughout the year. In Europe many nations have begun to take a particular interest in the study and control of rabies, and this killer disease may be carried by the raccoon dog. It was for this reason that the former protected status of the species in Finland was revoked. In the hunting season of 1984–85 a total of no fewer than 47,000 raccoon dogs were shot in Finland.

# Mink

## *Mustela vison*

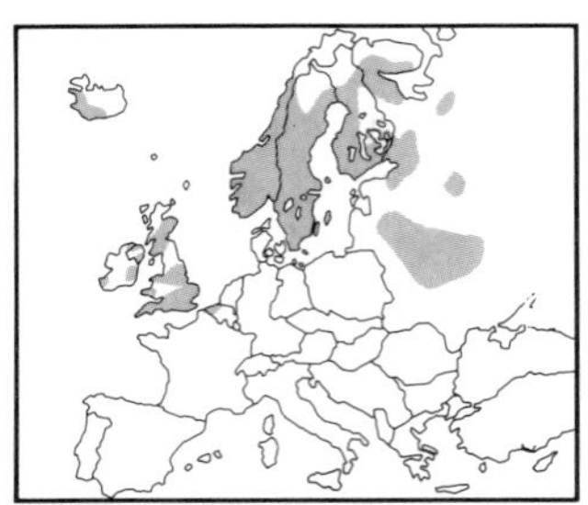

The mink is another of the mammals which did not voluntarily colonise Europe. The species is widely regarded as having virtually no value except commercially. Various other predatory members of the weasel family have been established in Europe for a very long time, and yet the feral mink has found an ecological niche where it faces little or no competition. On the contrary, many wild populations have flourished, often at the expense of many varieties of birds and small mammals.

The mink's arrival here from North America was brought about by the demand for mink fur in women's fashions, and especially in the mink coat trade. The mink is easy to breed and raise in captivity, and it can also be bred to produce pelts of the quality and consistency required by the fur trade. It has therefore become an ideal fur-farming proposition and a highly lucrative commercial activity. The mink's future as a commercially reared fur-bearing species was further confirmed when breeders and keepers found that it could be fed cheaply and conveniently on the waste products of the fishing industry and the abbatoir.

The mink is a native of North America. In Europe it occurs mainly in Fennoscandia, in the Baltic states, East Germany, the Netherlands, Spain, Iceland and the British Isles. It prefers aquatic environments, preferably freshwater ones, but it can also be found near the sea coast. Its diet comprises fish, frogs, small mammals and birds, the eggs and young of ground nesting birds, and some crustaceans and molluscs.

At the end of the 1920s feral mink which had escaped from fur farms were found at large in Sweden, and the same happened in the 1930s in Norway, Finland and Britain. Attempts were made to exterminate these feral populations, but they have never been entirely successful. However, in the U.S.S.R. the mink was deliberately introduced and encouraged as a feral species, because of the value of its fur. Nevertheless, the quality of wild mink pelts are generally regarded as inferior to those of farmed mink, and therefore of lower commercial value, and for this reason it is regarded throughout western Europe as an unwelcome pest species. In Finland 71,000 feral mink were killed during the 1984–85 hunting season, and more than 30,000 were accounted for in Sweden during the same period. In Britain and Ireland feral mink are present in virtually all the major river catchments, and their numbers are best controlled by intensive campaigns of cage trapping, to prevent serious damage to game fisheries and stocks of wild gamebirds and farm animals.

## Characteristics

Feral mink are usually dark brown, although individual trapped specimens may exhibit considerable variations in pelage. The bristles or guard-hairs are usually a brownish-black colour, while the underfur which is not normally seen is a greyish-brown. The underfur of the polecat, by contrast, is a yellowish-white. Another contrast between these two species is that, in the case of the mink, the lighter colouration on the head is confined to the front part of the lower jaw. White-speckled and light grey-coloured mink can occur, due to a mixture of colourations from fur-farm escapees. The mink's tail is 18–23 cm long and rather bushy, and there is a rudimentary webbing between the toes. The eyes are small and shiny, with a dark brown iris, while the dark ears protrude slightly from the fur of the head.

The male feral mink is about 40%–50% larger than the female, and he can weigh up to 1.4 kg, while the female seldom weighs more than 1.1 kg, although average weights for both are rather lower than this. Some males have been recorded at weights up to 2 kg, and some well-fed farm minks can weigh still more. Young mink are fully grown at the age of four months. Mink have 34 teeth which grow according to the following pattern:

$$\frac{3\quad 1\quad 3\quad 1}{3\quad 1\quad 3\quad 2}$$

The mink is a lithe animal, which moves sinuously and is mainly nocturnal in its habits. On occasion it may wander distances up to 10 km in a night, but it does not usually travel more than 5 km. Mink are good swimmers, and only the top of the head shows when an animal is in the water. During the day mink lie up and rest in cover. Just before mating male mink range across wide areas and traverse the territories of a number of females, and one male's territory is therefore very much larger than that of a female. When families of mink disperse the males travel furthest, but a young male will only hold a territory half the size of that of a mature male.

The mink has relatively poor eyesight, especially under water, but its sense of smell is very well developed. It uses secretions of scent from its anal glands to mark its territory.

Some experts regard the North American mink as the same species as the central European polecat (*Lustela lutreola*). If this were the case, the two populations would form the subspecies *Mustela l. lutreola* and *M. l. vison*. The fact that hybrids are produced when these two animals occur in the same area tends to support this theory.

## Reproduction

Mink are sexually mature at 10 months old, and they breed when they are one year old. Like most other members of the marten family, they undergo a period of delayed implantation, which almost doubles the actual period of gestation. Mating usually takes place in late winter or early spring, during which time the female may be in season up to four times. Mink are promiscuous, and a male will mate with several females, and these in their turn will mate with several other males. The female takes sole responsibility for the young.

The litter of 3–8 (occasionally 10) young, called kits, is born between March and May in a nest or burrow, usually close to water. They are born naked and blind, and weigh about 10 grams. The female, which has eight teats, suckles the kits until they are 6–8 weeks old, and they become independent several months later. Mink have a potentially very rapid rate of reproduction, but there is also a high mortality rate, especially among the young. Mink are not believed to live beyond the age of ten years at most.

# Pheasant

## *Phasianus colchicus*

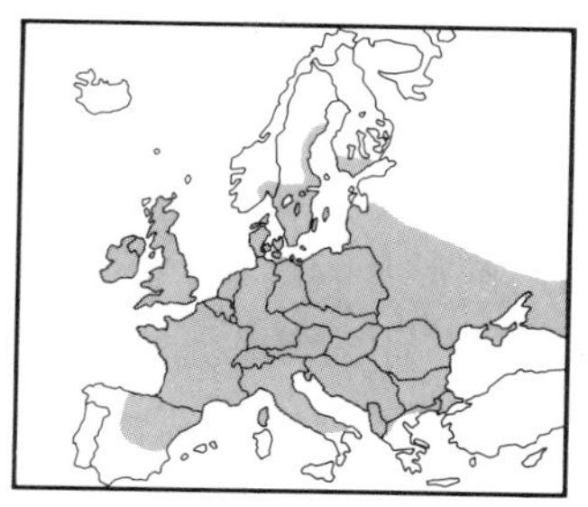

In almost every European language, this medium-sized game bird has the same name, although the spelling may differ slightly from country to country. It is generally agreed that its Latin name derives from the name of the river Phasis and of the region of Colchis, both of which are near the Black Sea in what is now the Soviet Republic of Georgia. This also tends to give us a clue to the fact that the pheasant is not a native of western Europe, but has been introduced from its original haunts further to the east and in Asia. However, some experts are interested in trying to discover whether there was at one time a natural population of pheasants in south-eastern Europe, which may have colonised naturally by spreading from further east.

The fact that the pheasant is referred to in some European countries as the "hunting pheasant" hints at the species' importance as a sporting quarry. In many areas it is now the commonest of the gallinaceous gamebirds. The *Phasianidae* or pheasant family comprises no less than 183 species, including the various species of pheasants and also the quails and the partridges. These gallinaceous gamebirds should be distinguished from the family of woodland grouse, which includes species like the capercaillie and the black grouse. (However, it should be noted in passing that the pheasant has been known to hybridise not only with other types of pheasant, but also with the black grouse and the domestic fowl.)

The feral populations of pheasants in Europe and the British Isles are almost all the result of interbreeding among various races and sub-species of pheasants. For example, in Asia there are a number of different and distinct local races of ring-necked pheasant, as well as the so-called "black necked" pheasants and the green pheasants of east Asia and Japan. All of these types have been thoroughly mixed by random cross-breeding in game farms and pheasant rearing enterprises.

### Characteristics

The sporting wild pheasant, which has been introduced not only to Europe but to many other countries as well, is an easy bird to recognise, as may be seen from the illustration. Both sexes have long, pointed tails, although the cock bird's tail is very much the longer of the two. Furthermore, the cock and hen birds have very different plumages. Pheasant chicks and young birds are very similar to the female bird at first, but the young males begin to moult into their distinctive male plumage at the age of 6–8 weeks. Among populations of wild pheasants the young males will have moulted completely by October–November.

Cock pheasants are about 18%–25% heavier than the hen birds, and they also have a greater wingspan. Typically, a cock pheasant's wing will measure in the range 230–267 mm, while that of the female will be 218–237 mm. The adult body weight will be in the range 1.1–1.5 kg. By the time they are only two weeks old, the cock chicks will already weigh slightly more than the young hen chicks. The adult cock pheasant has a prominent spur on the lower part of the tarsus.

As is common among gallinaceous birds, young pheasants fledge very quickly and some may be able to make very short flights at the age of 2–3 weeks. The adult pheasant flies with a rapid and powerful wingbeat, which often alternates with a gliding flight. When alarmed they can rocket upwards in a near-vertical flight for up to 10 metres. Otherwise, they fly on a straight and level course, and alight again after flying a relatively short distance. They do not undertake long-distance flights.

As early as their first autumn the young cock birds are able to crow like the adult males, and it is characteristic of the cock pheasant to crow loudly and to drum loudly with his short, rounded wings.

## Reproduction

The pheasant becomes sexually mature before it is one year old. Breeding males secure individual territories, which they defend and proclaim by crowing and wing-drumming. A breeding cock bird will claim and hold a number of breeding females within his territory, and these females form his harem, which he defends against the intrusion of other males. Even in autumn and winter it is not uncommon to see a single mature male pheasant with a number of attendant females. It is believed that hen pheasants tend to live longer than cock birds, but where densities are high the female mortality increases, and the rate of brood losses is also greater. Among populations of wild pheasants in their natural surroundings, and where there is virtually no sporting shooting of pheasants, the ratio of hens to cocks is believed to be equal, but in areas where there is significant shooting pressure there is often a surplus of females, owing to selective shooting of cock pheasants.

During the breeding season the females gradually move away after mating and establish their individual nests, although the dominant cock bird will continue to guard his females, especially when they are feeding in the open. The nest is usually a simple depression in the ground, often in quite dense cover. The eggs are laid in the period May–June, and a typical clutch will comprise 10–12 greyish-green eggs, measuring 42–49 mm x 32–38 mm and weighing 30–33 grams each. The female incubates these alone for approximately 24 days, although many clutches of eggs fall victim to predators, including corvids and stoats, and others may be crushed by farm machinery. If a clutch is lost the female will often lay a second clutch to replace it. If the first clutch hatches and fledges successfully, there will be no second clutch, and there are no recorded instances of pheasants rearing two broods in one year.

Pheasant chicks are well developed at birth, with a good covering of down, and they are lively and active as soon as they are dry, following the female and capable of picking up small items of food. At birth they weigh 17–26 grams, and they are nidifugous — i.e. as soon as all the eggs have hatched the young are ready to leave the nest. At two weeks old they will weigh 50 grams and after four weeks they will average 750 grams. They follow the mother bird for the first few weeks, after which they become progressively more independent. They communicate by high pitched *peep*-ing calls, especially if they have become separated from the rest of the family, or have become cold. The hen bird will respond to these calls by gathering the chicks together and brooding them. She is also capable of warning her chicks if danger threatens, so that they will hide or freeze and remain still.

## Enemies and environment

In its European range the pheasant encounters few competitors for food or breeding habitat. However, there are a number of birds and mammals which are significant predators of the pheasant. Members of the crow family are the principal predators of the eggs of pheasants, and studies tend to indicate that it is wandering, non-territorial crows which tend to find the early-laid pheasant eggs, which are often insufficiently hidden in sparse early vegetation. Later laid clutches, and second clutches, tend to have a better success rate, because the vegetation is higher and denser, affording more cover for the eggs, and also because the chicks hatch out when the weather is warmer and when there is more insect food for them to eat.

In continental Europe, goshawks can be serious predators of pheasants. There is one instance on record of a single female goshawk which caught and killed an entire wild population of 17 pheasants as a result of only one month of steady hunting. Goshawks and also sparrowhawks are attracted to places where unnaturally large numbers of reared pheasant poults have been released by sportsmen. It has also been shown that where pheasants become diseased they also become easier prey for goshawks. In one Swedish study area, where the birds were affected by avian tuberculosis, it was found that 25% of the pheasants fell victims to goshawks. The normal rate of predation by goshawks in the same area was only 8%.

## Adaptations and behaviour

The pheasant is a gregarious bird, which is active by day. Only during the breeding season do the individual hen birds seek solitude to nest and incubate their eggs. Non-territorial females may also lead a solitary life in the breeding season. Each territorial breeding male has his harem of females, which gradually disperses as egg-laying begins. Flocks of pheasants occur at different times of the year, and especially in winter, and individual birds may move from one flock to another at various times.

Breeding hen pheasants live with their families of chicks during the summer. In winter, however, mixed flocks of hens and cocks can often be seen, especially where supplementary feeding places are established. However, it is usual to see large flocks of hen birds and rather smaller flocks composed of cocks.

When they are laying claim to a territory the cock birds act very aggressively, and a territorial cock will attack any other male who enters his territory and seeks to claim it for himself. Antagonistic males will often face one another bill-to-bill, with raised throat feathers, raised tail feathers and trailing wings. Territorial cocks will also crow vigorously and repeatedly at one another, often when they are considerable distances apart. The lie of the land, the numbers of females and the availability of food and nesting habitat will influence the size of the cock bird's territory, and between 15% and 40% of males may be non-breeding and therefore non-territorial in any single year.

Studies of breeding pheasants in the rich farmlands of southern Sweden revealed a population density of 4.9 males per $km^2$ in May. Further north on the Swedish plains the density was counted at 1.1–1.9 per $km^2$, while further north still the density was only 0.3–0.7 males per $km^2$.

In Denmark studies were carried out of the movements of reared birds which were released into the wild. It was found that 97% of these birds were recovered within 4 km of their release point.

Released young pheasants were found to be more sedentary than mature wild birds, and the males wandered further than the females. However, none of the released birds in this study was found further than 17 km away (males) and 10 km (females).

The pheasant is generally a sedentary bird in Europe and in its native Asian haunts. However, one exception to this appears to be the pheasants of north-east Asia, Manchuria and Korea, where pheasant migrations have been observed.

## Occurrence and diet

The pheasant is a hardy and adaptable bird. Its natural habitat in central and eastern Asia is in open countryside, on grasslands with some shrubby vegetation, in open woodlands and in rushy marshes and along the margins of rivers and lakes. It also occurs on cultivated land. Although the pheasant is primarily a bird of the lowlands, in China it has been recorded as occurring at altitudes up to 4,000 metres, while in the Alps it can occur up to 800–1,000 metres. Much of its Asiatic habitat is characterised by hot summers and cold winters, but in southern Asia it often occurs in areas with a Mediterranean type of climate. Thus it seems clear that the pheasant has had little difficulty in adapting itself to the many alien areas to which it has been introduced outside its natural range.

Pheasants feed on vegetation of all kinds, and their diet also contains a proportion of animal material, principally insects, larvae and other small invertebrates. It is important that the adults, and especially the females, should be well fed and in good physical condition before they embark upon nesting and egg-laying. Various research studies have also shown that, after the eggs have hatched, the chicks feed eagerly on available insects and that a protein-rich diet of invertebrates may comprise up to 60%–80% of their food intake in the first three weeks of life. The plant food that pheasants eat vary with the seasons and with local availability. They will eat young buds, green leaves, roots, seeds and cereal grains, acorns and beechmast, and various types of berries and wild fruits. In winter, seeds, acorns and cereals are important, as well as green parts of plants. The latter are also an important component in their diet early in the year, while grain is most abundant in the aftermath of the harvest.

Ready availability of all these foods contributes to the size of the population and the success of breeding. Monoculture farming, early autumn ploughing and the use of insecticides and weedkillers have all reduced the amount of natural foods available to pheasants in many areas of Europe.

## Lifespan and mortality

The wild pheasant is unlikely to live for more than 7–8 years. Mortality among young pheasant chicks can be very high in some areas, and the first three months of the young bird's life is the most critical period. First year mortality has been estimated at 80%–85%, while the rate among older birds is slightly lower, at 60%–70% annually.

Sporting shooting is believed to be the largest single cause of mortality among adult pheasants in the autumn and winter. In the 1940s and 1950s many local populations of pheasants were seriously depleted by the use of methyl-mercury agrochemicals. Pheasants are also prone to tuberculosis, which can cause emaciation and death, or renders the birds weak and less able to avoid predators. Foxes, mink and stoats are important mammal predators, while goshawks and crows are the principal avian predators of the birds and their eggs. Gapeworm is a parasitic disease which chiefly affects reared pheasants in release pens.

## Distribution and subspecies

The common pheasant is to be found widely in Europe and eastwards across Asia as far as the Pacific coast. Pheasants also occur in some southern Asian areas, such as northern Iran and Afghanistan. While the familiar type of pheasant is common in the British Isles, it is not found in the islands of the Japanese archipelago, where its place is taken by the Japanese green pheasant (*P. versicolor*). This species has also been introduced to Europe and the British Isles, where it has interbred with other types of pheasants, resulting in the thoroughly mixed genetic make up of most European pheasants.

Across Asia some 30 local subspecies of pheasants have been identified, belonging to four or five general groups. In western Asia and closest to Europe are the pheasants of the nominate form, *Phasianus colchicus colchicus*, in which the male birds lack the conspicuous white neck collar of the so-called ring-neck pheasants. The latter belong to the eastern Asian populations of the *torquatus* group, and the white collar is also found in the birds of the *mongolicus* group.

## The pheasant in Britain

Pheasants may have been introduced to Britain by the Romans, and there is some archaeological evidence to support this. However, pheasants did not become well established as wild birds in Britain until the Middle Ages, and their spread and increase in numbers has been largely related to their importance as a sporting quarry for the shooter. The development of firearms and the ability to shoot fast-flying birds on the wing led to a greater appreciation of the pheasant, which was protected and fostered as an important sporting species. By the eighteenth century British sportsmen regularly shot pheasants by beating out areas of woodland in winter, after the best of the autumn partridge shooting was over.

The development of the breech-loading shotgun conferred a much greater degree of firepower on the sportsman, and pheasants were carefully conserved for sport. The rearing and releasing of pheasants also developed, both as a means of augmenting the wild stock of birds and also to introduce pheasants to sporting estates where they did not already exist. By the 1890s some sporting estates were rearing and releasing on a very large scale, and teams of eight guns were sometimes able to achieve daily bags of hundreds, and even thousands, of pheasants. During the twentieth century the partridge has steadily declined in numbers, and its former position as the premier gamebird of lowland Britain has been taken over by the pheasant. In the 1980s it was estimated that not less than 12 million pheasants were reared and released annually on British sporting estates, and the numbers released appear to be continuing to increase as shooting becomes more popular.

There is some evidence to show that the British populations of truly wild pheasants have been adversely affected by recent trends in agriculture, especially the widespread cultivation of cereal monocultures, the enlargement of fields and the use of pesticide and insecticide sprays. Annually the wild population is hugely augmented in autumn and winter by the release of millions of reared birds, but sportsmen and conservationists are increasingly interested in developing management techniques which will benefit wild breeding pheasants. The principal challenges appear to be to ensure that there is adequate nesting habitat for wild pheasants; that they are given some measure of protection from predation, especially in spring and early summer; and that the modified use of agrochemicals may help promote more abundant insects upon which the young chicks can feed in the first weeks of life.

# Canada goose

## *Branta canadensis*

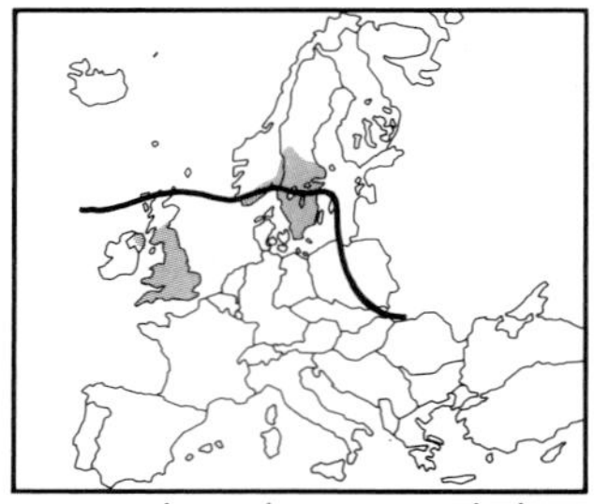

The Canada goose has been an inhabitant of Europe for some time, and there are records from as early as 1678 of the species' occurrence in England. The first wild-living geese were probably escapees from zoos and private waterfowl collections. The climates of the British Isles, southern Scandinavia and much of western Europe are very suitable for this species, and the result has been the rapid increase of Canada goose numbers and a considerable spread of the species. By 1953 Canada geese were known to nest in no fewer than 95 locations in the British Isles, and the total population was estimated at not less than 300–400 breeding pairs. Next in importance was Sweden, with 150 breeding pairs, and in 1957 it was reckoned that there was a non-breeding population of a further 1,200–1,500 birds. The period since the 1950s has seen the biggest expansion of Canada goose numbers and distribution.

The Canada goose has escaped or been deliberately released at a number of places in Europe, including Norway, but with varying degrees of success. After the introduction of Canada geese to Norway in the late 1950s, the population had risen to about 500 individuals by 1975. Finland, the Netherlands and West Germany also have populations of Canada geese, but the principal population explosions have taken place in Sweden, where there were an estimated 5,000 breeding pairs by the late 1970s; and in Britain, where the population was estimated at 20,000 breeding pairs by 1980.

### Characteristics

The Canada goose is numerous in its country of origin, North America. There it occurs in no fewer than 12 subspecies, several of which are quite distinctively different from the others. The largest subspecies, which was thought at one time to be extinct, can weigh up to 6–7 kg (mature males), while the smallest subspecies typically weigh not more than 2 kg. However, all the various subspecies are distinguished by one shared characteristic, a white cheek-patch and a black neck. The plumage is greyish-brown in some subspecies and reddish-brown in others.

The Canada geese now found in Europe are descendants of birds which came from north-eastern America, and they belong to the nominate form, Branta c. canadensis. This is one of the larger subspecies, which breeds in Quebec, Labrador and Newfoundland,

and winters along the Atlantic coast of the United States. It is also believed that some specimens of another inland subspecies, Branta c. interior, which breeds mainly around Hudson Bay, may have contributed to the European population.

The Canada goose is the largest goose species found in Europe. The male is the larger, with a wing length of 476-518 mm, while that of the female is in the range 445-495 mm. The maximum body weight for a mature male is about 5.8 kg. The goose's comparatively long legs and short bill are black.

Like other geese, the Canada goose is gregarious, and it forms flocks which may number up to several hundred individuals. Canada geese are active by day, grazing on salt marshes and also on inland fields. Although it gets a considerable amount of its food from the water, it also feeds readily on grassy meadows, on cereal stubble fields and on clover. In lakes it eats reed stalks and water plants such as Water Milfoil. It will reach deep under the surface to feed, while in coastal areas it will feed on various sea-grasses.

## Reproduction

The Canada goose, like many species of geese and swans, often pairs for many years, and the mates remain together all the year round. The family bonds are strong, and the goslings will stay with their parents during their first autumn and winter. A breeding pair will return to a nesting site chosen by the female, and females appear to be very faithful to their places of birth, while the males are more prone to wander. The female is responsible for building the nest and incubating the eggs, although the male may also occasionally take charge of the eggs while the clutch is being laid, for incubation does not begin in earnest until the last egg has been laid.

A typical clutch will comprise 5–6 yellowish-white eggs, although clutches of up to 11 eggs have been recorded. The eggs measure 79–99 mm x 53–63 mm, and weigh 220–240 grams. The younger breeding females may lay only 3–4 eggs. The Canada goose prefers to nest on islands, which may be small, low islets or larger rocky islands. They nest in a shallow depression in soft ground, which is lined with dry vegetation, seaweed, reeds and so on. The birds' down is also used to line the nests. On high ground the nest may be placed quite high up, giving the birds a good view of the surroundings.

Incubation takes 28 days, and the goslings are fledged after a further 6–7 weeks. As downy chicks, the goslings are coloured a mottled shade of yellowish-green, but as fledglings they are rather similar in appearance to the adults, but paler and with shorter necks.

Geese with their young eventually gather together into flocks, and in densely populated breeding areas these flocks may comprise 100 or more birds. At the same time, the non-breeding birds form their own flocks. Where there is a high density of breeding birds the pairs of geese will tolerate others quite close to them, and several females may form what seems like a scattered breeding colony with their nests only a few metres apart. However, the breeding territories are usually much larger, and there may be a high degree of aggression between neighbouring nesting pairs.

Occasionally a pair without young will try to attach themselves to another family. The breeding gander will often try to drive them off by flying at them with his neck outstretched, his bill open and cackling noisily. The female will keep watch over her goslings, standing upright with an upstretched neck. After his display of aggression the male will return to his mate and their young and will assume the same upright posture as the female, beating his wings as a final signal.

Breeding success is usually high, and 70%–90% of the eggs laid are hatched. Studies of a population of breeding Canada geese on freshwater lakes in western Sweden showed that each pair produced 4.7 eggs and reared 3.8 to fledging stage. Geese nesting along the sea shore were found to lay an average of 5.4 eggs, rearing 5.2 goslings to fledging.

## Migration

North American Canada geese are accustomed to making long distance migrations when the winter weather forces them away from their breeding grounds, and those that breed in the far north usually migrate directly southwards. In Europe some Swedish bred Canada geese which were leg-banded at the breeding sites were recovered in central Europe, and some as far south as Italy. Otherwise, they appear to gather at winter quarters in southern Sweden and in the Baltic Sea. Others continue south-westwards to the shores of the North Sea, and therefore none of the Swedish birds has been found to migrate very far. Canada geese in the British Isles tend to be very sedentary, and only rarely have they been found to migrate across the English Channel. Local movements in winter are very much influenced by the weather.

In any one year a high proportion of Europe's Canada geese are non-breeding birds, and like ducks these birds moult in summer and temporarily lose their powers of flight. Moulting birds gather together at specific moulting sites, and some geese are beginning to develop a pattern of pre-moulting migration flights.

## Distribution and numbers

As we have noted, Canada geese exist in many parts of north-western Europe. The two biggest populations are to be found in Britain and Sweden, each with over 30,000 birds. In Norway the population has been estimated at about 1,000 breeding birds, while in Finland there are only about 100 breeding pairs. 20–30 pairs breed in Denmark, while small numbers have been recorded as breeding in West Germany, Belgium and the Netherlands.

## Relations to other species

When Canada geese were first introduced into Europe, fears were expressed about the impact of this alien species upon native species such as the greylag goose and various duck species. However, experience has subsequently shown that Canada geese and greylag geese can breed side by side without difficulty. Nor have the two species hybridised to any significant extent, although some isolated instances have been recorded. The two species are quite different, and belong to the Branta and Anser families respectively.

Canada geese coexist quite readily with mute swans, although there have been some recorded instances of competition at certain nest sites. At others, however, the two species have nested side by side in perfect harmony. It is not uncommon for aggressive nesting male mute swans to drive away whole flocks of non-breeding Canada geese from their breeding places. Non-breeding geese and swans can coexist quite happily.

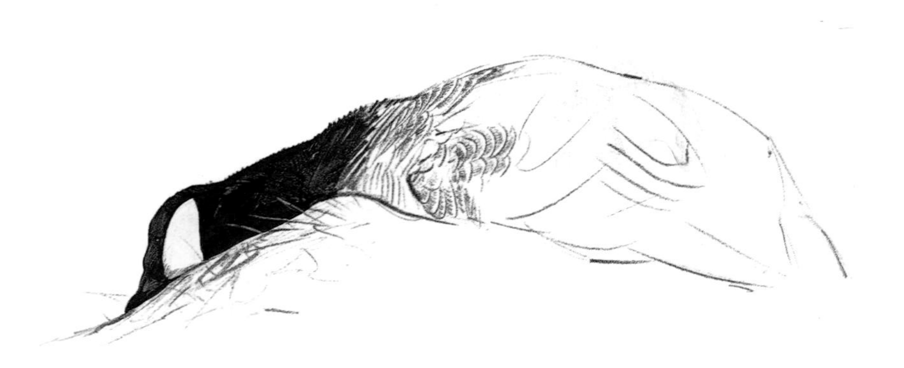

# Walrus

## *Odobenus rosmarus*

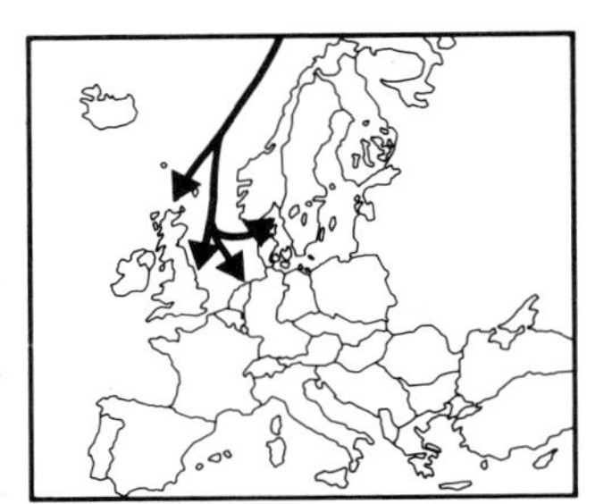

During the seventeenth century there was a good deal of discussion and argument about the "Wallr Oss" or "Ross marus", as the animal was referred to in an early Dutch treatise entitled "A Description of the Nature of Four-Footed Animals", published in Amsterdam in 1660. The author, Jonston, placed a fin above the walrus' foreleg and equipped it with a pair of ordinary hind legs in addition to its flipper-shaped hind legs. At that period, even whales and other large sea mammals were considered to be types of fish.

In 1685 there was published an important account of the capture of a walrus in the Arctic Ocean, written by Haquin Spegel, the chaplain to the royal court of Sweden. He described how Dutch seamen would capture "a fish" called the sea-ox or the whale-horse, which was large and powerful, but with a mouth so small that one could not insert a fist into it. The sea-ox was also said to be a great lover of music, even though he himself makes "a horrible, frightening noise".

A later writer, B. M. Keilhau, wrote in 1831 about walrus hunting on Bear Island between Norway and Spitzbergen, and described these animals as capable of lying lazily for several days without moving, and as having massive and clumsy looking bodies, which made them little more than "a mere excuse of an animal". Linnaeus, the great Swedish naturalist, recorded in the 1750s that the Russians believed that the walrus was capable of going deep down under the surface of the earth, like a giant mole.

## Characteristics

A bull walrus can reach a weight of 1.5 tonnes, and a cow up to one tonne. They reach a maximum length of 4.5 metres, and a newly born walrus calf weighs 40–60 kg. The skin of old and young walruses is heavy, baggy and wrinkled, and it is almost completely hairless, although there are bushy whiskers on the animal's upper lip.

The walrus's peculiar appearance is mainly due to the enormously enlarged "swollen" upper lip and the protruding tusks, which are present in both males and females. The tusks are in fact the canine teeth of the upper jaw, while the lower jaw lacks canine teeth and the animal has no real molars. The walrus therefore has a very limited number of teeth, and these are positioned according to the pattern:

$$\frac{1 \quad 1 \quad 3 \quad 0}{0 \quad 1 \quad 3 \quad 0}$$

The female walruses reach sexual maturity at 5–6 years old, and thereafter they give birth every second or third year. They cannot breed in consecutive years because the gestation period lasts for a full year. The calves remain with their mothers until they are two years old, and they are suckled for the first year. Walruses feed by eating mussels which they scrape off the rocks with their prominent tusks.

The walrus is a very sociable and gregarious animal, especially during the mating season and when the females give birth. They also remain together when they migrate, and groups of walruses often move together along the southernmost fringes of the Arctic ice, where they can be found on large ice floes as well as along the shoreline.

## Distribution

The walrus is a marine mammal which inhabits shallow waters close to sheltered coastlines and bays. It belongs to the large family of seals, but it is the only representative of the *Odobenus* family. It occurs all around the Arctic Ocean and it can also be found in sub-Arctic waters. The walrus, and especially the North Atlantic walrus (*O. r. rosmarus*), has been heavily hunted for centuries. However, it is still to be found throughout its range, which lies between the Hudson Bay area of Canada and Franz Josef Land to the north of Siberia. The populations of walrus which were most affected by hunting are those found around Bear Island and Spitzbergen, which occasionally migrated southwards towards northern Norway. A similar but distinctive subspecies of walrus, *O. r. divergens*, is found in the northern Pacific Ocean.

In 1776 the Nordic writer Magnus Orrelius said that the Spitzbergen walruses had become very wary as a result of hunting, and that they would have several animals alert and on guard while the rest of the herd slept. There is a record from 1608 of no fewer than 900 walruses being killed on Bear Island in a period of seven hours, and as recently as 1924 it was still possible for one hunter to kill 700 walruses in the course of one winter's hunting. There is also an account of two sailing vessels visiting the islands to the south of Spitzbergen in August 1852, where they saw a herd of walruses estimated at between 3,000 and 4,000 animals. A huge slaughter began, and in one day the crews of these two ships killed 900 walruses.

As a consequence of such severe hunting pressure the North Atlantic walrus population of several hundred thousand animals was reduced to only a few thousand, with 100 or so left in Greenland. The populations of Bear Island and Spitzbergen were completely wiped out, but in recent years a small colony of 10–20 animals has reappeared on Spitzbergen.

In recent times an increasing number of walruses have been sighted along the Norwegian coast, and it is believed this is associated with the gradual recolonisation of Spitzbergen. Additional sightings have also been reported off the coasts of Denmark, northern Germany and England. There have been six instances of walruses off the Swedish coast, one of which was caught in 1927 and is preserved in the Gothenberg Natural History Museum.

# Aurochs

## *Bos primigenius*

The wild ox is the predecessor of our domestic cattle. For long periods in European history the two were competitors for grazing and space, and as domesticated cattle were grazing in the fields, wild oxen would emerge from the forests to mate with cows in season. For the early farmers this was as undesirable as the interbreeding of wild boars with domestic pigs, which also took place.

It has always been believed that the last of Europe's truly wild indigenous oxen were to be found in a Polish royal game reserve at Jaktarow Forest, about 50 km south-west of Warsaw, where in 1564 a total of 38 animals were recorded. By 1599 the herd had dwindled to 24 animals and by 1603 only four remained. The last survivor, a cow, died in 1620 (some accounts say 1627) and one of its horns was presented to King Sigismund of Sweden, and is preserved in the Royal Armoury in Stockholm. Although that was the end of that particular herd, the species actually managed to survive for a further 150 years in the area around the Irtysh and Ob rivers, in what is now Russia.

In the 1930s two German biologists, Heinz and Lutz Heck, carried out an interesting experiment aimed at recreating the wild ox by selective breeding from primitive species of cattle. Their theory was that if wild animals could be domesticated and improved through selective breeding, the reverse should also be true. They experimented with breeding from Scottish Highland cattle and southern European bullfighting stock, and eventually produced an animal that had certain similarities to the wild oxen which can be seen in early illustrations. Specimens were released in the Bielawieza Forest in Poland, which had been the place of origin of the European bison. Unfortunately, during and after the Second World War these remarkable wild-living cattle disappeared completely, apart from a few animals which had been preserved in zoos.

### Characteristics

A physical description of the wild ox can be pieced together from drawings and paintings, and also from the many skeletons that have been found. The original wild ox apparently had a straight back and massively powerful forequarters and chest with high withers. The legs were thick, heavy and comparatively short, while the animal's hide was covered with short, smooth hair, with a prominent tuft of hair at the tip of the tail. The bull was blackish in colour with a forelock of rough curls, while the cow was a more reddish-brown in colour. Both bulls and cows had powerful horns that bent forward and curved inwards, with upward pointing tips. In contrast to the horns of the wild European bison, the base of the wild ox's horns was placed far back on the skull and parallel with the flat rear surface of the skull. The bulls probably weighed up to one tonne, and the cows about 25% less. The bulls were very aggressive, and it is believed that the wild oxen lived in herds comprising several mature cows, with their heifers and calves.

The record which has survived of the 38 animals that had lived in the Jaktarow Forest in 1569 shows that there were 22 adult cows, 5 calves and 11 bulls, 8 of which were fully grown. These animals were kept in enclosures, but it is now believed that the animals had a suitable genetic make-up to enable them to live naturally in the wild.

### Reproduction

The rutting season for the wild ox was in the period August–September, and mature bulls would fight heated battles for possession of cows in season. The gestation period lasted for about nine months, which meant that the calves were born in May–June, when there was an ample supply of good grazing for adults and young alike. Like our domesticated cattle today, the wild ox female usually produced a single calf, although twins were not unknown. The calves suckled for about six months and followed their mothers until just before the next calf was born.

The last surviving European wild ox, a female, is said to have been 30 years old, and it is presumed this was about the maximum natural lifespan. Cross-sections of fossil teeth have revealed ages up to 20–25 years. One of the bulls mentioned later in this chapter was probably 18–20 years old when it died, and it weighed approximately 1,000 kg. As cows were believed to reach sexual maturity at 2–3 years of age, it can be concluded that the lifetime reproductive capacity of both males and females was quite high.

## Distribution

The wild ox was widely distributed in Europe and Britain, and was to be found across a wide range from the Atlantic to the Pacific Ocean. In Europe remains have been found as far north as central Sweden, and carbon-14 dating of fossil remains has shown that the species was present in these northern areas as long ago as 9–11,000 years, and other evidence has shown that wild oxen were present in southern Sweden and Denmark during the Stone Age and the Iron Age. In many parts of Europe the wild oxen were exterminated in the thirteenth and fourteenth centuries, but as late as the 1600s some herds were kept in hunting reserves and enclosed areas.

The wild ox and the European bison were both woodland animals with a special fondness for mixed forests with glades, swampy meadows and water courses. The Jaktarow Forest in Poland is the same type of woodland as the European bison's Bielawieza Forest — a forest of oak, maple, alder and pine. The low-growing vegetation included an abundance of herbs and bushes. It is possible that there was some degree of competition between the European bison and the wild ox, especially as the latter was aggressive and belligerent.

## Ancient hunting

Every living thing must have nourishment. For our Stone Age ancestors hunting for large game must have been especially attractive as a successful hunt would result in an ample food supply for a large number of people, and it supplied other necessary by-products for survival as well.

9,500 years ago a wild ox bull died in a lake in Denmark, near Vig. It had apparently drowned there after having been pursued and wounded by hunters. Two fragments of flint were found in its carcase, and one wound had had time to heal, while the other had not. Three flint arrowheads were found among its ribs. It is possible to picture what had happened, for it seems likely that the bull had attacked the primitively armed hunters. A confrontation had taken place which was not dissimilar to that of the modern fighting bull, its distant cousin, and the picadors of a Spanish bullfight. A memory of the ancient hunts is retained in the spectacle of the bullfight.

About 8,500 years ago another wild bull died at a spot about three km away, in another lake. Its carcase was peppered with splinters of flint from the shafts of hunting arrows. Most of these flint fragments were found in the thigh, the back and the chest, and it seems clear that the same hunting methods had remained in practice for at least 1,000 years in the same area. The remains of another wild bull dating from about 5,000 years ago were found in southern Sweden in a peat bog which had formerly been a freshwater lake. The complete skeleton was uncovered, and among the remains were two arrowheads of the type used by Stone Age man. The remains of a Stone Age village were found about 2 km away.

These three discoveries have certain features in common. All were of bulls, and all three had died at the water's edge. Is it possible that the wild ox bulls, weighing 25% more than the cows, were very difficult to kill, and that therefore they were more likely to get away severely wounded? Did the hunted wild ox go in search of water, just as a hunted deer does today?

## The wild ox in Northern Europe

Many wild ox skeletons have been found in the peat bogs of southern Sweden, and similar remains have been found in the more northerly regions of Sweden, although they are not as numerous. It may be that there were fewer oxen in the northern parts. Another possibility is that the soil conditions in the south are more conducive to the preservation of bones. Also, peat bogs have been excavated to extract peat as fuel, and this has tended to bring any buried remains to light. But it seems reasonable to suppose that, like most animals, the wild ox was more scarce towards the northern limits of its range.

Pollen analysis and archaeological dating have been used to date various findings, and more recently the carbon-14 dating method has been employed. All the evidence indicates that the wild ox was present in central and southern Scandinavia from 11,000 years ago up to about 4,000 years ago.

The great naturalist Linnaeus had little to say about the wild ox, although the last remaining specimens were still alive in his day. In his celebrated lectures at Uppsala University in the 1740s he indicated that domesticated cattle and oxen were the same species as the wild cattle of old, and that they were the direct descendants of the wild cattle which were capable of tackling a fully grown bear and killing it with their horns. Linnaeus made no attempt to describe these wild cattle in detail, and this was left to the naturalist Bojanus who gave a scientific description of the wild ox in 1827.

The name "urus" or "aurochs" is thought to be of ancient Germanic origins. It has often been supposed that the term meant "the original ox", from the word "ur", meaning "original". However, it is now thought that the name has nothing to do with the fact that these were old or ancient cattle, but that the term derives from the ancient words "Ur" or "Or", meaning "forest". Thus the term denotes the wild ox as being the "forest ox" or the "woodland ox", and it is known to have been primarily an animal of the ancient woodlands and mixed forests of Europe.

# Musk ox

## *Ovibos moschatus*

The musk ox belongs to the same genus as the goat, the *Caprinae*, and it may come as a surprise to learn that it is actually more closely related to the goats than to animals like the bison or the yak. The musk ox is so called because it secretes a musky substance in glands under its eyes. It is an animal which is especially adapted to life in extreme Arctic conditions. It has a shaggy, woolly coat with two layers of hair, and the legs are short. Its tail is very short, and its eyes barely protrude from its head, all of which are adaptations to make life easier in the cold of the far north.

### Characteristics

The musk ox is a short animal, with short, stout legs, a short, thick neck, a large head and large horns. The horns curl downwards from the centre of the skull where they are rather flat, and the tips of the horns point forwards and upwards at the side of the head. The shoulders are sturdy and hunched, almost as if the animal had a hump. This hunched look is formed by outgrowths of the vertebrae and by the large muscles of the neck and throat, all of which support the enormous head. The hide is dark brown and shaggy in appearance, but the guard-hairs are long and soft, while the underfur is thick and soft. The coat is fully developed in the animal's third year, and the bulls have a thicker mane than the cows.

The cow is also smaller than the bull, but both have horns. A large bull can weigh up to and over 400 kg. Both sexes are fully mature between the ages of four and six years, but the bulls mature more slowly than the cows. The distance between the tips of the horns can be as much as 80 cm for the largest musk oxen bulls. The horns grow until the animal is about six years old, when they begin to darken in colour.

Musk oxen form small herds, but they will also live as solitary animals. The herds have a well developed defensive instinct, and they form a circle when they are attacked or if danger threatens. Both males and females are very aggressive, and are capable of charging much faster than their appearance might suggest. The winter herds comprise 20–30 animals, with the young protected in the middle of the group. The herds can move about quite a lot, although they will use established grazing grounds for long periods of time. During the summer the winter herds break up into smaller groups. Whatever the size of the herd, musk oxen will form a united front and stand up to any aggressor, including man. In the past, when musk oxen were hunted by Eskimos, this fearlessness made them a relatively easy quarry to hunt and kill. The result of this was that musk oxen became locally extinct or suffered considerable reductions in numbers.

In its natural habitat the musk ox chooses its summer grazing grounds close to water courses in low-lying areas which are rich in fresh greenery from bushes and herbs. In winter they tend to avoid the low ground, as it is often more thickly covered by snow than the open moors and upland ridges.

The musk ox feeds on the sparse, low growing vegetation of the tundra — willow, grasses and a variety of herbs. In winter it will forage in bare spots where wind has blown the snow clear, which is often the case on slopes and ridges, and it will also use its feet to scrape out vegetation from beneath the snow. In their natural haunts there are four months of complete darkness in the winter. Musk oxen are believed to live for up to 20 years in the wild.

The animals can be threatened by starvation, which can occur when the feeding grounds are covered by hard crusts of snow or by ice, and some widespread mortality by starvation has been noted. The musk ox normally builds up large reserves of fat before the onset of winter, and its thick and shaggy coat — shaggy because it is not shed regularly — prevents heat loss. However, it also limits the summer ranges to the coldest and most northerly areas. During winter musk oxen will live for several months in temperatures at or below minus 30°C.

### Reproduction

Musk oxen are considered to be sexually mature at the age of 3–4 years. The rut occurs in late summer and early autumn, from July until as late as September, and during this time the bulls battle among themselves for access to the breeding cows. Their eye glands release a profuse secretion, and sparring bulls will square up to one another at a distance of about 10 metres, and then rush at one another at full speed, with their heads lowered. They collide with great force and much noise. The flat horns, which almost grow together across the top of the head, forming a thick boss of horn, take the brunt of the impact. The procedure can be repeated again and again until, after perhaps ten collisions, the contest is decided by a final clash. It is surprising that the bulls' skulls are able to withstand the tremendous buffetings. The most hard-headed and strongest bulls gradually gather together a harem of cows while the unsuccessful challenger slinks away. Young, non-breeding bulls form their own herds, while old males may live a quite solitary life.

The gestation period is a little less than 10 months, and the single calf is born between April and June. In very exceptional cases twins may be born. Cows usually only calf every other year, and the intervals between calvings are determined by the age of the cow and the availability of food. A newborn calf will weigh 7–8 kg, and although the tundra is still covered with snow when it is born, the newborn calf is capable of following its mother within hours of being born, and the two will rejoin the herd after a few days. For the first few weeks the calf is suckled for up to five minutes at a time, and it continues to suckle up to the age of one year. The first signs of horns become visible when the youngster is only a month old. The young calves are playful and often form small social groups. They become independent by the time they are about two years old.

### Distribution

The musk ox was at one time very widespread in Europe and Asia, and a number of musk ox bones have been found in Sweden, Norway and Denmark. In continental Europe musk ox remains from the time of the last Ice Age have been found as far south as France and Hungary. The musk ox became extinct in Europe some 3-4,000 years ago, and its present range is limited to the northern tundra of Canada and Alaska. During this century it has also been introduced to suitable environments in north-western Greenland and in Spitzbergen. It is also believed to exist on Wrangel Island off the coast of Siberia. The total world population of the musk ox has been estimated at about 50,000 individuals.

killed by hunting each year. Elsewhere in Europe this species is protected.

There have been various attempts to account for the general decline of the great bustard. On the Brandenburg heaths in East Germany and on the Hungarian steppes there were some 3,000 and 7,000 birds respectively in the 1940s, but almost all have since disappeared. These areas have been affected by various changes in land use, including the cultivation of the heathlands, the spread of the human population and the widespread use of agricultural chemicals. In addition, the selective hunting of the large, mature cock birds is believed to have compounded the environmental damage and seriously reduced breeding success. In various parts of Europe bustard reserves have been created, and efforts have been made to provide supplementary feeding places in winter.

# Whitetail deer

## *Odocoileus virginianus*

The whitetail deer is originally from North America, and there its distribution range extends over much of the continent, from central Canada to the Caribbean Sea and parts of Central America, with the exception of the Pacific coastal regions and the desert areas of Nevada, Utah and Arizona.

This deer is mainly a woodland animal, and it has a preference for luxuriant, damp, marshy ground and areas of brush. It has adapted well to the presence of man, and can often be seen in the national parks and even along quite densely populated valleys, where it is drawn to the vegetation.

The whitetail deer is relatively large, and a mature male can weigh up to 180 kg, while the doe is unlikely to exceed 113 kg. Only the bucks have antlers, and the record span of a whitetail buck's antlers from North America is 85 cm. The pelage is reddish-brown in summer and greyish in winter. The muzzle is black, the chin and throat grey-white. The belly is white, as is the animal's rump patch, and also the underside of the tail, which is very conspicuous when the animal runs with its tail held erect. The tail is long — up to 35 cm — and bushy. The young calves are reddish-brown and richly dappled with white spots.

The whitetail's diet comprises grass and herbaceous plants, in addition to acorns and various forms of mushrooms and fungi. It often browses on bushes, branches and small trees, where it finds fresh leaves and tender twigs especially attractive.

Like many other deer, the whitetail forms herds in winter, and sometimes as many as 20–30 deer can be seen together. In summer females with their calves remain together, and the older bucks tend to live solitary and wandering lives. The whitetail deer moves quickly, with a galloping, jumping gait which may mean that they spring up to several metres in the air.

The whitetail deer ruts in the period November–February, and at this time the bucks can be heard grunting. Gestation lasts for 6½ months, and twin fawns are usual, although triplets and single calves are also known to occur. Only a few hours after they are born the fawns can stand on their own and after a day or two they are able to follow their mother. They suckle for about four months, but remain with their mother until the next calving time. The young does are in season in the autumn of their second year and normally give birth to their first fawns when they are two years old, and breed annually thereafter.

The whitetail deer was introduced into the wild in Finland in 1939, after a buck and four does had been introduced into a deer park in 1934, where two fawns were born. These deer formed the basis of the whitetail deer population which has subsequently spread widely in Finland. By 1949 the numbers were estimated at 100 animals, and by 1960 this had increased to 1,000. Tens of thousands of whitetail deer are now estimated to live in a distribution area that includes most of Finland. These deer have also emigrated from Finland to Sweden and Russia, but as yet they have not moved into Norway. In Sweden they have been regarded as undesirable immigrants, and it is permitted to shoot them at any time of the year. In Finland 15,000 whitetail deer were shot in the 1980 hunting season.

# Polar bear

## *Thalarctos maritimus*

Of the seven species of bear which occur world-wide, the polar bear is the only one which has totally white fur. It is at its whitest in the summer, just after it has shed its winter coat. In late winter and spring the fur usually has a yellowish tinge, which is caused by rancid seal fat. The normal weight of adult polar bears is between 275 and 500 kg, and the heaviest animals are the mature males. By the age of eight years and over, when they are fully grown, they can attain weights up to 600 kg. The polar bear is the world's largest land predator. The females are usually fully grown at the age of four years, and it is unusual for them to weigh more than 300 kg.

The polar bear is a lone wanderer of the high Arctic tundra and the pack ice, and is active both by day and by night. Occasionally in winter small groups — up to 20 at most — will gather together in places where there is a plentiful supply of food. Once such place, which has become well known for its winter numbers of polar bears, is the village of Churchill on the west side of Hudson Bay in Manitoba. The females tend to be followed by their young, and the two sexes only meet at mating time in late spring and summer. Polar bears are persistent wanderers that prefer to live in areas of pack ice, and they can wander for distances up to 20 km in one day. When occasion demands they can also run at speeds of up to 30–40 km per hour. In many areas they are active all year round, while elsewhere they go into hibernation. Pregnant females always hibernate.

The polar bear is a predator, and it is equipped with 42 teeth, which is the largest number to be found among carnivores. The wolf and the fox, both members of the dog family, have the same number of teeth: the brown bear, however, only has 40 teeth. The polar bear's canine teeth are large and prominent, and like the animal's long claws they are excellent tools for catching and holding on to their principal prey, the seal.

Their diet varies with the seasons of the year, but seals — especially ringed seals — predominate. The bears wait for the seals at their breathing holes in the ice, and it occasionally digs them out from the ice caves in which they give birth. The polar bear also uses its ability to dive and swim under water to catch seabirds, including auks, eider ducks and long-tailed ducks, along the edge of the open water. As the berried plants of the tundra ripen — especially blueberries, cranberries and crowberries — the bears will feed greedily on them.

The daily food requirements vary with the size of the individual bear, but a seal weighing 30 kg may be eaten at one sitting, leaving only the bones.

## Reproduction

The female polar bear becomes sexually mature at the age of three years, and most give birth to their first young when they are 3–5 years old. The males mature somewhat later. The mating season is in March–June, when wandering males use their highly developed sense of smell to detect females in season. A mature male will often mate with several females in succession during the season, but some only mate with one partner. The development of the foetus is delayed, and the fertilised egg is not implanted into the wall of the uterus until more than three months after mating. At that time the fertilised egg is given the nutritional and hormonal stimulus necessary to initiate the development of the foetus.

The polar bear young are born in dens, which are dug in the pack ice, or more commonly in snow banks some distance away inland. The cubs weigh only 5–700 grams at birth, and they are blind and deaf at this stage. They grow very quickly, mainly because the mother's milk is very nutritious, containing about 30% fat.

The cubs, normally two but occasionally one in the case of females breeding for the first time, and sometimes triplets, are born in the period November–January. Northern populations give birth to fewer cubs than those further south, where the incidence of twins is higher. The young remain with the mother in the winter den until March–April, by which time they weigh 8–12 kg. By August their weight will have increased five times, and by one year old they will weigh 150 kg. The cubs follow their mother until they are two years old, and they suckle throughout this time. Occasionally the family ties are not dissolved until the cubs are almost three years old, and because of this the females only breed every 3–4 years.

The average lifespan of a polar bear is 25 years, and the record for longevity is held by one wild polar bear which was found to be still alive at the age of 33 years. Even though it is well known that polar bears are capable of breeding when they are quite old, the average female is unlikely to give birth to more than 4–8 cubs during her whole life. Under ideal conditions a female may produce up to 10–12 young.

## Distribution

The polar bear, like many other animals of the Arctic, has a circumpolar distribution, inhabiting all the areas around the Arctic Ocean. In North America it is found in a broad zone between 52°N and 88°N, and the southernmost area of its range includes the entire Hudson Bay region. In Europe the permanent distribution area is further north, on Spitzbergen and Franz Josef Land. Occasionally Canadian polar bears have been seen in Newfoundland and as far south as the Gulf of St. Lawrence. In Europe individual polar bears have been seen as far south as the Varanger peninsula in Norway.

During the Ice Age the polar bear had a much more extensive southerly distribution, and fossil evidence to show this has been found in southern and south-western Sweden.

Polar bears have been found 150 km inland from the coast, and more than 150 km. out to sea. They are excellent swimmers, and often make use of their swimming ability. They can dive easily to a depth of several metres, and can remain underwater for up to two minutes.

Like other animals which live permanently in the north, the polar bear develops a thicker fur before the onset of winter. A thick layer of fat protects against heat loss as well as providing reserves of nourishment. The thick layer of fat also protects the bears against the cold water of the Arctic Ocean. The 15 cm thick fur is a protection against the cutting chill of the wind, and it is also thought that the polar bear's black skin may help to minimise heat loss.

Changes in climate and in the availability of food have been major factors in the disappearance of the polar bear from much of the southern regions of its former range. More recently, human persecution has been a major cause of declining numbers. However, in the early 1970s an international agreement gave complete protection to the polar bear, and only certain Eskimo communities are now permitted to hunt polar bears. The Eskimos eat the meat and sell the pelts as decorations, or use them as clothing. An annual total of 800–900 bears are licensed to be shot annually, and two-thirds of these are shot in the Canadian Arctic. Approximately 20 polar bears are allowed to be shot by non-resident sportsmen in Canada.